Trends in Logic

Volume 36

For further volumes:
http://www.springer.com/series/6645

TRENDS IN LOGIC

Studia Logica Library

VOLUME 36

SCOPE OF THE SERIES

Trends in Logic is a bookseries covering essentially the same area as the journal *Studia Logica* – that is, contemporary formal logic and its applications and relations to other disciplines. These include artificial intelligence, informatics, cognitive science, philosophy of science, and the philosophy of language. However, this list is not exhaustive, moreover, the range of applications, comparisons and sources of inspiration is open and evolves over time.

Yaroslav Shramko · Heinrich Wansing

Truth and Falsehood

An Inquiry into Generalized Logical Values

Springer

Prof. Yaroslav Shramko
Department of Philosophy
State Pedagogical University
50086 Kryvyi Rih
Ukraine
e-mail: shramko@rocketmail.com

Prof. Heinrich Wansing
Department of Philosophy II
Ruhr-University Bochum
44780 Bochum
Germany
e-mail: Heinrich.Wansing@rub.de

ISBN 978-94-007-0906-5 e-ISBN 978-94-007-0907-2
DOI 10.1007/978-94-007-0907-2
Springer Dordrecht Heidelberg London New York

Cover design: eStudio Calamar, Berlin/Figueres

Printed on acid-free paper

Springer is part of Springer Science+Business Media (www.springer.com)

To our beloved spouses Natalia and Petra

Preface

This study is the result of a research enterprise that started in 2003 when Y.S. visited H.W. at Dresden University of Technology as a Friedrich Wilhelm Bessel awardee of the Alexander von Humboldt Foundation. Actually, Robert Meyer's remark that taking semantical values such as *both true and false and neither true nor false* seriously leads to madness is what triggered our interest in moving from Nuel Belnap and Michael Dunn's set of four truth values represented as the powerset of the set of classical truth values to its sixteen-element powerset **16**. We soon realized that Bob's provocative verdict, inspiring as it proved to be, should not inhibit a generalization of Belnap and Dunn's approach. The outcome of our initial co-operation was published as [232, 233], and the central and fundamental structure of our investigations into generalized truth values became the trilattice $SIXTEEN_3$ defined on **16**.

Since we had generated quite a few problems and ideas for additional research, we decided to apply for a research grant in order to further investigate *The Logic of Generalized Truth Values*. Eventually, the project was funded by the German Research Council (DFG) under grant WA 936/6-1. We are very grateful to the DFG for its generous support, including a sabbatical for H.W. in the summer term 2007. Moreover, we are grateful to Ed Zalta for giving us the opportunity to write a survey on the notion of truth values in the Gestalt of an entry for the *Stanford Encyclopedia of Philosophy*, [238]. This helped us a lot in organizing our project.

As the project continued, it benefited further from a Humboldt connection. In 2008, Sergei Odintsov resumed his Humboldt research fellowship at TU Dresden in order to visit H.W. and to work on, among other things, the axiomatization of truth and falsity entailment in the trilattice $SIXTEEN_3$. From general algebraic results, we know that truth and falsity entailment in $SIXTEEN_3$ in the language with two versions of conjunction, disjunction, and negation (one for truth and the other for falsity) can be finitely axiomatized. The non-constructive existence proof, however, does not provide many clues for finding such an axiomatization. Sergei came up with axiom systems for truth and falsity entailment through extending the language by at least one implication connective. The matrix presentation of algebraic operations in $SIXTEEN_3$ employed in the development of these axiom systems gave rise to further investigations, in particular to the development of sequent calculi for truth and falsity entailment and the definition of intuitionistic variants of trilattice logics. Sergei's results are presented and discussed in Chap. 5. Also in 2008, Norihiro Kamide joined the Humboldt family by taking up a Humboldt research fellowship at TU Dresden. Norihiro contributed not only his expertise on sequent calculi resulting in the papers [141, 274] with H.W. but also investigated alternative semantics for trilattice logic. We are grateful to Norihiro for his willingness to collaborate and share his work with us in Chaps. 6 and 7.

The logics induced by the trilattice $SIXTEEN_3$ and its higher-valued extensions obtained by iterated powerset-formation are natural extensions of Nuel Belnap and Michael Dunn's logic of first-degree entailment, often referred to as Belnap's or Belnap and Dunn's useful four-valued logic. We therefore refer to these trilattice structures as *Belnap trilattices*. A trilattice of constructive truth values, isomorphic to $SIXTEEN_3$, was first presented in joint work by Y.S. Michael Dunn, and Tatsutoshi Takenaka [231]. In fact, the very idea of a truth value trilattice was developed in 1999–2000 during a Fulbright research stay of Y.S. with Michael Dunn at Indiana University. We are most grateful to Nuel Belnap and Michael Dunn for their insights and inspiration, which have been fundamental for our project. Moreover, H.W. is grateful to Nuel for joining him in a reply to critical comments on generalized truth values recently put forward by Didier Dubois, see [273].

In the course of the development of the present inquiry, several people (in addition to ones already mentioned) assisted us in various ways. First of all, we would like to thank Alexander Deck, who not only promptly provided us with all the literature we needed, but also programmed a validity tester for truth and falsity entailment in $SIXTEEN_3$ and prepared the website for our *2008 International Workshop on Truth Values*. Moreover, Christa Schröder, Konstantin Kleinichen, Andrea Kruse and Caroline Semmling helped us with the workshop. We would like to thank all the participants of this workshop for presenting their work and sharing ideas with us, the extremely sunny weather notwithstanding. In addition to the Humboldt Foundation, we would like to thank the German Society for Analytic Philosophy (GAP) and the Gesellschaft von Freunden und Förderern der Technischen Universität Dresden for their sponsorship of the workshop and Jacek Malinowski, the editor-in-chief of *Studia Logica*, for supporting the publication of two special issues of *Studia Logica* that evolved from the workshop, see [236, 237]. We are also grateful to the Alexander von Humboldt Foundation and the Gesellschaft von Freunden und Förderern der Technischen Universität Dresden for supporting Y.S.'s visit to Dresden University in 2010.

Furthermore, we would like to thank the audiences of various conferences and talks where we had an opportunity to present the results of our research project, including, in the case of H.W., *Logik und Wissen*, Darmstadt, June 2005; *The 49th Annual Meeting of the Australian Mathematical Society*, Perth, September 2005; *Trends in Logic IV. Towards Mathematical Philosophy*, Toruń, September 2006; *GAP.6, Philosophie—Grundlagen und Anwendungen*, Berlin, September 2006; *LOGICA 2007*, Hejnice, June 2007; *ASL Logic Colloquium 2007*, Wrocław,

July 2007; *CLE 30 YEARS/XV Brazilian Logic Conference/XIV Latin-American Symposium on Mathematical Logic*, Paraty, May 2008; *ICCL Summer School 2008*, Dresden, September 2008; *Lebenswelt und Wissenschaft. XXI. Deutscher Kongress für Philosophie*, Essen, September 2008; *Logics of Consequence: A Celebration of Nuel Belnap's Work in Philosophical Logic*, Pittsburgh, April 2009; *8th Smirnov Readings in Logic*, Moscow, June 2009; *LOGICA 2009*, Hejnice, June 2009; *GAP.7, Nachdenken und Vordenken—Herausforderungen an die Philosophie*, Bremen, September 2009; *Applications of Logic in Philosophy and Foundations of Mathematics 2010*, Szklarska Poręba, May 2010; talks at the Australian National University Canberra, September 2006; the University of Melbourne, September 2006; the Institute of Philosophy of the National Academy of Sciences, Kiev, May 2009, and at the University of Göttingen, May 2009. Y.S. presented lectures inter alia at the Institute of Logic and Philosophy of Science, Leipzig University, July 2004; the Institute of Philosophy, Humboldt University Berlin, November 2005; *7th Smirnov Readings in Logic*, Moscow, June 2007; *8th Smirnov Readings in Logic*, Moscow, June 2009; and *GAP.7, Nachdenken und Vordenken—Herausforderungen an die Philosophie*, Bremen, September 2009.

In the present book we, in a sense, summarize and systematize the results of our joint research project, many of which have been published during the last few years. More concretely, Chap. 1 contains material from our entry on Truth Values in the *Stanford Encyclopedia of Philosophy* [238]. Chapter 2 makes use of our analysis of the slingshot argument presented in [235]. In Chaps. 3 and 4 we recapitulate the motivating ideas, basic definitions, and results presented in [232] and [233]. As already mentioned, we have also greatly benefited from co-operation with Norihiro Kamide. Chapter 6 on sequent systems for trilattice logics is based on [141], and Chap. 7 on intuitionistic trilattice logics presents material from [273]. The idea of a harmonious many-valued logic, presented in Chap. 8, was first developed in [275]. The application of generalized truth values in the context of the discussion of Suszko's Thesis in Chap. 9 originates from our paper [276]. Moreover, this chapter also includes some basic results about the interconnections between various entailment relations reported in [234]. At this time we would also like to thank Springer-Verlag, Cambridge University Press, and Oxford University Press for their kind permission in letting us make use of the papers in question in this book.

Finally, we would like to express warm gratefulness to our families who have always been for us a constant source of understanding, support, and encouragement. We dedicate this book to our spouses, *Natalia Shramko* and *Petra Wansing*, and thank them for all their love and patience.

Kryvyi Rih and Bochum
September 2010

Yaroslav Shramko
Heinrich Wansing

Contents

Chapter 1
Truth Values

Abstract In this introductory chapter, we show how Gottlob Frege's notion of a truth value has become part of the standard philosophical and logical terminology. Nowadays, the notion of a truth value is an indispensable instrument of a functional analysis of language and realistic, model-theoretic approaches to logical semantics. Moreover, the idea of truth values has induced a radical rethinking of some central issues in the philosophy of logic, including: the categorial status of truth and falsehood, the theory of abstract objects, the subject-matter of logic and its ontological foundations, and the concept of a logical system. This chapter presents a general explication of the very idea of a truth value as well as an overview of the basic philosophical topics related to this concept.

1.1 The Idea of Truth Values

The notion of a truth value was explicitly introduced into logic and philosophy by Gottlob Frege—for the first time in [100], and most notably in his seminal paper [101]. Frege conceived this notion as a natural component of his language analysis where sentences, being saturated expressions, are interpreted as a special kind of names, which refer to (denote, designate, signify) a special kind of objects: truth values. Moreover, there are, according to Frege, only two such objects: *the True* (das Wahre) and *the False* (das Falsche):

> A sentence proper is a proper name, and its Bedeutung, if it has one, is a truth-value: the True or the False [21, p. 297].

It is worth noticing that Frege's own attitude towards truth values was expressly platonic. He considered them to be not simply "introduced" or "invented" but virtually "discovered" entities, regarding himself, in this respect, as acting just like a chemist discovering new chemical elements [108, p. 88]. Later on, we will address the topic of the nature of truth values at greater length. As yet we would

Y. Shramko and H. Wansing, *Truth and Falsehood*, Trends in Logic 36,
DOI: 10.1007/978-94-007-0907-2_1,

like only to remark that even if a realistic view on truth values has indeed merit, one need not be an extreme platonic realist to recognize the importance of the conception as a whole. The more so, as it is possible to present the conception of truth values in philosophically rather neutral terms, as, e.g., Nuel Belnap does by explaining its main semantic point:

> Truth values were put in play by Frege to be the *denotations* of sentences, in contrast with their *senses*. If I may use "**T**" and "**F**" as names of the two classical truth values, then the story is that the denotation of "snow is white" is **T** or **F** according as snow is or is not white. What a happy idea! [24, p. 306]

The idea turns out to be not only happy but very useful, too. It is surely among those new and revolutionary ideas at the turn of the twentieth century that have had a far reaching and manifold impact on the whole development of modern symbolic logic. Truth values provide an effective means to uniformly complete the formal apparatus of a functional analysis of language by generalizing the concept of a function and introducing a special kind of functions, namely propositional functions, or truth value functions, whose range of values consist of the set of truth values. Among the most typical representatives of propositional functions, one finds predicate expressions and logical connectives. As a result, one obtains a powerful tool for a conclusive implementation of the extensionality principle (also called the principle of compositionality), according to which the meaning of a complex expression is uniquely determined by the meanings of its components. On this basis, one can also discriminate between extensional and intensional contexts (cf. [48]) and advance further to the conception of intensional logics.

Moreover, the idea of truth values has induced a radical rethinking of some central issues in the philosophy of logic, including: the categorial status of truth, the theory of abstract objects, the subject-matter of logic and its ontological foundations, the concept of a logical system, the nature of logical notions, etc. Furthermore, truth values have been put to quite different uses in philosophy and logic, being characterized, for example, as:

- primitive abstract objects denoted by sentences in natural and formal languages,
- abstract entities hypostatized as the equivalence classes of sentences,
- what is aimed at in judgments,
- values indicating the degree of truth of sentences,
- entities that can be used to explain the vagueness of concepts,
- values that are preserved in valid inferences,
- values that convey information concerning a given proposition.

Depending on their particular use, truth values have been treated as unanalyzed, defined, unstructured, or structured entities.

In the subsequent sections of the present chapter, we first recall how truth values naturally arise from a certain approach to language analysis, then explicate them as abstract objects of some kind, and finally demonstrate the foundational role of truth values as determining the very subject-matter of logic as a whole.

1.2 Truth Values and the Functional Analysis of Language

The approach to language analysis developed by Frege essentially rests on a strict discrimination between two main kinds of expressions: proper names (singular terms) and functional expressions. Proper names designate (signify, denote, or refer to) singular objects, and functional expressions designate (signify, denote, or refer to) functions.[1] The name 'Ukraine',[2] for example, refers to a certain country, and the expression 'the capital of' denotes a one-place function from countries to cities, in particular, a function that maps the Ukraine to Kyiv (Kiev). Whereas names are "saturated" (complete) expressions, functional expressions are "unsaturated" (incomplete) and may be saturated by applying them to names, thus producing in this way new names. Similarly, the objects to which singular terms refer are saturated and the functions denoted by functional expression are unsaturated. Names to which a functional expression can be applied are called the *arguments* of this functional expression, and entities to which a function can be applied are called the *arguments* of this function. The object which serves as the reference for the name generated by an application of a functional expression to its arguments is called the *value* of the function for these arguments. Particularly, the above-mentioned functional expression 'the capital of' remains incomplete until applied to some name. An application of the function denoted by 'the capital of' to Ukraine (as an argument) returns Kyiv as the object denoted by the compound expression 'the capital of Ukraine' which, according to Frege, is a proper name of Kyiv. Note that Frege distinguishes between an n-place function f as an unsaturated entity that can be completed by and applied to arguments $a_1, \ldots, a_n$ and its

---

[1] It should be observed that there is no common agreement among researchers in various fields of philosophy of language, logic, and linguistics as to the unique term for naming the relation between expressions of a language and objects of an extralinguistic reality (the relation between names and their bearers or, more generally, between signs and what they stand for; i.e., the "word–world" relation). The situation here is sometimes even described (though rather polemically) as a "terminological chaos" [170]. Indeed, on the one hand, we find in the literature a number of terms such as 'designation', 'signification', 'denotation', 'reference', and some others which very often are taken to be synonymous and are used interchangeably not only in various writings by different authors but sometimes even in a work by one and the same author (see, e.g., [48, p. 97]). On the other hand, some authors consider it important and even necessary to draw subtle distinctions between these terms, as, e.g., in [63, pp. 309, 313], where 'reference' is interpreted as a genus of which 'denotation' and 'designation' are supposed to be different species. As we believe, this is mainly a question of a terminological convention depending on specific purposes set out in this or that investigation. In some circumstances, it may well be useful to take account of certain variations of a referential relation and to reflect this variety by means of a suitable terminological ramification. However, in many cases such a detailing is not needed at all, and then the identification of such expressions as 'designation', 'signification', 'denotation', and 'reference' seems to be absolutely justified. In what follows, we will line up with the latter tradition and will not differentiate between these terms, considering them to be synonyms.

[2] We use single quotation marks to refer to linguistic expressions and double quotation marks otherwise.

course of values, which can be seen as the set-theoretic representation of this function: the set $\{\langle a_1,\ldots,a_n,a\rangle \mid a = f(a_1,\ldots,a_n)\}$.

Pursuing this kind of analysis, one is very quickly confronted with two intricate problems. *First*, how should one treat declarative *sentences*? Should one perhaps separate them into a specific linguistic category distinct from the ones of names and function symbols? And *second*, how—from a functional point of view—should one deal with *predicate expressions* such as 'is a city', 'is tall', 'runs', 'is bigger than', 'loves', etc., which are used to denote classes of objects, properties of objects or relations between them and which can be combined with (applied to) singular terms to obtain sentences? If one considers predicates to be a kind of functional expressions, what sort of names are generated by applying predicates to their arguments, and what can serve as referents of these names, respectively values of these functions?

A uniform solution to both problems is obtained by introducing the notion of a *truth value*. Namely, by applying the criterion of "saturatedness" Frege provides a negative answer to the first of the above problems. Since sentences are a kind of complete entities, they should be treated as proper names, but names destined to denote some specific objects, namely the truth values: *the True* and *the False*. In this way one also obtains a solution to the second problem. Predicates are to be interpreted as some kind of functional expressions which, after being applied to these or those names, generate sentences referring to one of the two truth values. For example, if the predicate 'is a city' is applied to the name 'Kyiv', one gets the sentence 'Kyiv is a city', which designates *the True* (i.e., 'Kyiv is a city' *is true*). On the other hand, by using the name 'Mount Everest', one obtains the sentence 'Mount Everest is a city' which clearly designates *the False*, since 'Mount Everest is a city' *is false*.

Functions whose values are truth values are called *propositional functions*. Frege also referred to them as concepts (Begriffe). Typical examples of such functions (besides the ones denoted by predicates) are the functions denoted by propositional connectives. Negation, for example, can be interpreted as a unary function converting *the True* into *the False* and vice versa, and conjunction is a binary function that returns *the True* as a value when both its argument positions are filled in by *the True*, etc. Propositional functions mapping n-tuples of truth values into truth values are also called *truth-value functions*.

Thus Frege, in a first step, extended the familiar notion of a numerical function to functions on singular objects in general and, moreover, introduced a new kind of singular objects that can serve as arguments and values of functions on singular objects, the truth values. In a further step, he considered propositional functions taking functions as their arguments. The quantifier phrase 'every city', for example, can be applied to the predicate 'is a capital' to produce a sentence. The argument of the *second-order* function denoted by 'every city' is the *first-order* propositional function on singular objects denoted by 'is a capital'. The functional value denoted by the sentence 'Every city is a capital' is a truth value, the False.

Truth values thereby prove to be an extremely effective instrument for a logical and semantical analysis of language. But is the price we pay for such an effectiveness not too high? Do we not lapse here into a typical mistake by trying to explain unclear notions like "denotation" in even more problematic terms? Perhaps truth values are just nothing but artificial ad hoc constructions, a sort of twilight theoretical fictions? To clear up the situation we have to consider the nature of truth values and elucidate their conceptual location among other philosophical categories.

1.3 The Categorial Status of Truth and Falsehood

Truth values evidently have something to do with a general concept of truth. Therefore, it may seem rather tempting to try to incorporate considerations on truth values into the broader context of traditional truth-theories, such as correspondence, coherence, anti-realistic, or pragmatist conceptions of truth. Yet, it is unlikely that such attempts can give rise to any considerable success. Indeed, the immense fruitfulness of Frege's introduction of truth values into logic is, to a large extent, just due to its philosophical neutrality with respect to theories of truth. It does not commit one to any specific metaphysical doctrine of truth. In one significant respect, however, the idea of truth values contravenes traditional approaches to truth by bringing to the forefront the problem of its categorial classification.

In most of the established conceptions, truth is usually treated as a property. It is customary to talk about a "truth predicate" and its attribution to sentences, propositions, beliefs or the like. Such an understanding also corresponds to a routine linguistic practice, whereby one operates with the adjective 'true' and asserts, e.g., 'that 5 is a prime number is true'. By contrast with this apparently quite natural attitude, the suggestion to interpret truth as an object may seem very confusing, to say the least. Nevertheless, this suggestion is also equipped with a profound and strong motivation demonstrating that it is far from being just an oddity and has to be taken seriously (cf. [40]).

First, it should be noted that the view of truth as a property is not as natural as it appears on the face of it. Frege brought into play an argument to the effect that characterizing a sentence as *true* adds nothing new to the thought expressed, for 'It is true that 5 is a prime number' says exactly the same as just '5 is a prime number'. That is, the adjective 'true' is in a sense *redundant* and thus is not a real predicate expressing a real property such as the predicates 'white' or 'prime' which, on the contrary, cannot simply be eliminated from a sentence without an essential loss for the proposition expressed. In this case a superficial grammatical analogy is misleading. This idea gave an impetus to the deflationary conception of truth (advocated by Ramsey, Ayer, Quine, Horwich, and others).

However, even admitting the redundancy of truth as a property, Frege emphasizes its importance and indispensable role in some other respect. Namely, truth, accompanying every act of judgment as its ultimate goal, secures an objective *value of cognition* by arranging for every assertive sentence a transition from the level of sense (the thought expressed by a sentence) to the level of denotation (its truth value). This circumstance specifies the significance of taking truth as a particular object. As Tyler Burge explains:

> Normally, the point of using sentences, what "matters to us", is to claim truth for a thought. The object, in the sense of the point or *objective*, of sentence use was truth. It is illuminating therefore to see truth as an object [40, p. 120].

With certain modifications this can also be extended to the notion of falsity, which—in classical contexts—is usually expressed by an operation of negation (or rejection) of a sentence. 'It is false that 6 is a prime number' means accordingly nothing more than '6 *is not* a prime number'. Thus, if we have negation in our language (and we usually do), the property of "being false" is also easily eliminable from the context. And if we wish to operate with falsity as a non-trivial and useful notion, it should be codified as a specific truth value which is closely connected to negative judgments. The significance of this value is then primarily marked by its "negative role" as something that must be avoided in the course of a correct reasoning.

As it has been observed repeatedly in the literature (cf., e.g., [40, 213]), the stress Frege laid on the notion of a truth value was, to a great extent, pragmatically motivated. Besides an intended gain for his system of "Basic Laws" [103] reflected in enhanced technical clarity, simplicity, and unity, Frege also sought to substantiate in this way his view on logic as a theoretical discipline with truth as its main goal and primary subject-matter. Incidentally, Gabriel [111] demonstrated that in the latter respect Frege's ideas can be naturally linked up with a value-theoretical tradition in German philosophy of the second half of the 19th century. More specifically, Wilhelm Windelband, the founder and the principal representative of the Southwest school of Neo-Kantianism, was actually the first who employed the term 'truth value' ('Wahrheitswert') in his essay "What is Philosophy?" published in 1882 (see [283, p. 32]), i.e., nine years before [100], even if he was very far from treating a truth value as a value of a function.

Windelband defined philosophy as a "critical science about universal values". He considered philosophical statements to be not mere judgments but rather *assessments*, dealing with some fundamental values, *the value of truth* being one of the most important among them. This latter value is to be studied by logic as a special philosophical discipline. Thus, from a value-theoretical standpoint, the main task of philosophy, taken generally, is to establish the principles of logical, ethical and aesthetical assessments, and Windelband accordingly highlighted the triad of basic values: "true", "good", and "beautiful". Later this triad was literally taken up by Frege in [102] when he defined the subject-matter of logic (see below). Gabriel points out [110, p. 374] that this connection between logic and a value theory can be traced back to Hermann Lotze, whose seminars in Göttingen were attended by both Windelband and Frege.

The decisive move made by Frege was to bring together a philosophical and a mathematical understanding of values on the basis of a generalization of the notion of a function on numbers. While Frege may have been inspired by Windelband's use of the word 'value' (and even more concretely—'truth-value'), it is clear that he uses the word in its mathematical sense. If predicates are construed as a kind of functional expressions which, after being applied to singular terms as arguments, produce sentences, then the values of the corresponding functions must be references of sentences. Taking into account that the range of any function typically consists of objects, it is natural to conclude that references of sentences must be objects as well. And if one now just takes it that sentences refer to truth values (*the True* and *the False*), then it turns out that truth values are indeed objects, and it seems quite reasonable to generally explicate truth and falsity as objects and not as properties. As Frege explains:

> A statement contains no empty place, and therefore we must take its *Bedeutung* as an object. But this *Bedeutung* is a truth-value. Thus the two truth-values are objects [21, p. 140].

This explanation looks quite compelling but only as long as one agrees with its crucial point that sentences indeed refer to truth values. The point can be questioned, inasmuch as one could well suggest alternative (and at the first glance maybe even more suitable) candidates for possible references of sentences, which might appear rather plausible, e.g., facts, situations, or state of affairs. In such a way one would doubt that truth values are anything more than mere postulations. Such a questioning would undermine the very idea of truth values, and it would be therefore highly desirable to provide this idea not only with a pragmatical motivation but with strong theoretical justification. In the next chapter we will expose and critically examine a specific argument stemming from Frege's work and purporting to strictly *prove* that if sentences do have denotations at all, these cannot be but truth values.

1.4 The Ontological Background of Truth Values

If truth values are accepted and taken seriously as a special kind of objects, the obvious question as to the nature of these entities arises. The above characterization of truth values as objects is far too general and requires further specification. One approach is to qualify truth values as *abstract* objects. Note that Frege himself never used the word 'abstract' when describing truth values. Instead, he has a conception of so called "logical objects", truth values being the most fundamental (and primary) of them [105, p. 121]. Among the other logical objects Frege pays particular attention to are sets and numbers, thus emphasizing their logical nature (in accordance with his logicist view).

Church [53, p. 25], when considering truth values, explicitly attributes to them the property of being abstract. Since then it has been customary to label truth values as abstract objects, thus allocating them into the same category of entities as mathematical objects (numbers, classes, geometrical figures) and propositions. One may pose here an interesting question about the correlation between Fregean logical objects and abstract objects in the modern sense. Obviously, the universe of abstract objects is much broader than the universe of logical objects as Frege conceives them. The latter are construed as constituting an ontological foundation for logic, and hence for mathematics (pursuant to Frege's logicist program). Generally, the class of *abstracta* includes a wide diversity of platonic universals (such as redness, youngness, or geometrical forms) and not only those which are logically necessary. Nevertheless, it may safely be said that logical objects can be considered as paradigmatic cases of abstract entities or abstract objects in their purest form.

It should be noted that finding an adequate definition of abstract objects is a matter of considerable controversy. According to a common view, abstract entities lack spatiotemporal properties and relations, as opposed to concrete objects which exist in space and time [153, p. 515]. In this respect, truth values obviously *are* abstract as they clearly have nothing to do with physical spacetime. In a similar fashion, truth values fulfill another requirement often imposed upon abstract objects, namely the one of a causal inefficacy (see, e.g., [127, p. 7]). Here again, truth values are very much like numbers and geometrical figures: they have no causal power and make nothing happen.

Finally, it is of interest to consider how truth values can be introduced by applying so-called *abstraction principles*, which are used for supplying abstract objects with *criteria of identity*. The idea of this method of characterizing abstract objects is also largely due to Frege, who wrote:

> If the symbol *a* is to designate an object for us, then we must have a criterion that decides in all cases whether *b* is the same as *a*, even if it is not always in our power to apply this criterion [21, p. 109].

More precisely, one obtains a new object by abstracting it from some given kind of entities, by virtue of certain criteria of identity for this new (abstract) object. This abstraction is in terms of an equivalence relation defined on the given entities (see [289, p. 161]). The celebrated slogan by Quine "No entity without identity" [202, p. 23] is intended to express essentially the same understanding of an (abstract) object as an "item falling under a sortal concept which supplies a well-defined criterion of identity for its instances" [154, p. 619].

For truth values, such a criterion has been suggested by Dummett in [69, pp. 1–2] (cf. also [5, p. 2]), stating that for any two sentences p and q, the truth value of p is identical with the truth value of q if and only if p is (non-logically) equivalent with q. This idea can be formally explicated following the style of presentation in [154, p. 620]:

$$\forall p \forall q(\text{Sentence}(p) \ \& \ \text{Sentence}(q)) \Rightarrow (tv(p) = tv(q) \Leftrightarrow (p \leftrightarrow q)),$$

where $\&, \Rightarrow, \Leftrightarrow$, and $\forall$ stand, correspondingly, for 'and', 'if... then', 'if and only if' and 'for all' in the *metalanguage*, and $\leftrightarrow$ stands for some *object language* equivalence connective (biconditional).

Incidentally, Carnap [48, p. 26], when introducing truth values as extensions of sentences, is guided by essentially the same idea. Namely, he points out a strong analogy between extensions of predicators and truth values of sentences. Carnap considers a wide class of designating expressions ("designators") among which there are predicate expressions ("predicators"), functional expressions ("functors"), and some others. Applying the well-known technique of interpreting sentences as predicators of degree 0, he generalizes the fact that two predicators of degree n (say, P and Q) have the same extension if and only if $\forall x_1 \forall x_2 \ldots \forall x_n (Px_1 x_2 \ldots x_n \leftrightarrow Qx_1 x_2 \ldots x_n)$ holds. Then, analogously, two sentences (say, p and q), being interpreted as zero-degree predicators, must have the same extension if and only if $p \leftrightarrow q$ holds—that is, if and only if they are equivalent. And then, Carnap remarks, it seems quite natural to take truth values as extensions for sentences.

Note that here we have what Williamson [280, p. 145ff] has dubbed a "two-level criterion" of identity which employs a *functional dependency* between an introduced abstract object (in this case a truth value) and some other objects (sentences), see also [70, pp. 580–581; 154, p. 620]. More specifically, what is considered is the truth value *of* a sentence (or proposition, or the like). The criterion of identity for truth values is then formulated through the logical relation of equivalence holding between these other objects—sentences, propositions, or the like (with an explicit quantification over them).

Williamson contrasts this kind of criterion with a "one-level criterion" which should typically involve quantification over the introduced abstract objects themselves, defining identity by means of some other equivalence relation obtaining again between them.

For truth values the principle above can be reconstructed in "one-level"-style as follows (cf. [154, p. 623]):

$$\forall x \forall y ((TV(x) \ \& \ TV(y)) \Rightarrow (x = y \Leftrightarrow$$
$$\forall w \forall z ((\text{Sentence}(w) \ \& \ \text{Sentence}(z) \ \& \ Of(x, w) \ \& \ Of(y, z)) \Rightarrow (w \leftrightarrow z)))).$$

That is, truth values are identical if and only if any sentences of which they are truth values are equivalent.

It should also be remarked that the properties of the object language biconditional depend on the logical system in which the biconditional is employed. Biconditionals of different logics may have different logical properties, and it surely matters what kind of equivalence connective is used for defining truth values. This means that the concept of a truth value introduced by means of the identity criterion that involves a biconditional between sentences is also logic-relative. Thus, if '$\leftrightarrow$' stands for material equivalence, one obtains classical truth values, but if the intuitionistic biconditional is employed, one gets truth values of intuitionistic logic, etc. Taking into account the role truth values play in logic, such an outcome seems to be not at all unnatural.

1.5 Logic as the Science of Logical Values

In a late paper from 1918 [102] Frege claims that the word 'true' determines the subject-matter of logic in exactly the same way as the word 'beautiful' does for aesthetics and the word 'good' does for ethics. Thus, according to such a view, the proper task of logic consists, ultimately, of investigating "the laws of being true" [241, p. 86]. By doing so, logic is interested in truth as such, understood objectively, and not in what is merely taken to be true. Now, if one admits that truth is represented by a specific abstract object (the corresponding truth value), then logic in the first place has to explore the features of this object as well as its interrelations to other entities of various other kinds.

A prominent adherent of this conception was Jan Łukasiewicz. As he paradigmatically put it:

> All true propositions denote one and the same object, namely truth, and all false propositions denote one and the same object, namely falsehood. I consider truth and falsehood to be *singular* objects in the same sense as the number 2 or 4 is. ... Ontologically, truth has its analogue in being, and falsehood, in non-being. The objects denoted by propositions are called *logical values*. Truth is the positive, and falsehood is the negative logical value. ... Logic is the science of objects of a special kind, namely a science of *logical values* [158, p. 90].

This definition may seem rather unconventional, for logic is usually treated as the science of correct reasoning and valid inference. The latter understanding, however, calls for further justification. This becomes evident as soon as one asks *on what grounds* one should qualify this or that pattern of reasoning as correct or incorrect. Taking into account a quite common understanding that any valid inference should be based on *logical rules*, we arrive then at the question of how the logical rules are justified. This problem is of an undoubtedly foundational character, and in addressing it, various strategies are possible. Without going much into details, let us just shortly mention some of the most typical approaches, pointing out their crucial shortcomings.

1. *The psychologistic approach.* Logical rules essentially reflect the process of sound human thinking; more precisely they are based upon the so-called "laws of thought" and prescribe how we should think if we wish to think correctly.

   This strategy essentially turns logic into a branch of psychology. So conceived, logic becomes an empirical discipline depending on subjective and contingent circumstances concerned with the functioning of human minds. Psychologism was sharply criticized by Frege and Husserl, who provided various compelling arguments against it.

2. *The conventionalistic approach.* Logical rules are more or less voluntarily chosen conventions about correct reasoning, subject only to some formal requirements (restrictions) such as consistency, independence of rules, etc.
3. *The linguistic approach.* Logical rules are certain rules for operating with linguistic expressions. They represent specific regularities corresponding to structural features of the given linguistic system.

Both the conventionalistic and the linguistic strategy relativize logic too much with respect to arbitrarily adopted syntactic principles or "linguistic frameworks". In this way logic is actually deprived of foundations rather than provided with them.

4. *The transcendentalistic approach.* Logical rules represent some fundamental a priori structures of consciousness as such by means of which we synthesize our concepts and intuitions to acquire knowledge of the world as it is given in the process of apperception.

   This view is hardly compatible with the fact of the existence of many (non-classical) logical systems. In fact, a transcendentalist would be prone to insist that there is only one "right" logic and to consider the diversity of logical systems as a sort of deviation from the "normal situation".

All the approaches mentioned so far have evidently some "idealistic" (anti-realistic) flavor as they interlink logic with an activity of some subject, whether it be the intellectual or linguistic activity of a human being or the conscious activity of a transcendental subject in the Kantian sense. But if we seek to construe logic as a fully objective discipline, we might be dissatisfied with psychologistic, conventionalistic, purely linguistic or transcendentalistic answers to the foundational question raised above. In such a case it could be reasonable to take a look at another, namely ontological (realistic), strategy of justifying logical rules which seems to constitute quite a feasible alternative to idealistic approaches.

Let us take into account an established view according to which logical rules should at least guarantee that in a valid inference the conclusion(s) is (are) true if all the premises are true. Translating this demand into the Fregean terminology, it would mean that in the course of a correct inference, the possession of the truth value *the True* should be *preserved* from the premises to the conclusion(s). This demand is sometimes supplemented with an additional requirement that some of the premise must be false if the conclusion is false, which again would mean that the truth value *the False* should be preserved from the conclusion(s) to the premises. Thus, granting the realistic treatment of truth values adopted by Frege, the understanding of logic as the science of truth values does in fact provide logical rules with an ontological justification placing the roots of logic within a certain kind of entities (conceived objectively). The properties of these entities and relations between them (and maybe also some other entities) ultimately determine the characteristic features of logical rules.

These entities constitute a certain uniform domain which can be viewed as a subdomain of Frege's so-called "third realm" (the realm of the objective content of thoughts and generally abstract objects of various kinds, see [102], cf. [193] and also [41, p. 634].) Among the subdomains of this third realm one finds, e.g., the collection of mathematical objects (numbers, classes, etc.). The set of truth values may be regarded as forming another such subdomain, namely the one of *logical values*, and logic as a branch of science rests essentially on this *logical domain* ("logical realm") and on exploring its features and regularities.

1.6 Logical Structures

According to Frege, there are exactly two truth values, *the True* and *the False*. This opinion appears to be rather restrictive, and one may ask whether it is really indispensable for the concept of a truth value. One should observe that in elaborating this conception, Frege assumed specific requirements of his system of the *Begriffsschrift*, especially the principle of bivalence taken as a metatheoretical principle, viz. that there exist only two distinct logical values. On the object-language level, this principle finds its expression in the famous classical laws of excluded middle and non-contradiction. The further development of modern logic, however, has clearly demonstrated that classical logic is only one particular theory (although maybe a very distinctive one) among the vast variety of logical systems. In fact, the Fregean ontological interpretation of truth values depicts logical principles as a kind of ontological postulations, and as such they may well be modified or even abandoned. For example, by giving up the principle of bivalence, one is naturally led to the idea of postulating *many truth values*.

It was Łukasiewicz, who as early as 1918, proposed taking other logical values different from truth and falsehood seriously, see [155, 156]. Independently of Łukasiewicz, Emil Post, in his dissertation from 1920, published as [194], introduced m-valued truth tables, where m is any positive integer. Whereas Post's interest in *many-valued logic* (where "many" means "more than two") was almost exclusively mathematical, Łukasiewicz's motivation was philosophical. He contemplated the semantical value of sentences about the contingent future, as discussed in Aristotle's *De interpretatione*. Łukasiewicz introduced a third truth value and interpreted it as "possible". By generalizing this idea and also adopting the above understanding of the subject-matter of logic, one naturally arrives at a consideration of various collections of truth values as possible bases for different logical systems. It seems also quite plausible to assume that these collections have to be *organized* in one way or another, in other words, they must possess some sort of (inner) *structure*.

In this book, we take this assumption for granted and subsequently focus our consideration on two kinds of *logical structures* conceived as suitable representations of particular logical systems: *valuation systems* and *truth-value lattices*.

The notion of a valuation system has been developed, e.g., by Dummett in [68] (see also [70, 72, 216]). Consider a propositional language $\mathcal{L}$ built upon a non-empty set of atomic sentences *Atom* and a set of propositional connectives $\mathcal{C}$ (the set of sentences of $\mathcal{L}$ being the smallest set containing *Atom* and being closed under the connectives from $\mathcal{C}$). Then a *valuation system* **V** for the language $\mathcal{L}$ is a triple $\langle \mathcal{V}, \mathcal{D}, \mathcal{F} \rangle$, where $\mathcal{V}$ is a non-empty set with at least two elements, $\mathcal{D}$ is a non-empty proper subset of $\mathcal{V}$, and $\mathcal{F} = \{f_{c_1}, \ldots f_{c_m}\}$ is a set of functions such that f_i is an n-place function on $\mathcal{V}$ if c_i is an n-place connective. Intuitively, $\mathcal{V}$ is the set of truth values, $\mathscr{D}$ is the set of *designated* truth values, and $\mathscr{F}$ is the set of truth-value functions interpreting the elements of $\mathscr{C}$. If the set of truth values of a valuation system **V** has n elements, **V** is said to be n-valued. Any valuation system can be equipped with an assignment function which maps the set of atomic sentences into $\mathscr{V}$. Each assignment a relative to a valuation system **V** can be extended to all sentences of $\mathscr{L}$ by means of a valuation function v_a defined in accordance with the following conditions:

$$\forall p \in Atom, v_a(p) = a(p); \tag{1.1}$$

$$\forall c_i \in \mathscr{C}, v_a(c_i(A_1,\ldots,A_n)) = f_i(v_a(A_1),\ldots,v_a(A_n)). \tag{1.2}$$

Valuation systems are often referred to as *(logical) matrices*. Since the cardinality of $\mathscr{V}$ may be greater than 2, the notion of a valuation system provides a general and natural foundation not only for classical logic, but for various kinds of logical systems (particularly, many-valued ones).

In this way Fregean (classical) logic, with, say, the propositional connectives $\wedge, \vee, \rightarrow$ and $\sim$, can be presented as determined by a particular valuation system based on exactly two elements. Consider, for example, the valuation system $\mathbf{V}_{cl} = \langle \mathscr{V}_{cl}, \mathscr{D}_{cl}, \mathscr{F}_{cl} \rangle$, where $\mathscr{V}_{cl} = \{\{\varnothing\}, \varnothing\}$, $\mathscr{D}_{cl} = \{\{\varnothing\}\}$, and $\mathscr{F}_{cl} = \{f_\wedge, f_\vee, f_\rightarrow, f_\sim\}$, defined as follows:

Definition 1.1 For any $X, Y \in \mathscr{V}_{cl}$:

1. $f_\wedge(X,Y) = X \cap Y$;
2. $f_\vee(X,Y) = X \cup Y$;
3. $f_\rightarrow(X,Y) = (\mathscr{V}_{cl} - X) \cup Y$;
4. $f_\sim(X) = \mathscr{V}_{cl} - X$.

Then $\{\varnothing\}$ and $\varnothing$ can be understood as Frege's *the True* and *the False*, correspondingly. The following proposition shows that functions from $\mathscr{F}_{cl}$ determine exactly the propositional connectives of classical logic:

Proposition 1.1 *For any assignment a relative to* $\mathbf{V}_{cl}$, *and for any formulas A and B, the following holds:*

1. $v_a(A \wedge B) = \{\varnothing\}$ *iff* $v_a(A) = \{\varnothing\}$ and $v_a(B) = \{\varnothing\}$;
2. $v_a(A \vee B) = \{\varnothing\}$ *iff* $v_a(A) = \{\varnothing\}$ or $v_a(B) = \{\varnothing\}$;
3. $v_a(A \rightarrow B) = \{\varnothing\}$ *iff* $v_a(A) = \varnothing$ or $v_a(B) = \{\varnothing\}$;
4. $v_a(\sim A) = \{\varnothing\}$ *iff* $v_a(A) = \varnothing$.

The proof of this proposition is rather straightforward and is left to the reader.

The set $\mathscr{D}$ of designated values is of central importance for the notion of a valuation system. This set can be seen as a generalization of the classical truth value T (*the True*) in the sense that it determines many logical notions and thereby generalizes some of the roles played by Frege's *the True*, cf. the introductory remarks in Sect. 1.1 about uses of truth values. For example, the set of tautologies (logical laws) is directly specified by the given set of designated truth values: a sentence A is a *tautology* in a valuation system **V** iff for every assignment a relative to $\mathbf{V}$, $v_a(A) \in \mathcal{D}$. Another fundamental logical notion—the one of an entailment relation—can also be defined by referring to the set $\mathcal{D}$. For a given valuation system $\mathbf{V}$, a corresponding entailment relation ($\models_V$) is usually defined by postulating the preservation of designated values from the premises to the conclusion:

$$\Delta \models_V A \text{ iff } \forall a : (\forall B \in \Delta : v_a(B) \in \mathcal{D}) \Rightarrow v_a(A) \in \mathcal{D}. \tag{1.3}$$

A pair $\mathcal{M} = \langle \mathbf{V}, v_a \rangle$, where $\mathbf{V}$ is an (n-valued) valuation system and v_a a valuation in $\mathbf{V}$, may be called an (n-valued) *model* based on $\mathbf{V}$. Every model $\mathcal{M} = \langle \mathbf{V}, v_a \rangle$ comes with a corresponding entailment relation $\models_{\mathcal{M}}$ by defining $\Delta \models_{\mathcal{M}} A$ iff $(\forall B \in \Delta : v_a(B) \in \mathcal{D}) \Rightarrow v_a(A) \in \mathcal{D}$.

Suppose $\mathfrak{L}$ is a syntactically defined logical system $\mathfrak{L}$ with a consequence relation $\vdash_{\mathfrak{L}}$, specified as a relation between the power-set of $\mathcal{L}$ and $\mathcal{L}$. Then a valuational system $\mathbf{V}$ is said to be *strictly characteristic* for $\mathfrak{L}$ just in case $\Delta \models_V A$ iff $\Delta \vdash_{\mathfrak{L}} A$ (see [70, p. 431]). Conversely, one says that $\mathfrak{L}$ *is characterized* by $\mathbf{V}$. *By itself*, a valuation system is, properly speaking, *not* a logic. If, however, $\mathbf{V}$ strictly characterizes $\mathfrak{L}$, then $\mathbf{V}$ may be seen as a semantic basis for the logic $\mathfrak{L}$.

Another typical way of representing logical systems is by a *lattice-ordered set* of truth values. A partially-ordered set $(L, \leq)$ is called a *lattice* if and only if each two-element subset $\{a, b\}$ of L has an *infimum* (also called "meet"), denoted $a \sqcap b$, and a *supremum* (also called "join"), denoted $a \sqcup b$. Then, a *truth-value lattice* $(\mathcal{V}, \leq)$ can standardly be defined on some partially ordered set of truth values $\mathcal{V}$. The meet and join in a truth value lattice are normally seen as determining logical connectives of conjunction and disjunction, respectively. Other logical connectives can be defined through the truth-value lattice operations as well. More concretely, every truth-value lattice $(\mathcal{V}, \leq)$ can be equipped with a valuation function v mapping the set of atomic sentences of our language into $\mathcal{V}$, which can be extended to compound sentences by means of definitions of the following kind (for any n-place propositional connective c_i):

$$v(c_i(A_1, \ldots, A_n)) = o_i(v(A_1), \ldots, v(A_n)), \tag{1.4}$$

where o_i is some lattice operation on $(\mathcal{V}, \leq)$ or maybe a combination of several such operations.

Again, the cardinality of $\mathcal{V}$ may be greater than 2. The truth values of classical logic constitute just a specific lattice known as a two-element Boolean algebra, or as *the* two-element Boolean algebra, up to isomorphism. As an example of such a structure, consider the lattice $(\{\{\varnothing\}, \varnothing\}, \subseteq)$, where $\{\varnothing\}$ and $\varnothing$ (as above) stand for Frege's *the True* and *the False*, respectively. Clearly, we have $\varnothing \subseteq \{\varnothing\}$. The set-theoretical intersection and union play here the role of the lattice operations of meet and join. Moreover, one can define the function of ordering inversion $(-)$ such that $-\varnothing = \{\varnothing\}$ and $\varnothing = -\{\varnothing\}$ (in fact, the operation of lattice complement). Then we have the following definition of a valuation function for the classical propositional connectives:

Definition 1.2 For any formulas A and B:

1. $v(A \wedge B) = v(A) \cap v(B)$;
2. $v(A \vee B) = v(A) \cup v(B)$;
3. $v(A \to B) = -v(A) \cup v(B)$;
4. $v({\sim}A) = -v(A)$.

It is easy to see that Proposition 1.1 also holds for the valuation function so defined. Note that in a truth-value lattice $(\mathscr{V}, \leq)$, it is not necessary to distinguish the set of designated elements. Instead, one can effectively define the relation of logical entailment ($\models_{\mathscr{V}}$) with respect to $(\mathscr{V}, \leq)$ as expressing an agreement with the lattice ordering $\leq$:

$$\Delta \models_{\mathscr{V}} A \text{ iff } \forall v \sqcap \{v(B) \mid B \in \Delta\} \leq v(A). \tag{1.5}$$

In view of this definition, the ordering relation in a truth value lattice can be called *the logical order*. Having an entailment relation, one can easily introduce many other logical notions, e.g., the notion of a tautology, of a (semantic) contradiction, etc.

It should be noted that the truth values of a logical system do not always admit a lattice presentation. For example, truth values of any logic with non-commutative conjunction or disjunction (see, e.g. [20, 65, 190, 208, 212]) obviously do not constitute a lattice-ordered set with respect to these connectives. However, the class of logics whose truth values form lattices (cf. [210]) is rather extensive and includes many important logical systems. By developing a theory of generalized truth values, we will give special attention to various lattice-organized structures which can be defined on collections of such values.

1.7 Truth Values, Truth Degrees, and Vague Concepts

In many-valued logic, truth values are often called "truth degrees". The term 'truth degrees', used by Gottwald [124, 125] and many other authors, suggests that truth comes by degrees, and these degrees may be seen as truth values in an extended sense. The idea of truth as a graded notion has been applied to model vague predicates and to obtain a solution to the Sorites Paradox, the Paradox of the Heap. However, the success of applying many-valued logic to the problem of vagueness is highly controversial. Williamson [281, p. 97], for example, holds that the phenomenon of higher-order vagueness "makes most work on many-valued logic irrelevant to the problem of vagueness".

In any case, the vagueness of concepts has been much debated in philosophy and it was one of the major motivations for the development of *fuzzy logic*. In the 1960s, Zadeh [29] introduced the notion of a *fuzzy set*. A characteristic function of a set X is a mapping which is defined on a superset Y of X and which indicates membership of an element in X. The range of the characteristic function of a classical set X is the two-element set $\{0, 1\}$ (which may be seen as the set of classical truth values). The function assigns the value 1 to elements of X and the value 0 to all elements of Y not in X. A fuzzy set has a membership function ranging over the real interval [0, 1]. A vague predicate such as 'is much earlier than March 20th, 1963', 'is beautiful', or 'is a heap' may then be regarded as denoting a fuzzy set. The membership function g of the fuzzy set denoted by 'is much earlier than March 20th, 1963' thus assigns values (seen as truth degrees) from the interval [0, 1] to moments in time, for example g(1 p.m., August 1st, 2006) = 0, g(3 a.m., March 19th, 1963) = 0, g(9:16 a.m., April 9th, 1960) = 0.005, g(2 p.m., August 13th, 1943) = 0.05, g(7:02 a.m., December 2nd, 1278) = 1.

The application of continuum-valued logics to the Sorites Paradox has been suggested by Goguen [122]. The Sorites Paradox in its so-called conditional form is obtained by repeatedly applying *modus ponens* in arguments such as:

A collection of 100,000 grains of sand is a heap.
If a collection of 100,000 grains of sand is a heap, then a collection 99,999 grains of sand is a heap.
If a collection of 99,999 grains of sand is a heap, then a collection 99,998 grains of sand is a heap.

⋮

If a collection of two grains of sand is a heap, then a collection of one grain of sand is a heap.
Therefore: A collection of one grain of sand is a heap.

Whereas it seems that all premises are acceptable, because the first premise is true and one grain does not make a difference to a collection of grains being a heap or not, the conclusion is, of course, unacceptable. If the predicate 'is a heap' denotes a fuzzy set and the conditional is interpreted as implication in Łukasiewicz's continuum-valued logic, then the Sorites Paradox can be avoided. The truth-function $f_{\rightarrow}$ of Łukasiewicz's implication $\rightarrow$ is defined by stipulating that if $x \leq y$, then $f_{\rightarrow}(x, y) = 1$, and otherwise $f_{\rightarrow}(x, y) = 1 - (x - y)$. If, say, the truth value of the sentence 'A collection of 500 grains of sand is a heap' is 0.8 and the truth value of 'A collection of 499 grains of sand is a heap' is 0.7, then the truth value of the implication 'If a collection of 500 grains of sand is a heap, then a collection of 499 grains of sand is a heap' is 0.9. Moreover, if the acceptability of a statement is defined as having a value greater than j for $0 < j < 1$ and all the conditional premises of the Sorites Paradox do not fall below the value j, then *modus ponens* does not preserve acceptability, because the conclusion of the Sorites Argument, being evaluated as 0, is unacceptable.

Urquhart [257, p. 108] stresses "the extremely artificial nature of the attaching of precise numerical values to sentences like ... 'Picasso's *Guernica* is beautiful'". To overcome the problem of assigning precise values to predications of vague concepts, Zadeh [292] introduced *fuzzy truth values* as distinct from the numerical truth values in [0, 1], the former being fuzzy subsets of the set [0,1], understood as *true, very true, not very true*, etc.

The interpretation of continuum-valued logics in terms of fuzzy set theory has for some time been seen as defining the field of mathematical fuzzy logic. Haack [129] refers to such systems of mathematical fuzzy logic as "base logics" of fuzzy logic and reserves the term 'fuzzy logics' for systems in which the truth values themselves are fuzzy sets. Fuzzy logic in Zadeh's latter sense has been thoroughly criticized from a philosophical point of view by Haack [129] for its "methodological extravagances" and its linguistic incorrectness. Haack emphasizes that her criticisms of fuzzy logic do not apply to the base logics. Moreover, it should be pointed out that mathematical fuzzy logics are nowadays studied not in the first place as continuum-valued logics, but as many-valued logics related to residuated lattices, see [54, 112, 125, 130], whereas fuzzy logic in the broad sense is, to a large extent, concerned with certain engineering methods.

A fundamental concern about the semantical treatment of vague predicates is whether an adequate semantics should be truth-functional, that is, whether the truth value of a complex formula should depend functionally on the truth values of its subformulas. Whereas mathematical fuzzy logic is truth-functional, Williamson [281, p. 97] holds that "the nature of vagueness is not captured by any approach that generalizes truth-functionality". According to Williamson, the degree of truth of a conjunction, a disjunction, or a conditional just fails to be a function of the degrees of truth of vague component sentences. The sentences 'John is awake' and 'John is asleep', for example, may have the same degree of truth. By truth-functionality the sentences 'If John is awake, then John is awake' and 'If John is awake, then John is asleep' are alike in truth degree, indicating for Williamson the failure of degree-functionality.

One way of, in a certain sense, reasoning non-truthfunctionally about vagueness is supervaluationism. The method of supervaluations was developed by Mehlberg [172] and van Fraassen [97] and was later applied to vagueness by Fine [85], Keefe [146] and others.

Van Fraassen's aim was to develop a semantics for sentences containing non-denoting singular terms. Even if one grants that atomic sentences containing non-denoting singular terms and that some attributions of vague predicates are neither true nor false, it nevertheless seems natural not to preclude that compound sentences of a certain shape and containing non-denoting terms or vague predications *are* either true or false, e.g., sentences of the form 'If A, then A'. Supervaluational semantics provides a solution to this problem. A three-valued assignment a into $\{T, I, F\}$ may assign a truth-value gap (or rather the value I) to the vague sentence 'Picasso's *Guernica* is beautiful'. Any classical assignment a' that agrees with a whenever a assigns T or F may be seen as a precisification (or superassignment) of a. A sentence may then be said to be supertrue under assignment a if it is true under every precisification a' of a. Thus, if a is a three-valued assignment into $\{T, I, F\}$ and a' is a two-valued assignment into $\{T, F\}$ such that $a(p) = a'(p)$ if $a(p) \in \{T, F\}$, then a' is said to be a *superassignment* of a. It turns out that if a is an assignment extended to a valuation function v_a for the Kleene matrix $\mathfrak{K}_3$ defined in Sect. 9.5.1, then for every formula A in the language of $\mathfrak{K}_3$, $v_a(A) = v_{a'}(A)$ if $v_a(A) \in \{T, F\}$. Therefore, the function $v_{a'}$ may be called a *supervaluation* of v_a. A formula is then said to be *supertrue* under a valuation function v_a for $\mathfrak{K}_3$ if it is true under every supervaluation $v_{a'}$ of v_a, i.e., if $v_{a'}(A) = T$ for every supervaluation $v_{a'}$ of v_a. The property of being *superfalse* is defined analogously.

Since every supervaluation is a classical valuation, every classical tautology is supertrue under every valuation function in $\mathfrak{K}_3$. Supervaluationism is, however, not truth-functional with respect to supervalues. The supervalue of a disjunction, for example, does not depend on the supervalue of the disjuncts. Suppose $a(p) = I$. Then $a(\neg p) = I$ and $v_{a'}(p \vee \neg p) = T$ for every supervaluation $v_{a'}$ of v_a. Whereas $(p \vee \neg p)$ is thus supertrue under $v_a, p \vee p$ is *not*, because there are superassignments a' of a with $a'(p) = F$. An argument against the charge that supervaluationism requires a non-truth-functional semantics of the connectives can be found in [161] (cf. also other references given there).

Although the possession of supertruth is preserved from the premises to the conclusion(s) of valid inferences in supervaluationism, and although it might be tempting to consider supertruth an abstract object on its own, it seems that it has never been suggested to hypostatize supertruth in this way, comparable to Frege's *the True*. A sentence supertrue under a three-valued valuation v just takes the Fregean value *the True* under every supervaluation of v. The advice not to confuse supertruth with "real truth" can be found in [24].

Before we will embark on an elaboration of a theory of generalized truth values and lattice-organized structures defined on sets of generalized truth values, let us in the next chapter turn our attention to one remarkable argument in favor of the existence of truth values.

Chapter 2
Truth Values and the Slingshot Argument

Abstract The famous “slingshot argument” developed by Church, Gödel, Quine, and Davidson is often considered to be a formally strict proof of the Fregean conception that all true sentences, as well as all false ones, have one and the same denotation, namely their corresponding truth value: *the True* or *the False*. In this chapter, we present several versions of the slingshot argument and examine some possible ways of analyzing the argument by means of Roman Suszko’s non-Fregean logic. We show that the language of non-Fregean logic can serve as a useful tool for reconstructing the slingshot argument and formulate several embodiments of the argument in non-Fregean logics. In particular, a new version of the slingshot argument is presented, which can be circumvented neither by an appeal to a Russellian theory of definite descriptions nor by resorting to an analogous “Russellian” theory of λ-terms.

2.1 An Argument in Favor of Truth Values

There is a famous argument (or more precisely, a family of arguments) that is designed to provide a formally strict proof of the claim that all true sentences designate (denote, refer to) one and the same thing, as do all false sentences. These things are precisely the truth values: *the True* and *the False*. The argument has already been anticipated (implicitly at least) by Frege, who employs the principle of *substitutivity* of co-referential terms, according to which the reference of a complex singular term must remain unchanged when any of its sub-terms is replaced by an expression having the same reference. Frege asks:

> What else but the truth value could be found, that belongs quite generally to every sentence if the reference of its components is relevant, and remains unchanged by substitutions of the kind in question? [116, p. 64]

Y. Shramko and H. Wansing, *Truth and Falsehood*, Trends in Logic 36,
DOI: 10.1007/978-94-007-0907-2_2, © Springer Science+Business Media B.V. 2011

The idea underlying this question has been neatly reconstructed by Alonzo Church, who has explicitly formulated the corresponding argument in his review of Carnap's *Introduction to Semantics* [52]. In *Introduction to Mathematical Logic* [53] Church reconstructs the point of his proof by means of a rather informal line of reasoning. Other remarkable versions of the argument are those by Gödel [121] and Davidson [60], which make use of the formal apparatus of a theory of definite descriptions.

Stated generally, the pattern of the argument goes as follows (cf. [191]). One starts with a certain sentence, and then moves, step by step, to a completely different sentence. Every two sentences in any step designate presumably one and the same thing. Hence, the starting and the concluding sentences of the argument must have the same designation as well. But the only semantically significant thing they have in common seems to be their truth value. Thus, what any sentence designates is just its truth value.

Barwise and Perry [18] have dubbed this family of arguments "the slingshot", thus stressing its extraordinary simplicity and the minimality of presuppositions involved. Versions of the slingshot argument have been analyzed in detail by many authors, see, e.g., [66, 94, 171, 177, 185, 189, 191, 214, 263, 278], and especially the comprehensive study by Neale [178].

2.2 Reconstructing the Slingshot Arguments

2.2.1 Church's Slingshot

We first pose the argument as it was originally formulated by Church in [52]. Note that the slingshot argument in all its forms rests essentially on the assumption that every sentence normally has a designation. Another important assumption is the principle of *substitutivity* for co-referential terms in any expression:

(S) If we convert a linguistic expression into another expression by substituting any of its terms for a term with exactly the same designation, the resulting and the initial expressions also designate the same.

In particular, in any sentence, expressions with the same designation can be unrestrictedly substituted for one another without changing the designation of the sentence in which the substitution is made. This is actually just an instance of the compositionality principle mentioned in Sect. 1.1. Church formulates it in terms of "synonymy", saying that synonymous expressions are interchangeable, where two expressions are considered to be synonymous if and only if they have the same designation. Moreover, Church assumes (following Carnap, [47, p. 92]) that any two logically equivalent sentences are synonymous. He then employs a language with an abstraction operator λx such that for any sentence X, $\lambda x(X)$ means "the class of all x such that X".

Now let A and B be any true sentences. Consider the following five sentences:

C1. A
C2. $\lambda x(x = x \wedge \sim A) = \lambda x(x \neq x)$
C3. $\lambda x(x \neq x) = \lambda x(x \neq x)$
C4. $\lambda x(x = x \wedge \sim B) = \lambda x(x \neq x)$
C5. B

Church argues that all these sentences are synonymous. Indeed, C1 and C2 are logically equivalent, as well as C4 and C5. But C2, C3, and C4 are also synonymous by (S), since $\lambda x(x = x \wedge \sim A)$ and $\lambda x(x \neq x)$ have the same designation, namely the null class. But note again that A and B have been chosen absolutely arbitrarily, and the only (semantically relevant) feature they have in common is that they are both true. "Hence", Church concludes, "we have a means of showing any two true sentences to be synonymous. By a similar method any two false sentences can be shown to be synonymous. Therefore finally no possibility remains for the designata of sentences except that they be truth-values" [52, p. 300].

In [53, pp. 24–25] Church develops a "popularized" version of his argument which is by now rather widespread in the literature. Namely, he suggests to take a look at the following sequence of four sentences:

C1′. Sir Walter Scott is the author of *Waverley*.
C2′. Sir Walter Scott is the man who wrote 29 *Waverley* novels altogether.
C3′. The number, such that Sir Walter Scott is the man who wrote that many *Waverley* novels altogether, is 29.
C4′. The number of counties in Utah is 29.

Note that this sequence does not represent a logical inference (although it is not difficult to re-articulate the argument as a formal proof by applying suitable technical machinery). It is rather a number of conversion steps each producing co-referential sentences. It is claimed that C1′ and C2′ have the same designation by substitutivity, for the terms 'the author of *Waverley*' and 'the man who wrote 29 *Waverley* novels altogether' designate one and the same object, namely Walter Scott. And so have C3′ and C4′ because the number, such that Sir Walter Scott is the man who wrote that many *Waverley* novels altogether, is the same as the number of counties in Utah, namely 29. The step from C2′ to C3′ is justified by what Perry [191] calls *redistribution*: "rearrangement of the parts of a sentence does not effect what it designates, as long as the truth conditions remain the same". This principle may seem controversial, and Barwise and Perry [18] in fact reject it. (It must be said that they reject substitutivity as well.) Church himself argues that it is plausible to suppose that C2′, even if not synonymous with C3′, is at least so close to C3′ "so as to ensure its having the same denotation". If this is indeed the case, then C1′ and C4′ must have the same denotation (designation). But the only (semantically relevant) thing these sentences have in common is that both are true. Thus, taking it that there must be something which the sentences designate, one concludes that it is just their truth value. As Church remarks, a parallel example involving false sentences can be constructed in the same way (by considering, e.g., 'Sir Walter Scott is not the author of *Waverley*').

Church in [52] used his argument to demonstrate that sentences could not designate propositions as Carnap assumed in [47]. It seems that Carnap himself found the argument convincing enough: in his next book [48, p. 26] he *did* postulate truth values as "extensions" of sentences and, moreover, provided independent reasons for this.

2.2.2 Gödel's Slingshot

Gödel in [121, pp. 128–129] highlights the impact of various theories of descriptions on the problem of what sentences designate. According to him, if—in addition to substitutivity—we take the apparently obvious view that a descriptive phrase denotes the object described, then the conclusion that "all true sentences have the same signification (as well as all false ones)" is almost inevitable. Gödel hints at a "rigorous proof" of this claim by making use of some further assumptions. Let $(\iota x)(x = a \wedge Fx)$ stand for the definite description 'the x such that x is identical to a and x is F', and let for any sentence X, $[X]$ stand for what X designates. Then Gödel's assumptions can be articulated as follows:

(A1) $[Fa] = [a = (\iota x)(x = a \wedge Fx)]$.
(A2) Every sentence can be transformed into an equivalent sentence of the form Fa. (This assumption allows Gödel to expand his argument beyond the atomic sentences, cf. [177, p. 778].)

We will next reconstruct Gödel's proof in the form of an "official" logical inference. (This reconstruction stems essentially from [177, pp. 777–779, 789], although our formulation is somewhat different.) Making use of the introduced notation, we are going to prove that for any true sentences A and B, $[A] = [B]$. To do this, we will need an additional rule of inference governing the description operator:

$$\iota\text{-INTR: } \frac{A(x/a)}{a = (\iota x)(x = a \wedge A(x))}$$

where a is a singular term, $A(x)$ is a sentence containing at least one free occurrence of the variable x, and $A(x/a)$ is the result of replacing every occurrence of x in $A(x)$ with a.

Note the close connection between this rule and (A1). As Neale [177, p. 789] observes, ι-INTR (as well as its counterpart rule ι-ELIM, which goes in the opposite direction) must be a valid rule of inference in any (extensional) theory of descriptions (as it is in Russell's).

Another inference rule, which allows substitution of definite descriptions, is also taken from [177, p. 787]:

$$\iota\text{-SUB:}\quad \frac{(\iota x)\phi = (\iota x)\psi \quad A((\iota x)\phi)}{A((\iota x)\psi)} \quad \frac{(\iota x)\phi = a \quad A((\iota x)\phi)}{A(a)} \quad \frac{(\iota x)\phi = a \quad A(a)}{A((\iota x)\phi)}.$$

With this machinery at hand, suppose the sentences G1–G3 below are true.

G1. Fa
G2. $a \neq b$
G3. Gb

Then one can proceed as follows:

G4. $a = (\iota x)(x = a \wedge Fx)$ G1, ι-INTR
G5. $a = (\iota x)(x = a \wedge x \neq b)$ G2, ι-INTR
G6. $b = (\iota x)(x = a \wedge Gx)$ G3, ι-INTR
G7. $b = (\iota x)(x = b \wedge x \neq a)$ G2, ι-INTR
G8. $(\iota x)(x = a \wedge Fx) = (\iota x)(x = a \wedge x \neq b)$ G4, G5, ι-SUB
G9. $(\iota x)(x = b \wedge Gx) = (\iota x)(x = b \wedge x \neq a)$ G6, G7, ι-SUB
G10. $[Fa] = [a = (\iota x)(x = a \wedge Fx)]$ (A1)
G11. $[a \neq b] = [a = (\iota x)(x = a \wedge x \neq b)]$ (A1)
G12. $[Fa] = [a = (\iota x)(x = a \wedge x \neq b)]$ G8, G10, ι-SUB
G13. $[Fa] = [a \neq b]$ G11, G12, *Transitivity of* $=$
G14. $[Gb] = [b = (\iota x)(x = b \wedge Gx)]$ (A1)
G15. $[a \neq b] = [b = (\iota x)(x = b \wedge x \neq a)]$ (A1)
G16. $[Gb] = [b = (\iota x)(x = b \wedge x \neq a)]$ G9, G14, ι-SUB
G17. $[Gb] = [a \neq b]$ G15, G16, *Transitivity of* $=$
G18. $[Fa] = [Gb]$ G13, G17, *Transitivity of* $=$

The inference can be easily repeated if instead of G2 we take its negation $\neg(a \neq b)$. That is, in any case (in view of the assumption that sentences do designate) Fa and Gb must have one and the same designation. But Fa and Gb may be completely different sentences having nothing (of semantic relevance) in common, except that they are both true, as had been assumed. Then, taking into account the assumption (A2), we can easily extend this claim to any true sentences A and B.

The paper in which Gödel put forward his argument was published in a volume from the "Library of Living Philosophers" devoted to Bertrand Russell. As is well known, Russell held that any true sentence stands for a fact. In this case the argument above would demonstrate that all true sentences stand for one and the same fact, thus reducing Russell's view to an absurdity.

Therefore, the slingshot argument sometimes has been characterized as a "collapsing argument", for what it admittedly shows is that there are fewer entities of a given kind than one might suppose [177, p. 761]. In this sense the argument can be equally used to show that sentences do not designate situations, states of affairs or anything of this sort; any attempt to assume so will result in a breakdown of the class of supposed designata "into a class of just two entities (which might as well be called "Truth" and "Falsity")" [177, p. 761]. Another famous argument of this kind is the one by Quine (see [200, 201]), by which he intended to demonstrate that quantifying into modal contexts leads to a collapse of modality.

By the way, as Gödel, in contrast to Church and Davidson, offers only a rough outline of his argument, one can find in the literature several different reconstructions of Gödel's supposedly authentic proof, and it is not easy to decide which one of them reproduces Gödel's line of reasoning more accurately than the others. However, such reconstructions, being different in the assumptions involved in or in their technical implementation, all end up with the same conclusion. For example, if one comes back to Church and takes instead of (A1) a more general assumption:

(A3) If A and B are *logically equivalent*, then $[A] = [B]$,

then one obtains another, very simple (even simplified) version of the slingshot which—though not exactly Gödelian—is clearly inspired by Gödel's ideas. Moreover, this version (see, e.g., [160]) is to some degree intermediate between Gödel's original argument and the ones developed by Church and Davidson.

Namely, let R and T be any true sentences and a be some singular term which has a designation. Then we have the following four sentences which all have the same designation:

S1. R
S2. $a = (\iota x)(x = a \wedge R)$
S3. $a = (\iota x)(x = a \wedge T)$
S4. T

Indeed, S1 and S2 are logically equivalent. The same holds for S3 and S4. Now, because $(\iota x)(x = a \wedge R)$ and $(\iota x)(x = a \wedge T)$ designate one and the same object, namely a, by substitutivity, S2 and S3 also designate the same object. Hence, S1 and S4 must have the same designation as well. Quod erat demonstrandum.

2.2.3 Davidson's Slingshot

Davidson used the "slingshot" to undermine the view that true sentences correspond to the facts. In [60, pp. 305–306] he explicitly enunciates assumptions needed for the argument: "that logically equivalent singular terms have the same reference; and that a singular term does not change its reference if a contained singular term is replaced by another with the same reference".[1]

[1] While talking about "logically equivalent singular terms", Davidson, like Frege, treats declarative sentences as singular terms.

Then Davidson considers any two sentences S and R alike in truth values and argues that the following four sentences must all have the same designation:

D1. S
D2. $(\iota x)(x = x \wedge S) = (\iota x)(x = x)$
D3. $(\iota x)(x = x \wedge R) = (\iota x)(x = x)$
D4. R

(The sentences D1, D2 as well as the sentences D3, D4 are pairwise logically equivalent and hence co-referential. And taking into account the co-referentiality of the terms $(\iota x)(x = x \wedge S)$ and $(\iota x)(x = x \wedge R)$, D2 and D3 are also co-referential.)

Thus, Davidson maintains, if we wish to claim that sentences stand for facts, we are forced to admit that all true sentences refer to one and the same fact, which Davidson [61], nimbly enough, calls *The Great Fact*. This conclusion is often employed to make a case against the correspondence theory of truth. The idea is that facts—when related to a sentence—appear non-localizable, and thus any true sentence seems to correspond to the whole universe rather than to some of its "parts". As it was suggested by Lewis [151, p. 242], a proposition refers not to some limited state of affairs, but to the "kind of *total* state of affairs we call a world". And further: "All *true* propositions have the same extension, namely, this actual world; and all *false* propositions have the same extension, namely, zero-extension." Such an understanding is eminently congenial to the Fregean account of truth values.

2.3 The Slingshot Argument and Non-Fregean Logic

In [288], Wójtowicz aims to reassess the slingshot argument by translating it into the formal language of Roman Suszko's non-Fregean logic [245] extended by the ι-operator (or some suitable abstraction operator). Interestingly, she interprets the results of her analysis in the sense that the slingshot argument is circular and presupposes what it intends to prove. In [235] we have examined this claim by Wójtowicz and have shown that it is untenable. Nevertheless, as we believe, the very idea of employing non-Fregean logic to the analysis of the slingshot has merit by itself, since the technical apparatus of non-Fregean logic provides a natural and suitable framework for formalizing and investigating various versions of the argument. It should also be noted that Wójcicki in [286] has also drawn attention to the possibility of analyzing Church's argument by means of Suszko's non-Fregean logic.

In addition to the vocabulary of classical first-order logic (with the identity predicate), the language of non-Fregean logic contains a binary *identity connective* $\equiv$. Intuitively, a formula $A \equiv B$ states that the sentences A and B denote (or describe) the same situation.[2] Note that if one uses the notation employed in Sect. 2.2.2, this can be expressed as $A \equiv B$ iff $[A] = [B]$.

Suszko was very much animated by the idea of "abolishing" the so-called *Fregean Axiom* (see [245]) which is formulated as follows:

$$(A \leftrightarrow B) \rightarrow (A \equiv B). \qquad (FA)$$

That is, if A and B are materially equivalent, then they describe the same situation. Taking into account the material equivalence of all true sentences, as well as of all false sentences, this amounts to the claim that there are but two situations, one standing for all true sentences ("the True"), and another for all false ones ("the False") (see [285, p. 327]). Suszko calls the principle that "all true (and similarly all false) sentences describe the same state of affairs, that is, they have a common referent" the *semantic version of the Fregean Axiom* (see [187, p. 21]).

Generally, a non-Fregean logic (**NFL**) is a logic which does not validate (FA). On the other hand, the slingshot argument can actually be considered as a reasoning just for the substantiation of (FA), either in its syntactic or in its semantic form. Taking into account a direct correlation between (FA) and the conclusion of the slingshot, it may even be said that the very crux of this argument is to demonstrate that the Fregean Axiom is in fact not an axiom but can be strictly *proved* (by making a few quite natural assumptions). Therefore, the idea to employ the framework of **NFL** for an analysis of these assumptions and the argument as a whole looks quite reasonable.

NFL can be axiomatized in various ways. For example, one obtains the system **PCI** (predicate calculus with the identity connective) by adding to classical first-order logic with the identity predicate the following axiom schemata:

Ax1. $A \equiv B$ (where A and B vary at most in bound variables)
Ax2. $(A \equiv B) \rightarrow (C[A] \equiv C[A/B])$
Ax3. $(A \equiv B) \rightarrow (A \leftrightarrow B)$
Ax4. $(x_1 = y_1 \wedge \ldots \wedge x_n = y_n) \rightarrow (Px_1, \ldots, x_n \equiv Py_1, \ldots, y_n)$
Ax5. $\forall x(A \equiv B) \rightarrow (QxA \equiv QxB)$ (where $Q \in \{\forall, \exists\}$)

Note that the schemata $A \equiv B \rightarrow B \equiv A$ and $((A \equiv B) \wedge (B \equiv C)) \rightarrow (A \equiv C)$ are provable with the help of Ax2 (see [285, p. 325]).

Wójcicki has shown that the only identities in **PCI** are so-called "trivial identities", i.e., expressions of the form $A \equiv B$, where A and B differ at most with

[2] This reading rests on an original reading proposed by Suszko: "The proposed reading of the formula $p \equiv q$ is: that p is the same as that q, or the situation that p is the same as the situation that q" [244, p. 8]. It explicitly involves the expression 'the situation that A', which is fully in accord with Suszko's approach to semantics in terms of situations. Belnap [25] has remarked in this connection that the expression 'the situation that A' seems not to be customary in English. It is more common in the slingshot literature to use the expression 'the fact that A'. Nevertheless, Belnap also considers it acceptable to use as equivalents such readings of $A \equiv B$ as "the fact that A is the same as the fact that B", "the proposition that A is the same as the proposition that B" and "the situation described by the sentence A is the same as the situation described by the sentence B".

respect to the form of bound variables. Therefore it is sometimes desirable to consider some strengthening of **PCI** to make the non-Fregean logic more interesting. For example, if we wish to validate the assumption (A3) above what is needed for some versions of the slingshot argument, we can consider system **WBQ**,[3] which is formulated as follows:

$$\mathbf{WBQ} = \mathbf{PCI} \cup \{A \equiv B : A \leftrightarrow B \text{ is a theorem of classical predicate logic}\}.$$

Moreover, formulas A and B of the additional identities $A \equiv B$ added to **PCI** should be $\equiv$−free. That is, **WBQ** is obtained by adding to **PCI** the following inference rule (CL) for all $\equiv$−free formulas A and B:

$$\frac{\vdash A \leftrightarrow B}{\vdash A \equiv B}$$

Note that (*FA*) is still not a theorem of **WBQ**. Clearly, (A3) is expressible in **WBQ** and in fact is provable in it. Ax4 corresponds to the principle of substitutivity of co-denoting terms. However, these principles are not enough to prove the Fregean Axiom.

Consider the "intermediate" (Gödel–Davidson) version of the slingshot argument S1–S4 above. It involves the following basic thesis concerning the definite description operator:

(i) A is logically equivalent to $a = (\iota x)(x = a \wedge A)$.

Thus, we have to add the ι-operator to the language of **WBQ** to make (i) true and to accept some axiom schemes governing this operator. If we have in mind some kind of referential definite descriptions, the following principle looks quite natural:

($\star$) $(\iota x)(A(x) \wedge x = a) = a \leftrightarrow A(x/a)$, where $A(x)$ is a formula perhaps containing free occurrences of x, and $A(x/a)$ is the result of replacing every occurrence of x in $A(x)$ by a.

Notice that if $A(x)$ contains no free occurrences of the variable x, then $A(x)$ (as well as $A(x/a)$) is just A, and ($\star$) "degenerates" into $(\iota x)(A \wedge x = a) = a \leftrightarrow A$. One might wonder whether the condition attached to ($\star$) can (and ought to) be strengthened by stipulating that $A(x)$ must *actually* contain *at least one* free occurrence of x (cf. the formulation of the ι-INTR rule above). However, as Dunn has remarked in [78, pp. 351–352] (concerning the λ-abstractor), such a restriction in effect would be an "empty gesture", because, e.g., of the following equivalence:

$$A \leftrightarrow A \wedge (A \vee B(x)).$$

3 As Omyła [186, p. 13] explains, 'W' stands for Wittgenstein, 'B' for Boolean algebra, and 'Q' for quantifiers.

This equivalence allows one to “dummy in” occurrences of variables and to derive easily the “degenerate case” as follows: $(\iota x)(A \wedge x = a) = a \leftrightarrow (\iota x)((A \wedge (A \vee B(x))) \wedge x = a) = a \leftrightarrow (A \wedge (A \vee B(a))) \leftrightarrow A$. But notice that it would be impossible to derive $(\bigstar)$ in full generality from its degenerate version taken as a primitive.

Let us label with $\mathbf{WBQ}^{\iota}$ the system obtained from **WBQ** by merely extending its language with the ι-operator. That is, ι is added to the language of **WBQ**, but no further principles for ι are assumed, and in particular $(\bigstar)$ is not presupposed. Then we can observe that $(\bigstar)$ is not equivalent to (FA) as the following theorem reveals:

Theorem 2.1 *$(FA) \rightarrow (\bigstar)$ is not a theorem of* $\mathbf{WBQ}^{\iota}$.

Proof It is enough to demonstrate that by adding (FA) to $\mathbf{WBQ}^{\iota}$, $(\bigstar)$ remains unprovable. And this is indeed the case, for $\mathbf{WBQ} + (FA)$ gives us classical logic, and $(\bigstar)$ is not provable even in certain not merely linguistic extensions of classical logic, for example within Russell’s theory of descriptions. All the more so it is not provable in classical logic to which the ι-operator is added only as a purely linguistic extension. (Note that $\mathbf{WBQ}^{\iota} + (FA) = \mathbf{WBQ} + (FA)^{\iota}$.) □

Let $\mathbf{L}+(\bigstar)$ stand for a non-Fregean logic **L** with the axiom scheme $(\bigstar)$, which is not weaker than $\mathbf{WBQ}^{\iota}$. Then, the following key theorem puts the slingshot argument into the context of **NFL**:

Theorem 2.2 *(FA) is provable in any non-Fregean logic $\mathbf{L}+(\bigstar)$ which is not weaker than* $\mathbf{WBQ}^{\iota}$.

Proof We first show that $(A \wedge B) \rightarrow (A \equiv B)$ is provable in $\mathbf{L}+(\bigstar)$. We use the Deduction Theorem:

1. $A \wedge B$ *(Assumption)*
2. A *($\wedge$-Elimination)*
3. B *($\wedge$-Elimination)*
4. $A \leftrightarrow a = (\iota x)(A \wedge (x = a))(\bigstar)$
5. $B \leftrightarrow a = (\iota x)(B \wedge (x = a))(\bigstar)$
6. $a = (\iota x)(A \wedge (x = a))$ (2, 4, *modus ponens (MP)*)
7. $a = (\iota x)(B \wedge (x = a))(3, 5, MP)$
8. $(\iota x)(A \wedge (x = a)) = (\iota x)(B \wedge (x = a))(6, 7, \textit{Transitivity of} =)$
9. $(\iota x)(A \wedge (x = a)) = (\iota x)(B \wedge (x = a)) \rightarrow a = (\iota x)(A \wedge (x = a)) \equiv a = (\iota x)(B \wedge (x = a))$ (Ax4)
10. $a = (\iota x)(A \wedge (x = a)) \equiv a = (\iota x)(B \wedge (x = a))(8, 9, MP)$
11. $A \equiv a = (\iota x)(A \wedge (x = a))(4, CL)$
12. $A \equiv a = (\iota x)(B \wedge (x = a))(10, 11, \textit{Transitivity of} \equiv)$
13. $B \equiv a = (\iota x)(B \wedge (x = a))(5, CL)$
14. $A \equiv B(12, 13, \textit{Transitivity of} \equiv)$

Next, we show that $(\sim A \wedge \sim B) \rightarrow (A \equiv B)$ is also provable in $\mathbf{L}+(\bigstar)$:

1. $\sim A \wedge \sim B$ *(Assumption)*
2. $\sim A$ (1, *$\wedge$-Elimination*)

3. $\sim B$ (1, $\wedge$-*Elimination*)
4. $\sim A \leftrightarrow a = (\iota x)(\sim A \wedge (x = a))$ ($\star$)
5. $\sim B \leftrightarrow a = (\iota x)(\sim B \wedge (x = a))$ ($\star$)
6. $a = (\iota x)(\sim A \wedge (x = a))$ (2, 4, *MP*)
7. $a = (\iota x)(\sim B \wedge (x = a))$ (3, 5, *MP*)
8. $(\iota x)(\sim A \wedge (x = a)) = (\iota x)(\sim B \wedge (x = a))$(6, 7, *Transitivity of* $=$)
9. $\sim A \equiv a = (\iota x)(\sim A \wedge (x = a))$ (4, *CL*)
10. $\sim B \equiv a = (\iota x)(\sim B \wedge (x = a))$ (5, *CL*)
11. $A \equiv \sim(a = (\iota x)(\sim A \wedge (x = a)))$ (9, Ax2)
12. $B \equiv \sim(a = (\iota x)(\sim B \wedge (x = a)))$ (10, Ax2)
13. $(\iota x)(\sim A \wedge (x = a)) = (\iota x)(\sim B \wedge (x = a)) \rightarrow \sim(a = (\iota x)(\sim A \wedge (x = a))) \equiv \sim(a = (\iota x)(\sim B \wedge (x = a)))$ (Ax4)
14. $\sim(a = (\iota x)(\sim A \wedge (x = a))) \equiv \sim(a = (\iota x)(\sim B \wedge (x = a)))$ (8, 13, *MP*)
15. $A \equiv B$ (11, 12, 14, *Transitivity of* $\equiv$)

Thus, we have both $\vdash (A \wedge B) \rightarrow (A \equiv B)$ and $\vdash (\sim A \wedge \sim B) \rightarrow (A \equiv B)$. Hence, $\vdash (A \wedge B) \vee (\sim A \wedge \sim B) \rightarrow (A \equiv B)$, and consequently, $\vdash (A \leftrightarrow B) \rightarrow (A \equiv B)$. □

This proof shows that it is sufficient to have ($\star$) for incorporating the slingshot argument into the context of a non-Fregean logic. We have then the following corollary:

Corollary 2.1 ($\star$) $\rightarrow$ (*FA*) *is a theorem of* $\mathbf{WBQ}^{\iota}$.

This corollary expresses exactly the idea already articulated by Gödel in [121] (see Sect. 2.2.2), namely that accepting a certain kind of definite descriptions (together with some other assumptions) is sufficient to get the Fregean conception of "the True" and "the False" as the only possible denotations for sentences. As we will see in the next section, the condition imposed on such a description operator allows further weakening.

2.4 Non-Fregean Logic and Definite Descriptions

Theorem 2.2 and Corollary 2.1 exhibit the minimal assumptions needed for obtaining a certain version of the slingshot argument. Moreover, it turns out that the technical apparatus of **NFL** provides a natural and suitable framework for formalizing and investigating other variations of the argument. For example, it allows to adequately represent the genuine assumption (A1) made by Gödel on an object language level. Indeed, Gödel's version of the slingshot does not presuppose the co-referentiality of logically equivalent sentences, and correspondingly, Gödel need not postulate, e.g., the *logical* equivalence of 'Socrates is wise' and 'Socrates is the object which is wise and is identical with Socrates'. Gödel merely admits that these expressions "mean the same thing" [121, p. 129], which Neale interprets in the sense that they "stand for the same fact" [177, p. 777]. As Suszko would presumably say, they denote (or describe) the same situation.

The language of a non-Fregean logic makes it possible to directly express the idea behind Gödel's assumption (A1). Let us define **PCI**+(•) as the system obtained from the system **PCI** (see Sect. 2.3) by extending its language with the ι-operator which is subject to the following scheme:

(•) $(\iota x)(A(x) \wedge x = a) = a \equiv A(x/a)$, where $A(x)$ is a formula perhaps containing free occurrences of x, and $A(x/a)$ is the result of replacing every occurrence of x in $A(x)$ with a.

This scheme, being very similar to (★), does not express the logical equivalence of two expressions, but only the identity of the corresponding situations. Moreover, (•) generalizes the Gödelian assumption (A1) in a twofold respect: first, it deals with arbitrary formulas (maybe containing x and a) and not only with atomic sentences, and second, it allows x and a not to occur in A at all.

Notice that (•) is derivable in $\mathbf{WBQ}^{\iota}$+(★) by applying the *CL*-rule to ★. Yet, the system **WBQ** might be considered problematic from a rigorous "non-Fregean standpoint", since its characteristic rule *CL* comes pretty near to the Fregean Axiom. So, is it maybe not the scheme for the ι-operator alone but also the rule *CL* that bears responsibility for the validity of the slingshot argument? Some researchers (see, e.g. [18, 220]) indeed reject the view that logically equivalent sentences must have the same denotation and hope to reject in this way the slingshot argument as well.

It is therefore worth noting that (•) allows the construction of a kind of slingshot argument even on the basis of the weakest non-Fregean logic **PCI**.

Theorem 2.3 *Within* **PCI**+(•) *denotations of any true sentences are identical, i.e., any two true sentences have one and the same denotation.*

Proof Assume that we have two arbitrary *true* sentences A and B. Then we can outline the following proof:

1. A *(Assumption)*
2. B *(Assumption)*
3. $A \equiv a = (\iota x)(A \wedge (x = a))$(•)
4. $B \equiv a = (\iota x)(B \wedge (x = a))$(•)
5. $A \leftrightarrow a = (\iota x)(A \wedge (x = a))$ (3, Ax3)
6. $B \leftrightarrow a = (\iota x)(B \wedge (x = a))$ (4, Ax3)
7. $a = (\iota x)(A \wedge (x = a))$ (1, 5, *MP*)
8. $a = (\iota x)(B \wedge (x = a))$ (2, 6, *MP*)
9. $(\iota x)(A \wedge (x = a)) = (\iota x)(B \wedge (x = a))$ (7, 8, *Transitivity of* =)
10. $(\iota x)(A \wedge (x = a)) = (\iota x)(B \wedge (x = a)) \rightarrow a = (\iota x)(A \wedge (x = a)) \equiv a = (\iota x)$ $(B \wedge (x = a))$ (Ax4)
11. $a = (\iota x)(A \wedge (x = a)) \equiv a = (\iota x)(B \wedge (x = a))$ (9, 10, *MP*)
12. $A \equiv a = (\iota x)(B \wedge (x = a))$ (3, 11, *Transitivity of* ≡)
13. $A \equiv B$ (4, 12, *Transitivity of* ≡) □

In terms of situations, Theorem 2.3 means that any two true sentences describe one and the same situation, and hence there is exactly one situation that stands for *all* true sentences. Above we called this result "*a kind of* slingshot argument", because it expresses merely one half of the argument, namely the one concerning the *true* sentences. Still, the trivializing effect even of this "positive slingshot" is rather tremendous. Note that (•) is the only kind of non-trivial identities within **PCI**+(•). At the same time, (•) represents the minimal principle for any theory of descriptions in which definite descriptions are supposed to be singular terms that denote objects possessing some characteristics. Thus, Theorem 2.3 tells us that extending even the weakest non-Fregean logic with the minimal theory of descriptions of a certain kind immediately stultifies any attempt to differentiate between situations described by true sentences in this logic. And if we take into account that situations described by true sentences are *facts*, then it becomes clear that **PCI**+(•) succumbs to the destructive force of *The Great Fact* argument advocated by Davidson.

As to the *false* sentences, things are more intricate. To show that within non-Fregean logic all false sentences are coreferential, we have to assume $\sim A$ and $\sim B$ and deduce from these assumptions $A \equiv B$. However, *mutatis mutandis* as in Theorem 2.3 we can deduce only $\sim A \equiv \sim B$, but then the deduction process stops. Within **PCI**+(•) we cannot proceed further to $A \equiv B$, since neither

$$(\sim A \equiv \sim B) \rightarrow (A \equiv B) \tag{N}$$

is a theorem of **PCI**, nor

$$\frac{\sim A \equiv \sim B}{A \equiv B}$$

is a derivable rule of inference in **PCI**. Note, however, that this rule, as well as an even more general one:

$$\frac{C[A] \equiv C[A/B]}{A \equiv B}$$

is admissible in **PCI** in the sense that being added to **PCI**, it does not enlarge the set of theorems, i.e., $A \equiv B$ is always a theorem of **PCI** whenever $C[A] \equiv C[A/B]$ is a theorem. Indeed, suppose that we have a proof of $C[A] \equiv C[A/B]$. By Wójcicki's result, $C[A]$ and $C[A/B]$ differ at most with respect to bound variables. Hence, also A and B differ at most with respect to bound variables and therefore, by Ax1, $(A \equiv B)$. Unfortunately, in the present case the mere admissibility does not help us further, since we are reasoning here from assumptions and not only from theorems.

Thus, in pure **PCI** if A and B are arbitrary *false* sentences, this only implies the identity of the corresponding *negative facts*,[4] but does not demand the identity of the situations described by A and B themselves.

---

[4] Note that Suszko, by developing his non-Fregean logic, aimed to incorporate some basic ideas of Ludwig Wittgenstein's "Tractatus Logico-Philosophicus" concerning the interrelations between language and the world. According to Suszko's interpretation of these ideas, the reality consists of *situations* which are denoted (or described) by sentences. True sentences stand for

To obtain the latter effect, we need a certain strengthening of **PCI**. Namely, consider system $\mathbf{PCI}^{\mathrm{N}} = \mathbf{PCI} + (N)$. The principle represented by (*N*) embodies a basic assumption concerning interrelations between the negation operator and sentential identity and is generally unproblematic from the point of view of **NFL**. First, this principle is provable in **WBQ**. Second, it is accepted in most of the theories based on non-Fregean logic. For example, S. Bloom and R. Suszko consider in [37, pp. 303–307] "a particular **SCI** theory" (where **SCI** is the propositional part of **PCI**). Among the theorems of this theory, we find $\sim\sim A \equiv A$, from which one easily obtains (*N*) by using Ax2. Suszko, in his pioneering paper [243], where non-Fregean logic has been initiated, constructs "the ontology of situations", which assumes the axiom

$$(Np \equiv Nq) \rightarrow (p \equiv q),$$

where *Np* can be read as "it is a negative fact that *p*" [243, p. 12]. This axiom duplicates (*N*) with negation taken as a sort of modality. Thus, $\mathbf{PCI}^{\mathrm{N}}$ represents some rather weak extension of **PCI** and lies clearly within the scope of non-Fregean logic. In any case, $\mathbf{PCI}^{\mathrm{N}}$ is much weaker than **WBQ**.

Now we can prove the following theorem[5]:

Theorem 2.4 *(FA) is provable in* $\mathbf{PCI}^{\mathrm{N}}+(\bullet)$.

Proof By Theorem 2.3 and using the Deduction Theorem, we have $\vdash (A \wedge B) \rightarrow (A \equiv B)$ and also $\vdash (\sim A \wedge \sim B) \rightarrow (\sim A \equiv \sim B)$. By (*N*), $\vdash (\sim A \wedge \sim B) \rightarrow (A \equiv B)$. Hence, $\vdash (A \wedge B) \vee (\sim A \wedge \sim B) \rightarrow (A \equiv B)$, and consequently, $\vdash (A \leftrightarrow B) \rightarrow (A \equiv B)$. □

It also turns out that the context of the underlying non-Fregean logic $\mathbf{PCI}^{\mathrm{N}}$ allows it to take seriously the restriction to an actual free occurrence of the variable *x* in the given formula. Let us consider the following principle, which is a weakening of $(\bullet)$:

$(\bullet')$ $(\iota x)(A(x) \wedge x = a) = a \equiv A(x/a)$, where $A(x)$ is a formula containing at least one free occurrence of *x*, and $A(x/a)$ is the result of replacing every occurrence of *x* in $A(x)$ by *a*,

---

(Footnote 4 continued)
facts. Facts are divided into the positive ones and the negative ones. "Situations which exist create *positive facts*, those which do not exist create *negative facts*" [285, p. 326], cf. [243, p. 8]. See also Proposition 2.06 of *Tractatus*: "We call the existence of states of affairs a positive fact, and their non-existence a negative fact". A false sentences (e.g., 'Snow is green') does not stand for a fact, and it denotes a situation which does not exist (that snow is green). There is also the corresponding negative fact (that snow is not green) denoted by the true sentence 'Snow is not green'. The non-existence of the situation described by the false sentence 'Snow is green' creates the negative fact described by the true sentence 'Snow is not green'.

5 In our paper [235] this theorem as well as Theorems 2.5 and 2.7–2.11, below were claimed to hold with respect to $\mathbf{PCI} + (\bullet)$. Unfortunately, we overlooked the fact that Ax2 holds only in one direction and we have to postulate additionally (*N*) to be able to obtain the desired results in full generality.

and let us define the corresponding system $\mathbf{PCI}^{\mathbf{N}}+(\bullet')$. Identities like $A \equiv A \wedge (A \vee B(x))$ are not theorems of $\mathbf{PCI}^{\mathbf{N}}$. Hence, in $\mathbf{PCI}^{\mathbf{N}}+(\bullet')$ it is impossible to introduce at will "dummy" free occurrences of variables. Nevertheless, the slingshot argument (in its Gödelian version) remains in force even in this system. Namely, consider the following variation of the Fregean Axiom:

$$(A(a) \leftrightarrow B(b)) \rightarrow (A(a) \equiv B(b)). \qquad (FA')$$

In (FA') we are still dealing with arbitrary sentences $A(a)$ and $B(b)$ with the mere restriction that these sentences must contain arbitrary individual constants a and b, respectively. Then the following theorem holds:

Theorem 2.5 *(FA') is provable in* $\mathbf{PCI}^{\mathbf{N}}+(\bullet')$.

Proof First, consider the following proof:

1. $A(a) \wedge B(b)$ (*Assumption*)
2. $a \neq b$ (*Assumption*)
3. $A(a)$ (1, *$\wedge$-Elimination*)
4. $B(b)$ (1, *$\wedge$-Elimination*)
5. $A(a) \equiv a = (\iota x)(A(x) \wedge x = a)$ $(\bullet')$
6. $B(b) \equiv b = (\iota x)(B(x) \wedge x = b)$ $(\bullet')$
7. $a \neq b \equiv a = (\iota x)(x \neq b \wedge x = a)$ $(\bullet')$
8. $a \neq b \equiv b = (\iota x)(a \neq x \wedge x = b)$ $(\bullet')$
9. $A(a) \rightarrow a = (\iota x)(A(x) \wedge x = a)$ (5, Ax3)
10. $B(b) \rightarrow b = (\iota x)(B(x) \wedge x = b)$ (6, Ax3)
11. $a \neq b \rightarrow a = (\iota x)(x \neq b \wedge x = a)$ (7, Ax3)
12. $a \neq b \rightarrow b = (\iota x)(a \neq x \wedge x = b)$ (8, Ax3)
13. $a = (\iota x)(A(x) \wedge x = a)$ (3, 9, *MP*)
14. $b = (\iota x)(B(x) \wedge x = b)$ (4, 10, *MP*)
15. $a = (\iota x)(x \neq b \wedge x = a)$ (2, 11, *MP*)
16. $b = (\iota x)(a \neq x \wedge x = b)$ (2, 12, *MP*)
17. $(\iota x)(A(x) \wedge x = a) = (\iota x)(x \neq b \wedge x = a)$ (13, 15, *Transitivity of* $=$)
18. $(\iota x)(B(x) \wedge x = b) = (\iota x)(a \neq x \wedge x = b)$ (14, 16, *Transitivity of* $=$)
19. $A(a) \equiv a = (\iota x)(x \neq b \wedge x = a)$ (5, 17, Ax4)
20. $A(a) \equiv a \neq b$ (7, 19, *Transitivity of* $\equiv$)
21. $B(b) \equiv b = (\iota x)(a \neq x \wedge x = b)$ (6, 18, Ax4)
22. $B(b) \equiv a \neq b$ (8, 21, *Transitivity of* $\equiv$)
23. $A(a) \equiv B(b)$ (20, 22, *Transitivity of* $\equiv$)

Clearly, if in this proof instead of Assumption 2 we take its negation $\sim(a \neq b)$, we get the same conclusion. Thus, this assumption can be eliminated, and by the Deduction Theorem, we get $\vdash A(a) \wedge B(b) \rightarrow A(a) \equiv B(b)$.

The proof can be repeated by taking $\sim A(x)$, $\sim B(x)$, $\sim A(a)$, $\sim B(b)$ instead of $A(x), B(x), A(a), B(b)$, and thus, by the Deduction Theorem, we have $\vdash (\sim A(a) \wedge \sim B(b)) \rightarrow (\sim A(a) \equiv \sim B(b))$. By Ax2 and (N), $\vdash (\sim A(a) \equiv \sim B(b)) \leftrightarrow (A(a) \equiv B(b))$, and hence, $\vdash (\sim A(a) \wedge \sim B(b)) \rightarrow (A(a) \equiv B(b))$.

Consequently, we obtain $\vdash (A(a) \wedge B(b)) \vee ({\sim}A(a) \wedge {\sim}B(b)) \rightarrow (A(a) \equiv B(b))$, and then, $\vdash (A(a) \leftrightarrow B(a)) \rightarrow (A(a) \equiv B(b))$. □

Suszko conceived his non-Fregean logic as a foundation for a theory of situations in a Wittgensteinean spirit. Since **PCI** is usually regarded the weakest (or the basic) system of non-Fregean predicate logic, and (*N*) clearly falls into a general pattern of this logic, Theorems 2.4 and 2.5 indicate that a theory of situations based on Suszko's non-Fregean logic, which also presupposes some very weak principle involving negation (represented by (*N*)), is in fact incompatible with any theory of descriptions in which 'Sir Walter Scott is the man who wrote *Waverley* and is identical with Walter Scott' means the same (and hence describes the same situation) as 'Sir Walter Scott wrote *Waverley*'.

2.5 Non-Fregean Logic and λ-Expressions

In this section we get back to the original idea by Church, and examine the effect of combining Suszko's non-Fregean logic with the λ-abstractor (λx). Indeed, it is possible to prove the analogues of Theorems 2.4 and 2.5 by using λ-terms instead of ι-terms. We first adopt a standard interpretation of a *lambda-term* $\lambda x A(x)$ as "the class of all x such that $A(x)$" [52, p. 299]. Then we can make some natural assumption (analogous to the corresponding principle taken for the ι-operator) about λ-terms and sentential, descriptive identity, which uses the λ-operator twice. Again, it is possible to consider two versions of this assumption:

- ($\circ$) $(\lambda x(A(x) \wedge x = a) = \lambda x(x = a)) \equiv A(x/a)$, where $A(x)$ is a formula *perhaps* containing free occurrences of x, and $A(x/a)$ is the result of replacing every occurrence of x in $A(x)$ with a.
- ($\circ'$) $(\lambda x(A(x) \wedge x = a) = \lambda x(x = a)) \equiv A(x/a)$, where $A(x)$ is a formula containing *at least one* free occurrence of x, and $A(x/a)$ is the result of replacing every occurrence of x in $A(x)$ with a.

We now consider the system **PCI**+($\circ$) in the first-order language with the λ-operator. Evidently the following theorem holds true:

Theorem 2.6 *In* **PCI**+($\circ$) *denotations of any true sentences are identical, i.e., any two true sentences have one and the same denotation.*

Furthermore, we consider the systems $\mathbf{PCI^N}$+($\circ$) and $\mathbf{PCI^N}$+($\circ'$) and easily obtain the following theorems:

Theorem 2.7 (*FA*) *is provable in* $\mathbf{PCI^N}$+($\circ$).

Theorem 2.8 (*FA*$'$) *is provable in* $\mathbf{PCI^N}$+($\circ'$).

The proofs of these theorems are completely analogous to the proofs of Theorems 2.3–2.5, respectively. Note that no specific assumptions about equality between λ-terms from the theory of λ-conversion are used in this formalized (λ-)versions of the slingshot argument.

Let us next add to the language of $\mathbf{PCI^N}$ the new predicate constant $\in$ and consider expressions of the form $a \in \lambda x A(x)$. This is fully in accord with the set-theoretical (extensional) understanding of the lambda terms accepted above. The expression $a \in \lambda x A(x)$ is then a formula, and one can adopt $(\lambda x A(x))a$ as an abbreviation for this formula. Clearly, the λ-expressions as such remain singular terms and can still be equated, that is, $\lambda x A(x) = \lambda x B(x)$ is a completely legitimate statement saying that the respective classes are the same (i.e., are co-extensional).

Let $A(x)$ now mean that A possibly contains some free occurrences of x. Thus, $A(x)$ may contain no occurrences of x at all; in such a case $A(x)$ is just A, and the corresponding lambda-expression $\lambda x A$ means simply "the class of all x such that A".

We then assume the following basic principle for $\lambda, \in$ and sentential identity (the $\equiv$-version of *λ-conversion* or *$\beta\eta$-equality*):

$$(\beta\eta) \quad (\lambda x A(x))a \equiv A(x/a).$$

In particular, if x does not occur in $A(x)$, then $(\lambda x A)a \equiv A$.

Clearly, if we allow an unrestricted applicability of λ-terms, then $(\beta\eta)$ turns out to be just an instance of the Comprehension Axiom, and, as a result, Russell's Paradox is easily derivable.[6] Therefore, to exclude expressions of the form $(\lambda x A(x))\lambda x A(x)$, $(\beta\eta)$ has to be suitably restricted as, e.g., in various versions of *typed lambda calculus*. This can be done in a way analogous to a system of simple types by introducing the notion of a *well-typed λ-expression*. We assume two kinds of types which can be assigned to expressions of our language: σ (the base type) and $\omega \mapsto \tau$ (a functional type, where ω and τ are any types). Then we define recursively:

Definition 2.1 (*Well-typed expressions*)

Base case. Any individual variable, individual constant, or well-formed formula with no occurrences of the λ-operator is well-typed and has type σ.

Abstraction formation. For every expression M of type τ and every variable x of type σ, the (well-formed) term $\lambda x M$ is well-typed and has type $\sigma \mapsto \tau$.

Application. If M is well-typed of type $\omega \mapsto \tau$ and N is well-typed of type ω, then $(M)N$ is well-typed and has type τ.

Extremal clause. No expression is well-typed unless it is obtained from either of the three clauses above.

Here we need not delve too much into the specifics of various typed lambda calculi; it is enough just to require that in $(\beta\eta)$ all expressions must be well-typed.

---

[6] Let $A(x)$ be $x \notin x$, and consider the term $t = \lambda x(x \notin x)$. Then, by $(\beta\eta)$, $t \in \lambda x(x \notin x) \equiv t \notin t$, and by definition of t, $t \in \lambda x(x \notin x) \equiv t \in t$. In pure "bracket notation": $(\lambda x(\sim(x)x))\lambda x(\sim(x)x) \equiv \sim(\lambda x(\sim(x)x))\lambda x(\sim(x)x)$.

We mark the principle subjected to this requirement as $(\beta\eta)^{\tau}$. Moreover, it is not necessary to define more exactly the kind of set-theory for the predicate $\in$ which is implicit in any formula of the form $(\lambda x A(x))a$. We again just assume that this predicate should be such that $(\beta\eta)^{\tau}$ holds for it.

Let system $\mathbf{PCI^N} + (\beta\eta)^{\tau}$ be obtained from $\mathbf{PCI^N}$ by extending its language with the $\in$-predicate and the λ-operator both subject to $(\beta\eta)^{\tau}$. Then we have the following theorem:

Theorem 2.9 *(FA) is provable in* $\mathbf{PCI^N} + (\beta\eta)^{\tau}$.

Proof We have the following outline of a proof:

1. $A \wedge B$ (*Assumption*)
2. A (1, $\wedge$-*Elimination*)
3. B (1, $\wedge$-*Elimination*)
4. $A \rightarrow \lambda x(\sim A) = \lambda x(x \neq x)(\mathbf{PCI^N} + (\beta\eta)^{\tau})$
5. $B \rightarrow \lambda x(\sim B) = \lambda x(x \neq x)(\mathbf{PCI^N} + (\beta\eta)^{\tau})$
6. $\lambda x(\sim A) = \lambda x(x \neq x)$ (2, 4, *MP*)
7. $\lambda x(\sim B) = \lambda x(x \neq x)$ (3, 5, *MP*)
8. $\lambda x(\sim A) = \lambda x(\sim B)$ (6, 7, *Transitivity of* $=$)
9. $\lambda x(\sim A) = \lambda x(\sim B) \rightarrow (\lambda x(\sim A))a \equiv (\lambda x(\sim B))a$ (Ax4)
10. $(\lambda x(\sim A))a \equiv (\lambda x(\sim B))a$ (8, 9, *MP*)
11. $\sim A \equiv \sim B$ (10, $(\beta\eta)^{\tau}$, *Transitivity of* $\equiv$)
12. $\sim A \equiv \sim B \leftrightarrow A \equiv B$ (Ax2, (*N*))
13. $A \equiv B$ (11, 12, *MP*)

Thus, by the Deduction Theorem, $\vdash (A \wedge B) \rightarrow (A \equiv B)$. *Mutatis mutandis* we get also $\vdash (\sim A \wedge \sim B) \rightarrow (A \equiv B)$. Hence, $\vdash (A \leftrightarrow B) \rightarrow (A \equiv B)$. □

Now one can remark (and even find it problematic) that $\mathbf{PCI^N} + (\beta\eta)^{\tau}$ is formulated not in a purely logical language but in fact represents an applied theory of the $\in$-predicate (some kind of set theory). For such a critical observer we propose considering another approach to the λ-operator, where it serves as a device which allows to manipulate arbitrary formulas into predicates. If we take $\lambda x A(x)$ as the characteristic function of the interpretation of the predicate $A(x)$, then $\lambda x A(x)$ may be treated not as a singular term which denotes a function, but as a predicate. If x is not among the free variables of A, then the interpretation of a predicate $\lambda x A(x)$ in a model is the individual domain of the model, if A is true in the model, and otherwise it is the empty set. In any case, an expression $(\lambda x A(x))a$ then is a formula. This kind of abstracting a predicate from an open formula is called *predicate abstraction* and was introduced into modal logic by Stalnaker and Thomason in the late 1960s, see [93, Chaps. 9 and 10], the references given there, and [78, Sect. 3].

The lambda-expression $\lambda x A(x)$ stands then for "the property of being (an x such that x is) A". This is a somewhat "intensional" reading which strictly parallels the set-theoretic interpretation taken above.[7] The formula $(\lambda x A(x))a$ can be read as "The property of being (an x such that x is) A applies to a" or "a possesses the property $\lambda x A(x)$". If $char_A$ is the characteristic function of the interpretation of $A(x)$, the formula may be understood as saying that $char_A(a) = 1$. Thus, the λ-operator is now treated as a predicate-forming operator and *not* as a term-forming operator.

The schema $(\beta\eta)$ can be taken as it stands without any typing machinery, for the very distinction between singular terms and predicates outlaws the unwanted expressions like $(\lambda x A(x))\lambda x A(x)$. The principle $(\beta\eta)$ may be read then as saying that 'The property of being (an x such that x is) A applies to a' describes the same situation as 'a is A'. Predicate abstraction enables scope distinctions in the presence of non-rigid singular terms. We here assume rigidly designating individual constants.

However, considering λ as a predicate-forming operator has as a result that the lambda-expressions so conceived cannot be equated in a first-order language. Such equating is crucial for constructing the "λ-version" of the slingshot argument presented in Theorems 2.7 and 2.8. Nevertheless, we may consider a second-order extension of $\mathbf{PCI^N}$.

Let the system $\mathbf{PCI^{N^2}} + (\beta\eta)$ be second-order $\mathbf{PCI^N}$ extended by the predicate abstracting λ-operator subject to $(\beta\eta)$, the second-order identity predicate $=^2$, and the following second-order version $\mathrm{Ax4}^2$ of Axiom Ax4:

$\mathrm{Ax4}^2$. $(X =^2 Y) \rightarrow (X(a) \equiv Y(a))$,

for second-order variables X, Y. Then we have the following theorem:

Theorem 2.10 *(FA) is provable in* $\mathbf{PCI^{N^2}} + (\beta\eta)$.

Proof The proof is *mutatis mutandis* as in Theorem 2.9 with the difference being that in steps 4–9 the main sign of the (first-order) identity $(=)$ has to be changed to the second-order one $(=^2)$, and the justifications of these steps have to be modified correspondingly. □

2.6 Non-Fregean Logic and Indefinite Descriptions

The construction of λ-terms suggests the possibility to consider an enrichment of $\mathbf{PCI^N}$ not with lambda-expressions, but with indefinite descriptions. Indeed, we may well dissociate 'the property of being (an x such that x is) A' and concentrate our attention on 'an x such that (x is) A' itself. We may also just think of replacing 'the x such that $A(x)$' with 'an x such that $A(x)$'. The replacing expressions can be formalized by means of some sort of operator for indefinite descriptions. One can

[7] By the way, Carnap apparently considers both interpretations on a par when he introduces "abstraction expressions ('$(\lambda x)(..x..)$', 'the property (or class) of those x which are such that $..x..$')" [48, p. 3].

find in the literature several different accounts of indefinite descriptions (for an overview of these approaches and the corresponding operators consult, e.g., [131]). One of the best formally elaborated theories of such an operator is the famous ε-formalism introduced by David Hilbert, see [9, 132].

Thus, if $A(x)$ is a formula possibly containing free occurrences of x, then $\varepsilon x A(x)$ is a singular term for "an x, such that (x is) A". If x does not occur in $A(x)$, then $A(x)$ is just A, and $\varepsilon x A$ means "an x, such that A".

Now it is natural to suppose that an object, such that it is A and is equal to a, is exactly a, if and only if a is A. In other words, in the case when we consider some concrete object $x = a$, we can accept for indefinite descriptions the principles strictly analogous to (•) and (•′) just by changing ι to ε. Consequently, one immediately obtains the versions of the slingshot argument as formulated in Theorems 2.4 and 2.5 by using ε-expressions instead of ι-expressions.

However, in the "indefinite analogues" of (•) and (•′) the operator of description turns out to be explicitly submitted to an additional condition of uniqueness which is in fact characteristic of definite descriptions. As to indefinite descriptions, the uniqueness of a described object can only be considered a particular case and cannot be taken as a general condition. Indeed, as it was observed by Russell, "the only thing that distinguishes 'the so-and-so' from 'a so-and-so' is the implication of uniqueness" [215, p. 176].

It could also be possible to construct a slingshot argument within non-Fregean logic enriched by a kind of (indefinite) description operator without the uniqueness condition, provided this operator possesses some peculiar features. Namely, consider the operator κ such that if $A(x)$ is a formula possibly containing free occurrences of x, then $\kappa x A(x)$ is a singular term. We first require that if $A(x)$ is a false sentence, then $\kappa x A(x)$ should mean exactly the same as $\kappa x(x \neq x)$ (call this feature *F-property*). And second we assume the following scheme:

$$(\kappa^{=}) \quad \kappa x A(x) = \kappa x B(x) \rightarrow \exists x(A(x) \equiv B(x)).$$

That is to say, if $\kappa x A(x)$ and $\kappa x B(x)$ denote the same object(s), then there exists at least one object x such that $A(x)$ and $B(x)$ denote the same situation.

Now let $\mathbf{PCI^N} + (\kappa^{=})$ be $\mathbf{PCI^N}$ enriched with the κ-operator subject to $(\kappa^{=})$. Then we have the following theorem:

Theorem 2.11 *(FA) is provable in* $\mathbf{PCI^N} + (\kappa^{=})$.

Proof Consider the following outline of a proof:

1. $A \wedge B$ (*Assumption*)
2. A (1, *∧-Elimination*)
3. B (1, *∧-Elimination*)
4. $A \rightarrow \kappa x(\sim A) = \kappa x(x \neq x)$ (*F-property*)
5. $B \rightarrow \kappa x(\sim B) = \kappa x(x \neq x)$ (*F-property*)
6. $\kappa x(\sim A) = \kappa x(x \neq x)$ (2, 4, *MP*)
7. $\kappa x(\sim B) = \kappa x(x \neq x)$ (3, 5, *MP*)

8. $\kappa x(\sim A) = \kappa x(\sim B)$ (6, 7, *Transitivity of* =)
9. $\kappa x(\sim A) = \kappa x(\sim B) \rightarrow \exists x(\sim A \equiv \sim B)$ ($\kappa^{=}$)
10. $\exists x(\sim A \equiv \sim B)$ (8, 9, *MP*)
11. $\sim A \equiv \sim B$ (10, $\exists$-*Elimination*)
12. $\sim A \equiv \sim B \leftrightarrow A \equiv B$ (Ax2, (*N*))
13. $A \equiv B(11, 12, MP)$

Again, by the Deduction Theorem, $\vdash (A \wedge B) \rightarrow (A \equiv B)$. *Mutatis mutandis* we get also $\vdash (\sim A \wedge \sim B) \rightarrow (A \equiv B)$. Hence, $\vdash (A \leftrightarrow B) \rightarrow (A \equiv B)$. □

Thus, extending **NFL** with an operator of the κ-type also exerts a destructive effect on the connective of sentential identity. However, a possible informal reading (as well as the precise semantical interpretation) of the kappa-operator remains somewhat unclear. It obviously cannot be Hilbert's epsilon-operator, since ($\kappa^{=}$) turns out to be implausible if we just change κ to ε. Taking the standard understanding of εx as "some (particular) x", it is clear that even if "*some* (particular) man that is ill" and "*some* (particular) man that is tall" happen to denote the same person, it would hardly be convincing to conclude on this ground that to be ill and to be tall constitute the same situation for anyone.

Nevertheless, in the literature it has been repeatedly stressed that an indefinite description can ambiguously stand not only for "some (particular)", but also for "any (whatever)" object. As George Wilson put it: "In fact, the form of words $A(n)$ ϕ is ψ seems to be generally ambiguous between ... *Something which is* a(n) ϕ is ψ... and ... *Anything which is* a(n) ϕ is ψ" [282, p. 50]. Wilson also argues that a pure quantificational interpretation of indefinite descriptions reveals certain difficulties as well.

By taking this into account, it seems more natural to interpret the κ-operator in the second sense, i.e., as some kind of "any"-phrase (from "κάθε" in Greek), which does not necessarily have a pure quantificational meaning. Under such an informal understanding, ($\kappa^{=}$) looks plausible enough. But many questions concerning this operator (most importantly, its precise syntactical and semantical elaboration) are still open.

2.7 Concluding Remarks

The slingshot argument has caused much controversy, especially on the part of fact-theorists and adherents of situations, states of affairs, or other fact-like entities, who try to discredit its probative force in one way or another. In particular, opposition by Barwise and Perry from the standpoint of their situation semantics has gained general attention.

In this respect the idea to use Suszko's non-Fregean logic for an analysis of the slingshot argument suggested by Wójcicki and Wójtowicz turned out to be rather fruitful. Whereas the formal reconstruction of the slingshot argument in the style of Neale (see Sect. 2.2.2) focuses on the inferential properties of the ι-operator, the reconstruction by means of **NFL** makes use of the properties of sentential, descriptive identity as axiomatized in **PCI**, $\mathbf{PCI^N}$ or **WBQ** extended by the ι-operator or the λ-operator (or even the κ-operator). In summary, it seems fair to say that despite a multifarious and sometimes sophisticated criticism, the slingshot argument presents a powerful and lucid justification of the view that sentences do signify truth values.

Nevertheless, this does not at all show that an ontology and semantics of situations (facts, states of affairs, etc.) is not worthy of investigation or is even technically infeasible, see, for example, [19, 285]. Still, one has to be very cautious when combining such theories with definite descriptions or lambda-expressions. As Wójtowicz [288, p. 190] emphasizes, "the description or abstraction operators must have their interpretation in such an ontology". Gödel [121] already explained that the problem posed by his version of the slingshot argument can be solved by adopting Russell's contextual theory of definite descriptions. In the case of the λ-abstractor, Wójtowicz suggests considering a Russellian theory of λ-terms in which an expression $A(\lambda x B(x))$ is regarded as an abbreviation of the sentence $\exists x \forall y((y \in x \leftrightarrow B(y)) \wedge A(x))$.[8] We can observe that this move in fact blocks the derivation from the proof of Theorems 2.7, 2.8, and 2.9 because we are no longer dealing with a substitution of singular terms licensed by axiom Ax4.

The version of the slingshot argument using predicate abstraction, however, is *not* blocked in this way because now we are using *λ-predicates* and, instead of an application of Ax4, we have an application of it's second-order version Ax4^2. In other words, the slingshot argument based on predicate abstraction cannot be circumvented by an appeal to a "Russellian" theory of *λ-terms*. In this sense $(\beta\eta)$, (N) and Ax4^2 pinpoint minimal assumptions about descriptive sentential identity needed to derive the Fregean Axiom.

8 See [288, Footnote 7], which contains a typographical mistake, however.

Chapter 3
Generalized Truth Values: From $FOUR_2$ to $SIXTEEN_3$

Abstract In the present chapter, we discuss the possibility of generalizing the very notion of a truth value by constructing truth values as complex units which may possess a ramified inner structure. We consider some approaches to truth values as structured entities and summarize this point in the notion of a *generalized truth value* conceived as a subset of some basic set of initial truth values of a lower degree. We are essentially led by the idea which is at the heart of Belnap and Dunn's useful four-valued logic, where the set $\mathbf{2} = \{T, F\}$ of classical truth values is generalized to the set $\mathbf{4} = \mathscr{P}(\mathbf{2}) = \{\varnothing, \{T\}, \{F\}, \{T, F\}\}$. We argue in favor of extending this process to the set $\mathbf{16} = \mathscr{P}(\mathbf{4})$. It turns out that this generalization is well-motivated and leads to a notion of a *truth value multilattice*. In particular, we proceed from the bilattice $FOUR_2$ with both an information and truth-and-falsity ordering to another algebraic structure, namely the trilattice $SIXTEEN_3$ with an information ordering together with a truth ordering *and* a (distinct) falsity ordering. We also consider another exemplification of essentially the same structure based on the set of truth values one can find in various constructive logics.

3.1 Truth Values as Structured Entities

One of the interesting peculiarities of the Fregean conception of truth values is that it provides a possibility of distinguishing between various "parts within truth values" [101, p. 30]. Although Frege, when pointing out to this possibility, immediately specifies that the word 'part' is used here "in a special sense", the basic idea seems nevertheless to be that truth values can be considered not as something amorphous but as possessing some inner structure. It is not quite clear how serious Frege is about this view, but it seems to suggest that truth values may

Y. Shramko and H. Wansing, *Truth and Falsehood*, Trends in Logic 36,
DOI: 10.1007/978-94-007-0907-2_3,

well be interpreted as complex, structured entities that can be divided into elementary constituents.

This idea finds its implementation in some approaches to semantic constructions for various non-classical logics, where truth values are represented as being made up from some primitive components. As an example, consider a formulation of a valuation system for intuitionistic logic based on so-called *Kripke frames*, cf. [216, pp. 21–25].

Let W be a non-empty set, $\leq$ be a reflexive and transitive relation defined on W (a pre-order), and w_0 be a distinguished element from W. Then we can define an intuitionistic valuation system $\mathbf{V}_W$ based on the frame $(W, \leq, w_0)$ in the following way:

$\mathbf{V}_W = \langle \mathscr{P}(W), \mathscr{D}_{w_0}, \mathscr{F}_W \rangle$, where $\mathscr{P}(W)$ is the power-set of W, $\mathscr{D}_{w_0} = \{X \subseteq W \mid w_0 \in X\}$, and $\mathscr{F}_W = \{f_\wedge, f_\vee, f_\rightarrow, f_\sim\}$, defined as follows:

Definition 3.1 For any $X, Y \in \mathscr{P}(W)$:

1. $f_\wedge(X, Y) = X \cap Y$;
2. $f_\vee(X, Y) = X \cup Y$;
3. $f_\rightarrow(X, Y) = \{x \in W \mid \forall y((x \leq y \text{ and } y \in X) \Rightarrow y \in Y)\}$;
4. $f_\sim(X) = \{x \in W \mid \forall y(x \leq y \Rightarrow y \notin X)\}$.

Now let any assignment function a relative to any $\mathbf{V}_W$ be subject to the following hereditary condition (for any $x, y \in W$):

$$x \in a(p) \quad \text{and} \quad x \leq y \Rightarrow y \in a(p). \tag{3.1}$$

Then we can prove the following proposition:

Proposition 3.1 *For any* $\mathbf{V}_W$, *for any assignment a relative to* $\mathbf{V}_W$, *and for any formulas A and B, the following holds:*

1. $x \in v_a(A \wedge B)$ *iff* $x \in v_a(A)$ *and* $x \in v_a(B)$;
2. $x \in v_a(A \vee B)$ *iff* $x \in v_a(A)$ *or* $x \in v_a(B)$;
3. $x \in v_a(A \rightarrow B)$ *iff* $\forall y((x \leq y$ *and* $y \in v_a(A)) \Rightarrow y \in v_a(B))$;
4. $x \in v_a(\sim A)$ *iff* $\forall y(x \leq y \Rightarrow y \notin v_a(A))$.

Moreover, condition (3.1) can easily be extended to any valuation v_a and to any formula A.

We say that sentence A is *intuitionistically true* in the given intuitionistic valuation system $\mathbf{V}_W$ based on the frame $(W, \leq, w_0)$ with respect to the given valuation v_a, if and only if $w_0 \in v_a(A)$. Otherwise A is said to be *intuitionistically false* (in the given intuitionistic valuation system $\mathbf{V}_W$ based on the frame $(W, \leq, w_0)$ with respect to the given valuation v_a). A sentence is called *valid* if and only if it is true in any intuitionistic valuation system $\mathbf{V}_W$ based on any frame $(W, \leq, w_0)$ by any valuation v_a. One can show that the set of all valid sentences so defined is exactly the set of theorems of intuitionistic propositional logic **IPL**. This fact does not, of course, contradict Gödel's observation [120] that **IPL** is not characterized by any finite matrix.

Informally, W can be understood as a set of states of some (intuitionistic) theory, and $\leq$ as the "earlier or simultaneous" relation between the states of the theory. Any subset of W can be seen as a proposition accepted exactly at the states belonging to this subset. Then any intuitionistic valuation system determines the corresponding truth values *intuitionistically true* (as the set of all propositions accepted in the distinguished state of this system) and *intuitionistically false* (as the set of all propositions which are not accepted in the distinguished state of this system). Clearly, the intuitionistic truth values so conceived are composed from some simpler elements, viz. the states and their sets, and as such they turn out to be complex units.

One can also recall in this connection the so-called "factor semantics" introduced by Alexander Karpenko for many-valued logic, where truth values are construed as ordered n-tuples of classical truth values (T–F sequences), see [144]. A valuation system for an n-valued Łukasiewicz logic can be formulated then as follows.

Let us have an algebraic system $A_s = (B^s, \cong, R, \neg^{\oplus}, \supset^{\oplus})$, where

$B^s = \{\langle a_1, \ldots, a_s\rangle \mid a_i \in \{T, F\}, 1 \leq i \leq s\}$;

$\alpha \cong \beta$ iff $\alpha^T = \beta^T$ ($\alpha, \beta \in B^s$, α^T is the number of occurrences of T in α);

$$R(\alpha,\beta) \text{ iff } \begin{cases} \forall k \leq s (a_k = T \Rightarrow b_k = T) & \text{if } \alpha^T \leq \beta^T (a_k \in \alpha, b_k \in \beta), \\ \forall k \leq s (b_k = T \Rightarrow a_k = T) & \text{otherwise;} \end{cases}$$

$\neg^{\oplus}\langle a_1, \ldots, a_s\rangle = \langle \neg^{+} a_1, \ldots, \neg^{+} a_s\rangle$;

$\langle a_1, \ldots, a_s\rangle \supset^{\oplus} \langle b_1, \ldots, b_s\rangle = \langle a_1 \supset^{+} b_1, \ldots, a_s \supset^{+} b_s\rangle$, $\neg^{+}, \supset^{+}$ being the standard Boolean operations of negation and material implication defined on $\{T, F\}$.

Then a *factor valuation system* $\mathbf{V}_{A_s}$ associated with the algebraic system A_s can be defined as a structure $\langle B_s/\cong, \{[T^s]\}, \{\neg, \rightarrow\}\rangle$, where $B_s/\cong$ is the factor set of B_s determined by $\cong$, $[T^s]$ is the equivalence class determined by $\langle T, T, \ldots T\rangle$ (s times), and for any $[\alpha], [\beta] \in B_s/\cong$: $\neg[\alpha] = [\neg^{\oplus}\alpha]$, $[\alpha] \rightarrow [\beta] = [\alpha' \supset^{\oplus} \beta']$, where $\alpha' \in [\alpha]$, $\beta' \in [\beta]$, and $R(\alpha', \beta')$.

One can show (see Theorem 1 in [144, p. 180]) that a factor valuation system $\mathbf{V}_{A_s}$ is strictly characteristic for $(s+1)$-valued Łukasiewicz logic. In this way one also obtains a certain intuitive understanding of "intermediate" truth values. For example, the value $3/5$ can be interpreted in a factor valuation system as a T–F sequence of length 5 with exactly three occurrences of T. Most importantly, the classical values T and F are used here as "building blocks" for non-classical truth values.

In what follows we adopt the fundamental idea of structuring truth values and show how one can effectively employ it for generalizing the very notion of a truth value and developing on this basis a universal method of "producing" truth values for various non-classical logics. It turns out that these non-classical (generalized) truth values can be conceived as resting essentially on the classical values *the True* and *the False* as some basic (fundamental) entities. Being so conceived the generalized truth values provide a natural justification for some possible limitations of certain classical principles, most notably the one of bivalence and non-contradiction.

3.2 Generalized Valuations, Four-Valued Logic and Bilattices

As it was remarked in Sect. 1.6, it is by no means unusual in modern symbolic logic to allow sentences to be sometimes *neither* true *nor* false. This is exactly the idea which goes back to Łukasiewicz and which has been implemented in various systems of many-valued logic, partial logic, and elsewhere. There is also a "dual" idea according to which some sentences can occasionally be viewed as having *both* of these truth values. In the last couple of decades it has been noticed repeatedly that it is sometimes quite *useful* to evaluate sentences as both true and false. This view has recently found numerous fruitful applications in paraconsistent logics, logics of information states and in computer science. Notably, this idea also can largely be traced back to Łukasiewicz, namely to his critique of the *Principle of Contradiction* by Aristotle in [159]. While discussing other forerunners of inconsistency-tolerant logics, one should mention the Russian logician Nikolai Vasiliev with his "imaginary logic" (see [260]).

It is quite telling that both these ideas were brought together and combined by considering the so-called "under-determined" and "over-determined" valuations in investigations of implication and entailment within the framework of *relevance logic*—a direction in modern non-classical logic initiated by Alan Anderson and Nuel Belnap, see [3]. More specifically, this strategy of semantic analysis was carefully elaborated in the 1960s and 1970s by J. Michael Dunn, who stressed that a sentence can be *rationally* considered to be not just true or just false, but also neither true nor false as well as both true and false.[1] This can be made explicit by developing a suitable valuation procedure that generalizes the notion of an ordinary, classical truth value function by allowing "under-determined" and "over-determined" valuations.

Consider the set $\mathbf{2} = \{F, T\}$ to be the usual set of classical truth values. A standard classical valuation v^2 (a two-valuation) is then a *function* from the set of sentences into **2**, thus ascribing to every sentence *one and only one* element from **2**, i.e., either truth or falsity. There can be then different ways of generalizing such a two-valued truth value function. For example, Dunn in [73, pp. 121–132], introduces a so-called "aboutness valuation" that ascribes to each sentence a pair (X_1, X_2) (called a "proposition surrogate", see also [75], p. 161), where X_1 represents "topics" the sentence gives definite information about, and X_2 represents topics the negation of the sentence gives definite information about (cf. [79, p. 36]). Let U be a "universe of discourse" such that $X_1 \subseteq U$ and $X_2 \subseteq U$. Unlike

[1] Dunn had already developed this approach in his doctoral dissertation [73] and then presented it in a number of conference talks and publications, most notably in [75] (see also [74]). The reader may consult [79, 80] for a comprehensive account and systematization of Dunn's (and other) work in this area (cf. also [261]). In the literature, the semantic strategy in question is sometimes called the "American Plan" as opposed to the so-called "Australian Plan". Both labels were brought into usage by Meyer [173] to contrast the *four-valued* approach of the "Americans" Dunn and Belnap to the *star semantics* of the "Australians" Routley and himself.

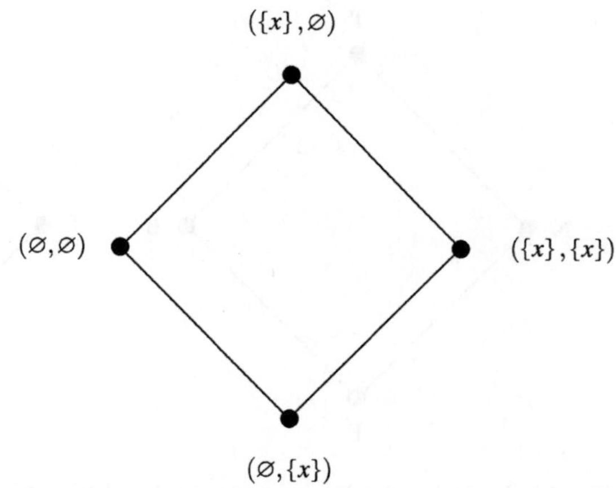

Fig. 3.1 The De Morgan lattice of proposition surrogates

a classical truth value function, the aboutness valuation needs to be neither disjoint ($X_1 \cap X_2 = \varnothing$) nor exhaustive ($X_1 \cup X_2 = U$), thus making "truth-value gaps" and "truth-value gluts" possible. If the universe of discourse consists of a single topic x, the usual ("normal") valuations can be represented as $(\{x\}, \varnothing)$ and $(\varnothing, \{x\})$—for truth and for falsity respectively, and under-determined and over-determined valuations as $(\varnothing, \varnothing)$ and $(\{x\}, \{x\})$. It appears that these valuations form a lattice, which is called the "De Morgan lattice of proposition surrogates" in [73, p. 129]. We present this lattice in Fig. 3.1.

The same idea is realized in [75] by interpreting a valuation not as a function but just as a *relation* connecting sentences of the language in use with elements from **2**. Such a valuation relates to each sentence either the value "true" (T) or "false" (F), or *neither* of these values (partial function), or *both* of them (non-functional relation). Dunn provides the non-standard valuations with an intuitive motivation employing the terminology of abstract (epistemic) "situations" that may well be incomplete or inconsistent (see [75, pp. 155–157]). However, the principle of functionality for valuations is explicitly given up by this approach, which can go against the grain of those who strictly adhere to Frege's functional analysis in general and to truth-value functions in particular.

Therefore, one may prefer an equivalent (although somehow more "ontological") way of grasping the same point by considering a valuation still a function, though not from sentences to *elements* of the set **2**, but from sentences to *subsets* of this set. This latter interpretation can be found in [75] as well (cf. also [80, p. 7]). Taking this method of generalizing classical truth value functions as paradigmatic, we call a function conceived in this way a *generalized truth value function.*[2]

Nuel D. Belnap (see [22, 23], reproduced in [4, § 81]) took Dunn's idea a step further. Namely, he explicitly considered the empty and the overcomplete subsets of the set $\{F,T\}$ new truth values in their own right. Using a highly heuristic interpretation of a truth value as information that "has been told to a computer",

[2] In [231, p. 762] valuations of this kind have been called *multivaluations.*

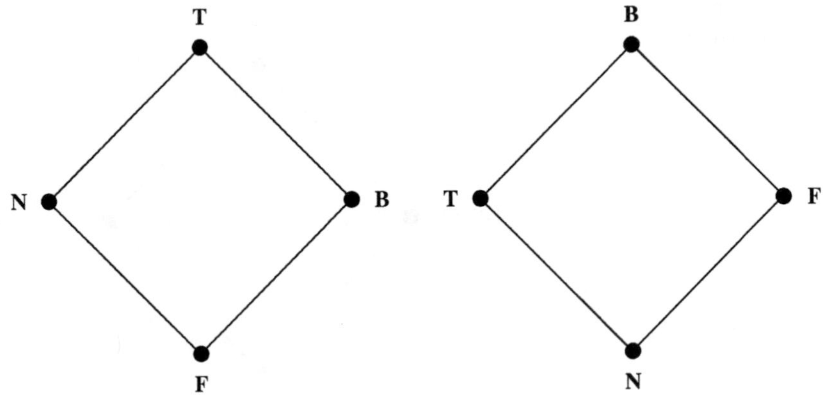

Fig. 3.2 Logical lattice **L4** and approximation lattice **A4**

he arrived at a "useful four-valued logic" of "how a computer should think" based on the following four "told truth values"[3]:

N = { }—none ("told neither falsity nor truth");
F = $\{F\}$—"plain" falsehood ("told only falsity");
T = $\{T\}$—"plain" truth ("told only truth");
B = $\{F, T\}$—both falsehood and truth ("told both falsity and truth").

Now the lattice from Fig. 3.1 can be represented as a lattice of Belnap's four truth values as in Fig. 3.2, left. Belnap in [22, p. 14] calls it "the logical lattice **L4**". This lattice is "logical" because the ordering on it is in effect a *logical order* with the usual truth-functional conjunction and disjunction as meet and join, respectively. Belnap's papers present another lattice, also based on the same elements. Belnap calls it the "approximation lattice **A4**" (Fig. 3.2, right) because its ordering can be naturally explicated as "approximates the information in." The idea of this lattice goes back to Scott (see, e.g., [219]) who considers various examples of an approximation order. Belnap remarks that **N** is the bottom of **A4** because it gives no information at all, whereas **B** is at the top because it gives too much (indeed inconsistent) information.

Belnap's four-valued logic has found numerous applications in various fields such as the theory of deductive databases, distributed logic programming, and other areas. Inspired by applications of logic in computer science and AI, Matthew Ginsberg [118, 119] introduced the notion of a *bilattice* and pointed out that Belnap's four truth values form the smallest non-trivial bilattice.[4] Roughly speaking, a bilattice is a non-empty set with *two* partial orderings, each

[3] The idea of the four truth values has been also expressed by Scott in [219], p. 170.

[4] Bilattices have been studied by many authors in various contexts, see, e.g., [7, 8, 10, 11, 115, 217], [86–92] and references therein.

Fig. 3.3 Bilattice $FOUR_2$

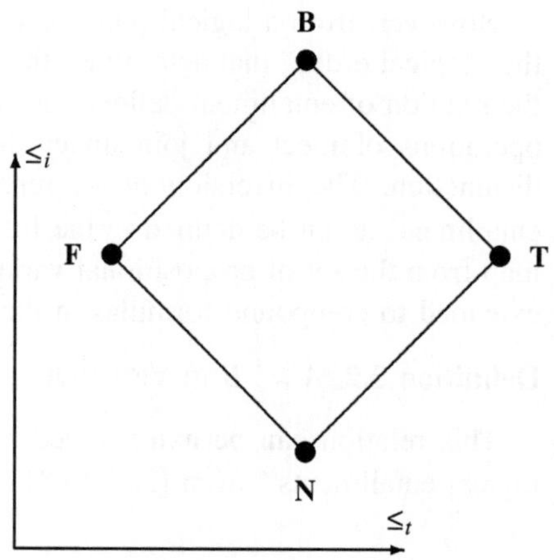

constituting its own lattice on this set. The truth value bilattice based on the set $\mathbf{4} = \{\mathbf{N}, \mathbf{F}, \mathbf{T}, \mathbf{B}\}$, which we will call $FOUR_2$ (the subscript '2' stands for '*bi*'), is presented by a double Hasse diagram in Fig. 3.3.

This diagram is placed into a two-dimensional coordinate plane, where the horizontal axis stands for a truth order ($\leq_t$) and the vertical axis stands for an information order ($\leq_i$).[5] These orderings represent an increase in truth and information, respectively. That is, $x \leq_t y$ means that y is "at least as true" as x, and $x \leq_i y$ means that y is "at least as informative" as x.

Let us formally define the ordering relations ($\leq_i$ and $\leq_t$) in $FOUR_2$. The definition of $\leq_i$ is very simple: for any $x \in \mathbf{4}$ we just put $x \leq_i y$ iff $x \subseteq y$. For $\leq_t$ the situation is less obvious. For each element of $\mathbf{4}$ we first define its "truth part" and its "falsity part" as follows:

$$x^t := \{z \in x \mid z = T\}; \quad x^f := \{z \in x \mid z = F\}.$$

Then we have: $x \leq_t y$ iff $x^t \subseteq y^t$ and $y^f \subseteq x^f$.[6] This definition clearly shows that within $FOUR_2$, $\leq_t$ is in fact not just a *truth* order but rather a *truth-and-falsity* order: by ordering the truth values we have to take into account not only the "truth-content" of each value but also its "falsity-content". An increase in truth-content automatically means a decrease in falsity-content (cf. [91, p. 480]). In other words, $\leq_t$ in $FOUR_2$ seems to presuppose that falsehood *by itself* is less true than truth, and thus one may suspect that truth and falsity in $FOUR_2$ are not entirely autonomous notions.

---

[5] The information order is sometimes referred to as a "knowledge order" (denoted by $\leq_k$), which is not quite accurate from a philosophical point of view when we are taking into account the classical definition of knowledge as justified *true* belief.

[6] As Fitting put it: "[W]e might call a truth value t_1 *less-true-or-more-false* than t_2 if t_1 contains *false* but t_2 doesn't, or t_2 contains *true* but t_1 doesn't" [89, p. 94].

However, from a logical point of view, just $\leq_t$ is most important. It is exactly the "logical order" that determines the properties of logical connectives as well as the relation of entailment defined on $FOUR_2$. Namely, as noted above, the lattice operations of meet and join under this order are just logical conjunction and disjunction. The inversion of $\leq_t$ represents a certain kind of negation. As to entailment, it can be defined in the following way. Let v^4 (a four-valuation) be a map from the set of propositional variables into $FOUR_2$, and let this valuation be extended to compound formulas in the usual way. Then we have:

Definition 3.2 $A \models^4 B$ iff $\forall v^4(v^4(A) \leq_t v^4(B))$.

This relation can be axiomatized by the consequence system $\mathbf{E}_{fde}$ of "tautological entailments" from [3, § 15.2] (also called *First-Degree Entailment*).

3.3 Taking Generalization Seriously: From Isolated Computers to Computer Networks

There is an interesting question concerning Dunn's and Belnap's four-valued semantics, namely: Why should we stop the "generalization procedure" just at the four-valued stage and not proceed further to considering, say, combinations of **T** and **B**, **N** and **B**, etc? If we can get a "useful four valued logic" by taking the power-set of **2**, why should one not consider a *sixteen-valued logic* based on $\mathcal{P}(\mathbf{4})$, the power-set of the set of Belnap's truth values obtained previously? Would such a logic be "useful"?[7] We believe that the above question is not only interesting but also important since answering it allows us to present the whole story of truth values in a greater generality and arrive, in this way, at a far reaching generalization of the very notion of a truth value itself, leading, for example, to an extension of the familiar set of propositional connectives and a broad conception of logical systems.

The "oral tradition" supplies the above question with one typical reply, arguing to the effect that any combination of Belnap's four truth values would be in a sense *superfluous*. The argument usually goes as follows. Consider, e.g., the combination $\mathbf{TB}(=\{\{T\},\{F,T\}\})$ of **T** and **B**. This new truth value would then mean "true and true-and-false". But a repetition of truths gives us no new information (is superfluous)! Thus, the meaning of **TB**, it is claimed, collapses into just "true-and-false", and in this way we simply obtain **B**. An analogous argument reduces **FB** to **B**, and it is not difficult to argue in a similar way that **FT** is, in fact, also **B**. Further, a combination of **N** with any other truth value seems to be superfluous as well, for unifying the empty set with any other set gives just this latter set. As a

[7] As R. Meyer put it: "[I]f we take seriously both true and false and neither true nor false separately, what is to prevent our taking them seriously conjunctively? As in 'It is both true and false and neither true nor false that snow is white'" [173, p. 19]. As we will argue below, Meyer's own answer to this question—"This way, in the end, lies madness" [173] appears a bit overhasty.

consequence one might conclude that any attempt to continue generalizing truth values beyond the four values introduced by Belnap should fail due to a collapse of any new truth value into one of the initial four.

However, a more careful examination shows that such a conclusion is not justified. First, recall that the proper interpretation of **T** is not simply "true" but "true-*only*" (and analogously for falsehood). The combination of "true-only" and "true-and-false", which we get in the new truth value **TB**, is not so trivial and, in any case, is not so easily reducible to "true-and-false" as the above argument seems to suggest. Second, one may notice that this argument works only under the implicit interpretation of the comma between elements in the notation for the new truth values as set-theoretical union and the identification of a set x with the singleton $\{x\}$. Only then, one would be able to conduct the suggested manipulation: $\{\{T\},\{F,T\}\} = \{\{T\} \cup \{F,T\}\} = \{T,F,T\} = \{F,T\}$, which is obviously incorrect. Indeed, $\{\{T\},\{F,T\}\}$ is, of course, distinct from $\mathbf{B} = \{T\} \cup \{F,T\}$, and therefore, it would be rather more natural to consider the truth value $\{\{T\},\{F,T\}\}$ an independent value in its own right. Similarly, $\{\varnothing,\{F,T\}\}$ is not the same as $\{F,T\}$, etc.

Thus, there are good reasons for taking the above mentioned generalizing procedure seriously and for considering a *second-order generalization*, which results from generating the powerset of the set $\{\mathbf{N},\mathbf{F},\mathbf{T},\mathbf{B}\}$.[8] In this way we obtain the following set **16** of *sixteen* generalized truth values (denotations for most values are obvious, and **A** stands for "all"):

1. $\mathbf{N} = \varnothing$	9. $\mathbf{FT} = \{\{F\},\{T\}\}$
2. $\boldsymbol{N} = \{\varnothing\}$	10. $\mathbf{FB} = \{\{F\},\{F,T\}\}$
3. $\boldsymbol{F} = \{\{F\}\}$	11. $\mathbf{TB} = \{\{T\},\{F,T\}\}$
4. $\boldsymbol{T} = \{\{T\}\}$	12. $\mathbf{NFT} = \{\varnothing,\{F\},\{T\}\}$
5. $\boldsymbol{B} = \{\{F,T\}\}$	13. $\mathbf{NFB} = \{\varnothing,\{F\},\{F,T\}\}$
6. $\mathbf{NF} = \{\varnothing,\{F\}\}$	14. $\mathbf{NTB} = \{\varnothing,\{T\},\{F,T\}\}$
7. $\mathbf{NT} = \{\varnothing,\{T\}\}$	15. $\mathbf{FTB} = \{\{F\},\{T\},\{F,T\}\}$
8. $\mathbf{NB} = \{\varnothing,\{F,T\}\}$	16. $\mathbf{A} = \{\varnothing,\{T\},\{F\},\{F,T\}\}$.

That **16** makes perfect sense can also be shown by employing Belnap's interpretation of truth values as information that could be told to a computer. One of the main motivations for Belnap's interpretation is the obvious observation that a computer can receive information from various (maybe independent) sources. Now a situation is overall possible when one source informs a computer that a sentence is true-only while another informant supplies (perhaps without being aware of this) inconsistent data, namely that the sentence is both true and false. And what if a computer has been simultaneously "told" that a sentence is true-only (informant 1), false-only (informant 2), both-true-and-false (informant 3), and neither-true-nor-false

[8] The idea to generalize Belnap's construction by considering truth values as subsets of a set containing more than two elements (T and F) has been also expressed by Karpenko in [145, p. 46].

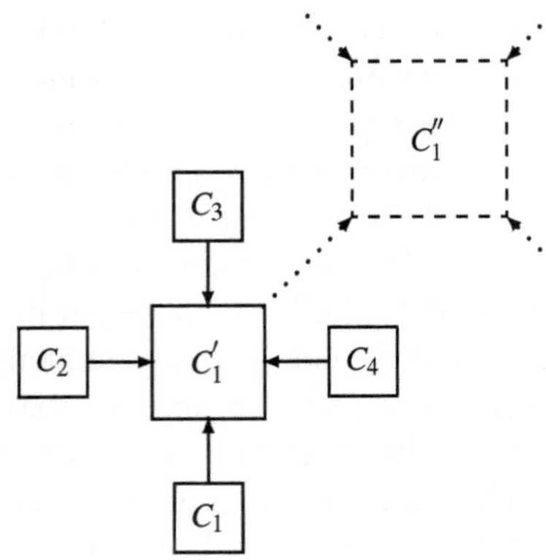

Fig. 3.4 A computer network

(informant 4)? Dunn and Hardegree [81, p. 277, our emphasis] seem to argue along the same lines when they remark that "there can be states of information that are inconsistent, incomplete, *or both*". Obviously, an information state that is both inconsistent *and* incomplete cannot be represented by any single value from **4**, whereas **16** offers the value **NB** to account for this situation.

In fact, Belnap's interpretation is perfectly well-suited when applied to a single computer. In addition, it presupposes that this computer receives information from *classical* sources, i.e. from sources which can operate exclusively with classical truth values (T, F). But what if the computer's informants behave nonclassically? Moreover, nowadays, it is quite rare for a computer to stay completely isolated and not connected (even from time to time) to other computers, thereby constituting a network. It appears that Belnap's interpretation cannot adequately be applied when we deal not just with one computer but with several interconnected computers, i.e., a *computer network* (conceived hierarchically).

Consider four Belnap computers[9] (C_1, C_2, C_3, C_4) connected to some *central computer* (C_1'), a server, to which they are supposed to supply information (Fig. 3.4). It is fairly clear that the logic of the server itself (so, the network as a whole) cannot remain four-valued any more. Suppose C_1 informs C_1' that a sentence is true only (has the value **T**), whereas C_2 supplies inconsistent information (the sentence is both true and false, i.e., has the value **B**). In this situation C_1' has received the information that the sentence simultaneously is true only as well as both true and false, in other words, it has a value not from **4**, but from $\mathcal{P}(\mathbf{4})$, namely the value $\mathbf{TB} = \{\{T\}, \{T, B\}\}$. If C_1' has been informed simultaneously by C_1 that a sentence is true-only, by C_2 that it is false-only, by C_3 that it is both-true-and-false, and by C_4 that it is neither-true-nor-false, then the value $\mathbf{A} = \{\varnothing, \{T\}, \{F\}, \{T, F\}\}$ is far from being a "madness" but is just an adequate value which should be ascribed to the sentence by C'. That is, the logic of C_1' has

[9] A "Belnap computer" is just a computer that uses Belnap's four-valued logic. Note also that it is not crucial for our example to have exactly *four* Belnap computers. There can well be more of them or fewer—even one would be enough. The main point is that it should be connected to some "higher" computer.

to be 16-valued, and to get an adequate tool for handling this situation we have to involve **16**. Incidentally, it is interesting to observe that if we wish to extend our network and connect our server to some "higher" computer (C_1''), then generalized truth values of the *third order* (the set $\mathscr{P}(\mathbf{16})$) come into question (and so on), which motivates a further generalization of the present construction.

Note that in **16**, each of Belnap's original four truth values from **4** underwent an important transformation by being promoted to the single element of some "higher set". We mark this by putting the corresponding truth values in **16** into italics. Obviously, $\mathbf{N} \in \boldsymbol{N}, \mathbf{F} \in \boldsymbol{F}, \mathbf{T} \in \boldsymbol{T}$ and $\mathbf{B} \in \boldsymbol{B}$. It is also interesting to observe the difference between **FT** and Belnap's **B**. We will discuss this difference and generally an intuitive motivation for "generalized truth values of higher order" in more detail later.

3.4 Generalized Truth Values and Multilattices

We can now summarize the main points of the above considerations. Having a basic set of some initial truth values X, we call the power-set of this basic set $\mathscr{P}(X)$ the set of *generalized truth values* based on X. A generalized truth value function then will be a function from the set of sentences of a given language into $\mathscr{P}(X)$. In this way one obtains a generalization of both the notion of a truth value *function* and the very notion of a *truth value* itself.

We can also employ the idea of a bilattice to introduce general logical structures suitable for dealing with generalized truth values—*multilattices*. A bilattice turns out to be then just a particular case of such a structure with *more than one* ordering relation by which a set of (generalized) truth values is to be organized.

Definition 3.3 An *n-dimensional multilattice* (or simply *n-lattice*) is a structure $\mathscr{M}_n = (S, \leq_1, \ldots, \leq_n)$ such that S is a non-empty set and $\leq_1, \ldots, \leq_n$ are partial orders defined on S such that $(S, \leq_1), \ldots, (S, \leq_n)$ are all distinct lattices.

The notion of a *multilattice* (cf. [268]) is a generalization of the notion of a bilattice along the lines proposed in [231] (cf. also [227, 228]), where a set of generalized truth values of constructive logics has been presented in the form of a *trilattice*. The trilattice in [231] is a direct extension of a bilattice structure with a *third* partial order ($\leq_c$), which represents there an increase in *constructivity*. However, the elements of the given set of generalized truth values may well be compared also with respect to other properties, and hence some additional partial orders may be defined on it, thus determining, depending on the purposes of the analysis, different multilattices.

Note that our notion of a multilattice is extremely general. We do not impose any additional conditions that interconnect individual lattices of a given multilattice. According to our definition, a bilattice ($n = 2$) is just a set with two lattice-forming partial orders on it. Strangely enough, there is no uniform and generally accepted definition of a bilattice in the literature. When Ginsberg first introduced this notion in [118], he defined it as a quintuple $(B, \wedge, \vee, \cdot, +)$ such that $(B, \wedge, \vee)$ and $(B, \cdot, +)$ are both lattices and each operation respects the lattice relations in the alternate lattice. (Fitting calls this latter condition the *interlacing condition*, i.e., a bilattice à la [118] is an interlaced bilattice in Fitting's sense.) But already in [119] we find a quite different definition according to which an operation of *negation* becomes a necessary element of *any* bilattice. This definition has been adopted in [7, 8, 88] and by some other authors. However, we believe that it is too restrictive, because "there are interesting bilattice-like structures that do not have a notion of negation" [88, p. 241]. Fitting in [87] introduces the notion of a *pre-bilattice* as just a non-empty set with two partial orderings, each giving this set the structure of a lattice. Then in [92] he defines a bilattice as a pre-bilattice with some "useful connections between orderings". But when exactly is a connection "useful"? Could we speak of "interesting" connections instead? And if we just omit the "usefulness requirement", then the definition is not very informative, for it is not difficult to introduce at any time *some* (maybe even trivial) kind of connection between given orderings. We therefore prefer to refer to Fitting's pre-bilattices just as two-lattices (or simply bilattices). What is really crucial here, is the number of different partial orderings defined on *one and the same* set. All other properties and conditions (including conditions that connect the ordering relations) can be specified later, thereby giving rise to different types of multilattices (and bilattices).

Following Fitting and others (see, e.g., [92]), we call an n-lattice *complete* iff all the lattices that constitute this multilattice are complete, *interlaced* iff each pair of meet and join is monotone (i.e., order preserving) with respect to each partial order of the multilattice, and *distributive* iff all $2\,(2n^2 - n)$ distributive laws connecting its meets and joins hold.

Consider any two distinct ordering relations, each defined on some non-empty set. We say that these relations are *mutually independent* with respect to these definitions (or are defined *mutually independently*) iff they are not inversions of each other and the only common terms that are used in both definitions, except for metalogical connectives and quantifiers, are the usual set theoretical terms (for the intuitive motivation for the idea of "mutual independence" see [231, pp. 782–783]). The following definition introduces an important class of multilattices and allows a reasonable reduction of the amount of partial orders in a multilattice to relations that are in a certain sense "non-trivial" (or "interesting").

Definition 3.4 A multilattice is called *proper* iff all its partial orders can be defined mutually independently.

There are also some other works in the literature which deal with the notion of a trilattice. Lakshmanan and Sadri introduced this notion in [150, p. 257]. They were directly motivated by enriching Ginsberg's bilattices with a third ordering relation, which they called a *precision ordering*. In this way they aim to construct a probabilistic calculus as a suitable framework for probabilistic deductive databases, thus dealing with an algebra defined on a set of *interval pairs* rather than on a set of generalized truth values. Another line of research concerning trilattices

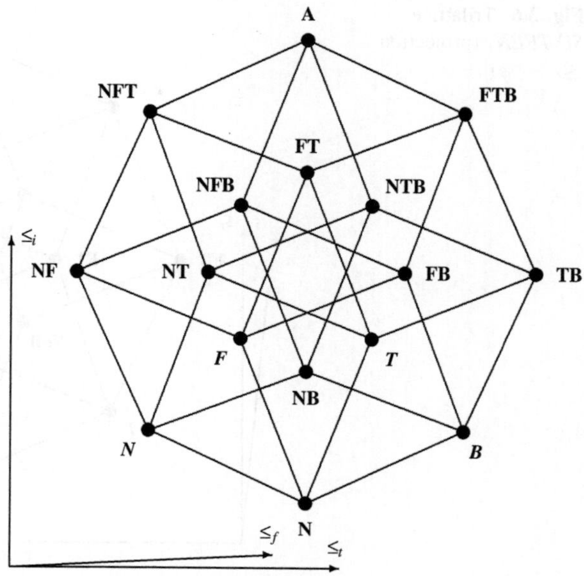

Fig. 3.5 Trilattice *SIXTEEN*$_3$ (projection $\leq_i - \leq_t$)

comes from *formal concept analysis*, a research area established and investigated by R. Wille, B. Ganter and their collaborators (see [33, 114, 262, 279]). This tradition seems to be developing totally independently of investigations in the field of bilattices and multi-valued logic.

3.5 The Trilattice of 16 Truth Values

Now, when we turn to the algebraic structure of **16**, it appears that it is generally possible here to discriminate between an increase in truth and a decrease in falsity and thus to define a truth order and a (non-)falsity order as distinct and in effect *mutually independent* relations. To do so, we have to redefine the sets x^t and x^f, and to explicitly introduce their complements. Thus, for every x in **16** we denote by x^t the subset of x that contains exactly those elements of x which in turn contain T as an element and by x^{-t}—the "truthless" subset of x:

$$x^t := \{y \in x \mid T \in y\}; \quad x^{-t} := \{y \in x \mid T \notin y\};$$

and analogously for falsity[10]:

[10] Note that we could introduce sets x^{-t} and x^{-f} for $FOUR_2$ as well, for example, as follows:

$$x^{-t} := \{z \in x \mid z \neq T\}; \quad x^{-f} := \{z \in x \mid z \neq F\}.$$

It turns out then, however, that $x^{-t} = x^f$ and $x^{-f} = x^t$, which once again confirms our observation that truth and falsity in $FOUR_2$ are still interdependent.

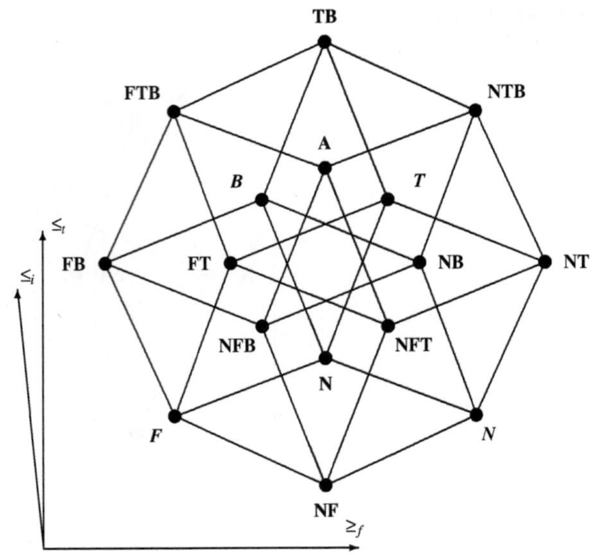

Fig. 3.6 Trilattice *SIXTEEN*$_3$ (projection $\leq_t - \leq_f$)

$$x^f := \{y \in x \mid F \in y\}; \quad x^{-f} := \{y \in x \mid F \notin y\}.$$

Then we define:

Definition 3.5 For every x, y in **16**:

1. $x \leq_i y$ iff $x \subseteq y$;
2. $x \leq_t y$ iff $x^t \subseteq y^t$ and $y^{-t} \subseteq x^{-t}$;
3. $x \leq_f y$ iff $x^f \subseteq y^f$ and $y^{-f} \subseteq x^{-f}$.

In this way, the algebraic structure of **16** is that of a proper three-lattice, a *trilattice* with *three* mutually independent partial orders that represent an increase in information, truth, and falsehood. This trilattice, which we for obvious reason call *SIXTEEN*$_3$, is presented by a triple Hasse diagram in Fig. 3.5 (cf. Fig. 3.5 in [231]). One can clearly observe in this diagram all three partial orderings. **A** and **N** are, respectively, the lattice top and bottom relative to $\leq_i$, **TB** and **NF** relative to $\leq_t$, and **FB** and **NT** relative to $\leq_f$. In accordance with the underlying interpretation, **A** and **N** are then the most and the least *informative* elements of **16**, **TB** and **NF** are the most and the least *true* of its elements, and **FB** and **NT** are the most and the least *false* elements. Note that the f-axis in Fig. 3.5 is drawn a bit approximately and gives only a rough idea about the third "dimension" of *SIXTEEN*$_3$. Another projection of *SIXTEEN*$_3$ (on the plain $\leq_t - \leq_f$) is represented in Fig. 3.6.

It is interesting to observe that *SIXTEEN*$_3$ has altogether exactly *eight* distinct partial orders that are not inversions of each other. In addition to the relations introduced in Definition 3.5 there are: a "truth-only" order (with ***T*** and **NFB** as top and bottom); a "falsity-only" order (with ***F*** and **NTB**); a "truth-only-and-falsity-only" order (with **FT** and **NB**); a "both-only" order (with ***B*** and **NFT**); a "none-only" order (with ***N*** and **FTB**). These five additional relations are all not independent and in fact "derivative" of the truth order and falsity order from Definition 3.5 in the sense that any of them can be defined through a certain combination of the functions $(\cdot)^t, (\cdot)^{-t}, (\cdot)^f$ and $(\cdot)^{-f}$ used in the definitions of $\leq_t$ and $\leq_f$.

One might consider the generalized truth values $\boldsymbol{B}, \boldsymbol{N}, \boldsymbol{F}$, and $\boldsymbol{T}$ from **16** as *analogues* of the truth values **B, N, F** and **T** from **4**.[11] However, some of these truth values from **16** and **4**, respectively, behave quite differently under the corresponding truth and information orderings in *SIXTEEN*$_3$ and in *FOUR*$_2$. For example, within *FOUR*$_2$, **B** is more and **N** is less informative than **F** and **T**, but in *SIXTEEN*$_3$, we see that $\boldsymbol{B}, \boldsymbol{N}, \boldsymbol{F}$, and $\boldsymbol{T}$ are situated on the same informational level. This latter fact might seem to violate some basic intuitive motivations. Recall that according to Belnap's interpretation, **B** carries more information than **T** because **B** stands for "told both true and false", whereas **T** stands for "told only true" (and analogously for other truth values). To simplify the exposition, in the remainder of this chapter, when discussing Belnap's truth values and their combinations, we will just say "true" and "false" instead of "told true" and "told false".

However, it turns out that the behavior of $\boldsymbol{B}, \boldsymbol{N}, \boldsymbol{F}$, and $\boldsymbol{T}$ in *SIXTEEN*$_3$ is much more natural than one might think on the face of it. First, notice that analogous values from **4** and from **16** are *not* the same, e.g., $\mathbf{T} = \{T\}$ but $\boldsymbol{T} = \{\{T\}\}$, etc. Moreover, we remark that under $\leq_i$ generalized truth values should be ordered exclusively by the amount of elements in the corresponding sets: the more elements a set has, the more informative it is. But in **16** the truth values $\boldsymbol{B}, \boldsymbol{N}, \boldsymbol{F}$, and $\boldsymbol{T}$ are *all* singletons (in contrast to **N, F, T,** and **B** in **4**), and hence they are all equally informative, precisely as *SIXTEEN*$_3$ suggests. Intuitively this means that—as one may easily observe—*any* of Belnap's initial truth values, and not only **B**, may be viewed as saying (explicitly or implicitly) *something* both about truth and falsehood, either in a positive or in a negative mode. To make this point explicit, the elements of **4** need a slightly different reading. To justify this reading, we remark that, e.g., **T**—"truth-only"—actually means nothing more than "true and *not false*". From a purely quantitative point of view, a negative piece of information is exactly of *the same* cash value as a positive one. In this way, we arrive at the following reinterpretation (or maybe more precise interpretation) of Belnap's four truth values:

N—"a sentence is not false and not true", [*non-falsehood, non-truth*];
F—"a sentence is false and not true", [*falsehood*, *non-truth*];
T—"a sentence is not false and true", [*non-falsehood*, *truth*];
B—"a sentence is false and true", [*falsehood*, *truth*].

Let us emphasize that this reading for elements of **4** could be explicated only within the higher-order construction **16**. Generally speaking, the key point of the semantic approach by Dunn and Belnap consists of taking the power set of some *basic set* of truth values and thereby obtaining a new set of generalized truth values, which

[11] Similarly, **T** and **F** from **4** can be viewed as analogues (or *representatives*) of the classical values T and F.

in turn should provide information *concerning this basic set* or, more specifically, information about the assignment of elements of the given base to a sentence. Thus, $\sigma(x) := \{x\}$ may naturally be viewed as an operation of "informatization". It creates a "piece of information" that refers to some "reality of a (one-step) lower order": the truth value $\{x\}$ is supposed to supply information just about x.[12]

It is now clearer what the difference between **N** and $\boldsymbol{N}$ consists of. The only feature of **N** is that it presents *no information* at all (relative to the truth values from the corresponding base). That is, in **4** the value **N** just gives *no* information concerning the classical truth values and in **16** concerning Belnap's truth values. But $\boldsymbol{N}$ in **16** is more expressive. Namely, it provides specific information saying that a sentence has been assigned Belnap's value **N**, which can be articulated by the *metastatement* that a sentence is neither classically true nor classically false.

The reader may also observe the difference between **FT** and **B**. Recall that Belnap's **B** is often interpreted as representing the idea of *paraconsistency*, the view that there are nontrivial contradictory theories. However, this interpretation makes sense only under some implicit linguistic convention, namely the assumption that truth and falsehood are, in effect, contradictory notions. The real, *logical* contradiction to truth, that does not depend on any assumption, is just *non*-truth, and a logical contradiction to falsehood is *non*-falsehood. Thus, **FT**—saying "false and not-true *as well as* not-false and true"—is not only more informative but seems to express the idea of a (nontrivial) contradictory truth value much better than **B** does.

Moreover, it appears that the second-order value **A** is inconsistent in an even stronger sense, stating that a sentence with this value is *not only* true-and-false, but also true and not-false, false and not-true, and neither-true-nor-false. Such a sentence takes all the values available at the level of first-order values, and it becomes clear that this idea of strengthening the notion of inconsistency can be extended to higher levels. If we define $\mathscr{P}_1(\mathbf{2}) := \mathscr{P}(\mathbf{2})$, and $\mathscr{P}_n(\mathbf{2}) := \mathscr{P}(\mathscr{P}_{n-1}(\mathbf{2}))$, then a sentence taking the value $\{x \mid x \in \mathscr{P}_n(\mathbf{2})\}$ is of maximal inconsistency of order $n+1$, while **B** can be seen to represent inconsistency of order 0.

It is also instructive to notice that **N, F, T** and **B** in $FOUR_2$ and $\boldsymbol{B}, \boldsymbol{N}, \boldsymbol{F}$, and $\boldsymbol{T}$ in $SIXTEEN_3$ are ordered differently under $\leq_t$ as well. Within $FOUR_2$, **T** is "more true" than **B**, and **N** is more true than **F**, whereas **N** and **B** are of the same truth level. But in $SIXTEEN_3$, $\boldsymbol{B}$ comes out more true than $\boldsymbol{N}$, while $\boldsymbol{T}$ and $\boldsymbol{B}$ as well as $\boldsymbol{N}$ and $\boldsymbol{F}$ are (pairwise) "equally true". This situation is entirely clear. As it was noted above, in $SIXTEEN_3$ (but not in $FOUR_2$) $\leq_t$ is a *pure truth relation*: it orders the truth values by exclusively taking into account what they say about *truth*, leaving any information about falsehood without attention.

[12] Thus, Belnap's informational interpretation of generalized truth values is not just an incidental *façon de parler*, but expresses the very essence of his construction. Therefore it is not by chance that this semantics has found so many fruitful applications in theoretical computer science and other areas related to information theory.

Clearly, meets and joins exist in *SIXTEEN*$_3$ for all three partial orders. We will use $\sqcap$ and $\sqcup$ with the appropriate subscripts for these operations under the corresponding ordering relations. *SIXTEEN*$_3$ appears then as the structure $(\mathbf{16}, \sqcap_i, \sqcup_i, \sqcap_t, \sqcup_t, \sqcap_f, \sqcup_f)$.

Now we would like to highlight one crucial point. It was already mentioned above that in *FOUR*$_2$ the partial order $\leq_t$ is often called (and, in fact, identified with) a "logical order", for it is supposed to determine the properties of logical connectives and the entailment relation. As we have seen, within *FOUR*$_2$ actually *two* ordering relations—one for truth and one for falsehood—are merged into one logical order. However, when proceeding one step higher to *SIXTEEN*$_3$, it turns out that an increase in truth does not necessarily mean a decrease in falsehood (and *vice versa*) any more. Hence, within *SIXTEEN*$_3$ the logical order explicitly splits into two distinct relations: the truth order $\leq_t$ and the falsity order $\leq_f$. To display both orders in a precise way, we present in Fig. 3.6, another projection of *SIXTEEN*$_3$ (namely the projection on the plain $\leq_t - \leq_f$). Here the falsity order is inverted because for defining central logical notions, we will be interested in decreasing (rather than increasing) falsehood.

Important properties of $\sqcap_t$ and $\sqcup_t$, as well as $\sqcap_f$ and $\sqcup_f$ are summarized in the following, directly checkable propositions:

Proposition 3.2 *For any x, y in SIXTEEN*$_3$:

1. $\mathbf{T} \in x \sqcap_t y \Leftrightarrow \mathbf{T} \in x$ *and* $\mathbf{T} \in y$;
   $\mathbf{B} \in x \sqcap_t y \Leftrightarrow \mathbf{B} \in x$ *and* $\mathbf{B} \in y$;
   $\mathbf{F} \in x \sqcap_t y \Leftrightarrow \mathbf{F} \in x$ *or* $\mathbf{F} \in y$;
   $\mathbf{N} \in x \sqcap_t y \Leftrightarrow \mathbf{N} \in x$ *or* $\mathbf{N} \in y$;
2. $\mathbf{T} \in x \sqcup_t y \Leftrightarrow \mathbf{T} \in x$ *or* $\mathbf{T} \in y$;
   $\mathbf{B} \in x \sqcup_t y \Leftrightarrow \mathbf{B} \in x$ *or* $\mathbf{B} \in y$;
   $\mathbf{F} \in x \sqcup_t y \Leftrightarrow \mathbf{F} \in x$ *and* $\mathbf{F} \in y$;
   $\mathbf{N} \in x \sqcup_t y \Leftrightarrow \mathbf{N} \in x$ *and* $\mathbf{N} \in y$.
3. $\mathbf{T} \in x \sqcup_f y \Leftrightarrow \mathbf{T} \in x$ *and* $\mathbf{T} \in y$;
   $\mathbf{N} \in x \sqcup_f y \Leftrightarrow \mathbf{N} \in x$ *and* $\mathbf{N} \in y$;
   $\mathbf{F} \in x \sqcup_f y \Leftrightarrow \mathbf{F} \in x$ *or* $\mathbf{F} \in y$;
   $\mathbf{B} \in x \sqcup_f y \Leftrightarrow \mathbf{B} \in x$ *or* $\mathbf{B} \in y$;
4. $\mathbf{T} \in x \sqcap_f y \Leftrightarrow \mathbf{T} \in x$ *or* $\mathbf{T} \in y$;
   $\mathbf{N} \in x \sqcap_f y \Leftrightarrow \mathbf{N} \in x$ *or* $\mathbf{N} \in y$;
   $\mathbf{F} \in x \sqcap_f y \Leftrightarrow \mathbf{F} \in x$ *and* $\mathbf{F} \in y$;
   $\mathbf{B} \in x \sqcap_f y \Leftrightarrow \mathbf{B} \in x$ *and* $\mathbf{B} \in y$.

In bilattices, a logical negation is usually defined as an operation that inverts the truth order only, leaving the information order unchanged. Fitting [87] also considers an operation of *conflation* that inverts $\leq_i$ while $\leq_t$ remains as it is. For trilattices this point has been generalized in [231], where several unary operations have been introduced under the general label of *inversion*. The idea is that an inversion can invert some partial orders in a trilattice, possibly leaving the other(s) without change. We then have the following definition:

Table 3.1 Inversions in *SIXTEEN*$_3$

a	$-_t a$	$-_f a$	$-_i a$	$-_{tf} a$	$-_{ti} a$	$-_{fi} a$	$-_{tfi} a$
N	**N**	**N**	**A**	**N**	**A**	**A**	**A**
N	***T***	***F***	**NFT**	***B***	**NTB**	**NFB**	**FTB**
F	***B***	***N***	**NFB**	***T***	**FTB**	**NFT**	**NTB**
T	***N***	***B***	**NTB**	***F***	**NFT**	**FTB**	**NFB**
B	***F***	***T***	**FTB**	***N***	**NFB**	**NTB**	**NFT**
NF	**TB**	**NF**	**NF**	**TB**	**TB**	**NF**	**TB**
NT	**NT**	**FB**	**NT**	**FB**	**NT**	**FB**	**FB**
FT	**NB**	**NB**	**NB**	**FT**	**FT**	**FT**	**NB**
NB	**FT**	**FT**	**FT**	**NB**	**NB**	**NB**	**FT**
FB	**FB**	**NT**	**FB**	**NT**	**FB**	**NT**	**NT**
TB	**NF**	**TB**	**TB**	**NF**	**NF**	**TB**	**NF**
NFT	**NTB**	**NFB**	***N***	**FTB**	***T***	***F***	***B***
NFB	**FTB**	**NFT**	***F***	**NTB**	***B***	***N***	***T***
NTB	**NFT**	**FTB**	***T***	**NFB**	**N**	***B***	***F***
FTB	**NFB**	**NTB**	***B***	**NFT**	***F***	***T***	***N***
A	**A**	**A**	**N**	**A**	**N**	**N**	**N**

Definition 3.6 Let $\mathscr{T}$ be a trilattice with three partial orders $\leq_t$, $\leq_f$, and $\leq_i$. Then we can introduce the following kinds of unary operations on $\mathscr{T}$ with the following properties:

1. t-inversion$(-_t)$:
(a) $a \leq_t b \Rightarrow -_t b \leq_t -_t a$;
(b) $a \leq_f b \Rightarrow -_t a \leq_f -_t b$;
(c) $a \leq_i b \Rightarrow -_t a \leq_i -_t b$;
(d) $-_t -_t a = a$.

2. f-inversion$(-_f)$:
(a) $a \leq_t b \Rightarrow -_f a \leq_t -_f b$;
(b) $a \leq_f b \Rightarrow -_f b \leq_f -_f a$;
(c) $a \leq_i b \Rightarrow -_f a \leq_i -_f b$;
(d) $-_f -_f a = a$.

3. i-inversion$(-_i)$:
(a) $a \leq_t b \Rightarrow -_i a \leq_t -_i b$;
(b) $a \leq_f b \Rightarrow -_i a \leq_f -_i b$;
(c) $a \leq_i b \Rightarrow -_i b \leq_i -_i a$;
(d) $-_i -_i a = a$.

4. tf-inversion$(-_{tf})$:
(a) $a \leq_t b \Rightarrow -_{tf} b \leq_t -_{tf} a$;
(b) $a \leq_f b \Rightarrow -_{tf} b \leq_f -_{tf} a$;
(c) $a \leq_i b \Rightarrow -_{tf} a \leq_i -_{tf} b$;
(d) $-_{tf} -_{tf} a = a$.

5. ti-inversion$(-_{ti})$:
(a) $a \leq_t b \Rightarrow -_{ti} b \leq_t -_{ti} a$;
(b) $a \leq_f b \Rightarrow -_{ti} a \leq_f -_{ti} b$;
(c) $a \leq_i b \Rightarrow -_{ti} b \leq_i -_{ti} a$;
(d) $-_{ti} -_{ti} a = a$.

6. fi-inversion$(-_{fi})$:
(a) $a \leq_t b \Rightarrow -_{if} a \leq_t -_{if} b$;
(b) $a \leq_f b \Rightarrow -_{if} b \leq_f -_{if} a$;
(c) $a \leq_i b \Rightarrow -_{if} b \leq_i -_{if} a$;
(d) $-_{if} -_{if} a = a$.

7. tfi-inversion$(-_{tfi})$:
(a) $a \leq_t b \Rightarrow -_{tif} b \leq_t -_{tif} a$;
(b) $a \leq_f b \Rightarrow -_{tif} b \leq_f -_{tif} a$;
(c) $a \leq_i b \Rightarrow -_{tif} b \leq_i -_{tif} a$;
(d) $-_{tif} -_{tif} a = a$.

In *SIXTEEN*$_3$ such inversion operations can be defined as shown in Table 3.1.

A routine calculation over this table immediately gives us the following proposition:

Proposition 3.3 *For any x in SIXTEEN*$_3$:

1. $-_t-_fx = -_f-_tx = -_{tf}x$;
2. $-_t-_ix = -_i-_tx = -_{ti}x$;
3. $-_f-_ix = -_i-_fx = -_{fi}x$;
4. $-_t-_f-_ix = -_f-_t-_ix = -_t-_i-_fx = -_f-_i-_tx =$
$-_i-_t-_fx = -_i-_f-_tx = -_{tf}-_ix = -_{ti}-_fx =$
$-_{fi}-_tx = -_i-_{tf}x = -_f-_{ti}x = -_t-_{fi}x = -_{tfi}x.$

Our main concern will naturally be focused on *t*-inversion, *f*-inversion, and *tf*-inversion as the most obvious candidates for representing an object-language negation. The following proposition highlights some key features of these operations that will be employed in the further analysis.

Proposition 3.4 *For any x in SIXTEEN*$_3$:

1. $\mathbf{T} \in -_tx \Leftrightarrow \mathbf{N} \in x$;	2. $\mathbf{T} \in -_fx \Leftrightarrow \mathbf{B} \in x$;	3. $\mathbf{T} \in -_{tf}x \Leftrightarrow \mathbf{F} \in x$;
$\mathbf{N} \in -_tx \Leftrightarrow \mathbf{T} \in x$;	$\mathbf{B} \in -_fx \Leftrightarrow \mathbf{T} \in x$;	$\mathbf{B} \in -_{tf}x \Leftrightarrow \mathbf{N} \in x$;
$\mathbf{F} \in -_tx \Leftrightarrow \mathbf{B} \in x$;	$\mathbf{F} \in -_fx \Leftrightarrow \mathbf{N} \in x$;	$\mathbf{F} \in -_{tf}x \Leftrightarrow \mathbf{T} \in x$;
$\mathbf{B} \in -_tx \Leftrightarrow \mathbf{F} \in x$;	$\mathbf{N} \in -_fx \Leftrightarrow \mathbf{F} \in x$;	$\mathbf{N} \in -_{tf}x \Leftrightarrow \mathbf{B} \in x$.

Note that $\sqcap_t, \sqcup_t$ and $-_t$ are now not the only algebraic operations that naturally correspond to logical conjunction, disjunction, and negation; $\sqcup_f, \sqcap_f$ and $-_f$ (or even $-_{tf}$) may play this role as well. Taking into account the fact that $x \sqcap_t y \neq x \sqcup_f y, x \sqcup_t y \neq x \sqcap_f y$ and $-_tx \neq -_fx$, we can state that both logical orders bring into existence "parallel" and, in fact, *distinct* logical connectives.

Thus, it seems rather natural to explore the possibility of a unified approach to all of these operations within a joint logical framework. To determine such a framework syntactically, we consider (in the most general case) the language $\mathscr{L}_{tf}$ that comprises $\wedge_t, \vee_t, \sim_t, \wedge_f, \vee_f$, and $\sim_f$ as propositional connectives. As to the semantics, let v^{16} (a 16-valuation) be a map from the set of propositional variables into *SIXTEEN*$_3$, and let us define:

Definition 3.7 For any $\mathscr{L}_{tf}$-formulas A and B:

1. $v^{16}(A \wedge_t B) = v^{16}(A) \sqcap_t v^{16}(B)$;
2. $v^{16}(A \vee_t B) = v^{16}(A) \sqcup_t v^{16}(B)$;
3. $v^{16}(\sim_t A) = -_t v^{16}(A)$;
4. $v^{16}(A \wedge_f B) = v^{16}(A) \sqcup_f v^{16}(B)$;
5. $v^{16}(A \vee_f B) = v^{16}(A) \sqcap_f v^{16}(B)$;
6. $v^{16}(\sim_f A) = -_f v^{16}(A)$.

This definition naturally extends a 16-valuation v^{16} to a valuation of compound formulas, thereby providing a natural machinery for an evaluation of arbitrary formulas from $\mathscr{L}_{tf}$.[13] In this way, *SIXTEEN*$_3$ allows a nontrivial coexistence of pairs of different (although analogous) logical connectives without collapsing them into each other. It may be helpful to think of $\wedge_t$, $\vee_t$, $\sim_t$ in terms of the *presence of truth* and to treat $\wedge_f$, $\vee_f$, $\sim_f$ as essentially highlighting the *absence of falsity* in the sense that, e.g., $\mathbf{T} \in x \sqcap_t y$ iff ($\mathbf{T} \in x$ and $\mathbf{T} \in y$); and $\mathbf{F} \notin x \sqcup_f y$ iff ($\mathbf{F} \notin x$ and $\mathbf{F} \notin y$). See also the corresponding discussion in Sect. 8.2.

Incidentally, one may notice that *SIXTEEN*$_3$ in a way "improves" some perhaps disputable aspects of *FOUR*$_2$. It has been observed that the account of the truth functions applied to the "nonstandard" truth values (**N** and **B**) in *FOUR*$_2$ looks a bit "puzzling" or even "odd" (see [4, pp. 516–518]). Indeed, intuitively it is not so evident why we should get $\mathbf{N} \wedge \mathbf{B} = \mathbf{F}$ and $\mathbf{N} \vee \mathbf{B} = \mathbf{T}$. *SIXTEEN*$_3$ offers in fact a quite different account of gaps and gluts. Their conjunction, for example, can produce again either a gap or a glut, depending on whether we wish to stress the presence of truth (using $\wedge_t$) or the absence of falsity (using $\wedge_f$).

3.6 Another Example of a Trilattice: Truth Values in Constructive Logics

In the concluding section of this chapter, we present another implementation of a trilattice isomorphic to *SIXTEEN*$_3$ based on what has been called in [231] a "truth values space of constructive logics". As it has been observed in [231, p. 767], the constructive character of such logics as, e.g., intuitionistic or minimal logic, finds its expression in what can be dubbed a 'constructive understanding of truth' which these logics adopt. A sentence is said to be constructively true if and only if it is proved. A natural counterpart of this notion of constructive truth is the notion of constructive falsity interpreted as refutation: a sentence is constructively false if and only if it is

[13] Note that according to Propositions 3.2 and 3.4, any 16-valuation for an arbitrary formula A can be unambiguously modeled by a certain combination of the expressions $\mathbf{N} \in v^{16}(A), \mathbf{F} \in v^{16}(A), \mathbf{T} \in v^{16}(A), \mathbf{B} \in v^{16}(A)$ and their negations. For example $v^{16}(A) = \mathbf{NT}$ is representable as $\mathbf{N} \in v^{16}(A), \mathbf{F} \notin v^{16}(A), \mathbf{T} \in v^{16}(A)$ and $\mathbf{B} \notin v^{16}(A)$, etc. This will greatly simplify the whole semantic exposition, see Chap. 5.

refuted. In terms of truth values, this means an acceptance of the corresponding constructive truth values: *the constructive truth* and *the constructive falsehood*.

Not any constructive logic obligatorily makes use of both these truth values. It may well be the case that, e.g., only the notion of truth employed in this or that logic is constructive, whereas falsity is taken in a non-constructive sense. Thus, the above mentioned pair of constructive truth values can be naturally complemented by two non-constructive ones, and in this way we arrive at the following four basic truth values for a truth value space of constructive logics considered in [231, p. 774]:

T—constructive truth (a sentence is *proved*);
F—constructive falsehood (a sentence is *refuted*);
t—non-constructive truth (a sentence is not refuted, i.e., it is *acceptable*);
f—non-constructive falsehood (a sentence is not proved, i.e., it is *rejectable*).

Intuitionistic logic can be interpreted as based on the pair of truth values $\langle \mathbf{T}, \mathbf{f} \rangle$. Another constructive logic, which involves both the constructive truth and the constructive falsehood $\langle \mathbf{T}, \mathbf{F} \rangle$, is David Nelson's logic of constructible falsity [180]. One can formulate dual intuitionistic logic as a purely falsificationistic logic in which only falsity is constructive, and truth is not: $\langle \mathbf{t}, \mathbf{F} \rangle$, see [229]. Dual intuitionistic logic has been considered by many authors, see, for example, [123, 203, 255, 271].

Now it is possible to obtain on the basis of the above four truth values the set of *generalized constructive truth values* by considering all the combinations of elements of the basic set[14]:

$$\{\{\}, \{\mathbf{T}\}, \{\mathbf{F}\}, \{\mathbf{t}\}, \{\mathbf{f}\}, \{\mathbf{T},\mathbf{F}\}, \{\mathbf{T},\mathbf{t}\}, \{\mathbf{T},\mathbf{f}\}, \{\mathbf{F},\mathbf{t}\}, \{\mathbf{F},\mathbf{f}\}, \{\mathbf{t},\mathbf{f}\},$$
$$\{\mathbf{T},\mathbf{F},\mathbf{t}\}, \{\mathbf{T},\mathbf{F},\mathbf{f}\}, \{\mathbf{T},\mathbf{t},\mathbf{f}\}, \{\mathbf{F},\mathbf{t},\mathbf{f}\}, \{\mathbf{T},\mathbf{F},\mathbf{t},\mathbf{f}\}\}.$$

One can straightforwardly define on this set an information order $\leq_i$ as the subset relation. Moreover, a logical order $\leq_l$ as a unified truth-and-falsity order can also be introduced. Namely, for every generalized constructive truth value x let us mark by x^{Tt} (x^{Ff}) the part of x that contains exactly those of the values **T** or **t** (**F** or **f**) that are in x. Then, it is possible to define: $x \leq_l y \Leftrightarrow x^{Tt} \subseteq y^{Tt}$ and $y^{Ff} \subseteq x^{Ff}$. One can also observe that as soon as we deal with generalized truth values of *constructive* logics, we have to take into account another important property that essentially characterizes each such truth value, the property of *constructivity*. Indeed, each of these truth values possesses not only some degree of information and truth but also a certain degree of constructivity. This means that we have here

[14] An approach to a generalization of truth values in intuitionistic logic and formulating on this base the notion of a *relevant intuitionistic entailment* has been developed in [222–226]. Dunn in [80] presents the results of a similar generalization of Nelson's logic. The construction below arises from a unification of both approaches into a joint framework. An intuitionistic (constructive) relevant logic has also been developed by some other authors, Garrel Pottinger and Neil Tennant among them, see, for example, [195, 249, 250].

Fig. 3.7 Trilattice *SIXTEEN*$_3^c$ (projection $\leq_i - \leq_t$)

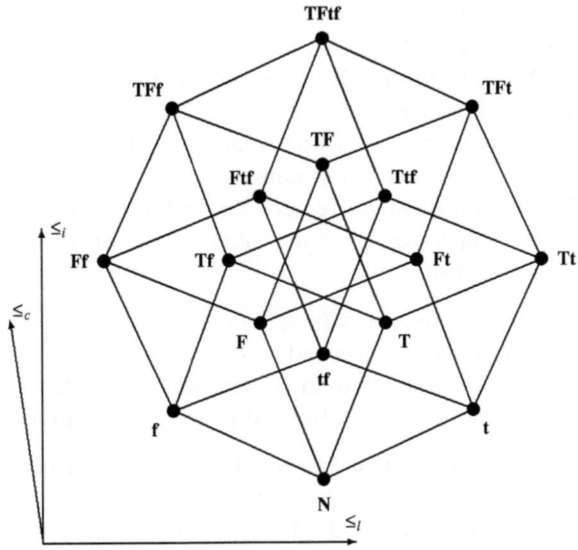

another partial ordering ($\leq_c$) which represents an increase in constructivity (or a decrease in non-constructivity). For every x, the part of x that contains exactly those of the values **T** or **F** (**t** or **f**) that are in x is marked by $x^{TF}(x^{tf})$. Then: $x \leq_c y \Leftrightarrow x^{TF} \subseteq y^{TF}$ and $y^{tf} \subseteq x^{tf}$.

We again omit brackets and commas in our notation for generalized truth values, and write **N** instead of $\{\}$. Each of the three partial orderings defined above form a complete lattice. **TFtf** and **N** are respectively the lattice top and bottom relative to $\leq_i$, **Tt** and **Ff** the bounds relative to $\leq_t$, and **TF** and **tf** relative to $\leq_c$. Indeed, **TFtf** and **N** are the most and the least *informative* elements in the set of sixteen generalized constructive truth values, **Tt** and **Ff** are the most and the least *true* (the least and the most *false*) of its elements, and **TF** and **tf** are the most and the least *constructive* elements. We denote the resulting trilattice as *SIXTEEN*$_3^c$ and present it by a triple Hasse diagram as in Fig. 3.7.

In this book, we will not delve further into an investigation of *SIXTEEN*$_3^c$. Our purpose in presenting it here is to give an interesting (and historically the first) exemplification of a sixteen-valued trilattice of truth values. We believe that such an investigation may constitute the subject of a separate research project on generalized constructive logics and their semantic foundations. In Chap. 7 we will present another approach to a generalization of intuitionistic logic on the base of *SIXTEEN*$_3$.

Chapter 4
Generalized Truth Values: $SIXTEEN_3$ and Beyond

Abstract In this chapter we define two principal entailment relations on $SIXTEEN_3$, one determined by the truth order and another determined by the falsity order. It turns out that the logics generated separately by the algebraic operations under the truth order and under the falsity order in $SIXTEEN_3$ coincide with the logic of $FOUR_2$, namely *first-degree entailment*. In the present setting, however, it becomes rather natural to consider also logical systems in the language obtained by combining the vocabulary of the logic of the truth order and the falsity order. Some important fragments of these combined logics are axiomatized. We also consider iterated powerset formation applied to **4** and introduce Belnap trilattices. These structures again give rise to relations of truth entailment and falsity entailment. It is observed that the logic of truth and the logic of falsity for every Belnap trilattice is one and the same, namely again, first-degree entailment. Finally, we consider Priest's notion of hyper-contradiction and examine some approaches to generalizations of Priest's logic of Paradox and Kleene's logic of uncertainty.

4.1 Entailment Relations on $SIXTEEN_3$

Returning back to $SIXTEEN_3$ we may notice—following Belnap [4, p. 518]—that at this point we have a nice algebraic structure, but we still do not have a *logic*. To get a full-fledged logic, a mere lattice of truth values is not enough (no matter how beautiful it is)—this lattice also has to be equipped with a suitable *entailment relation*. The canonical way to do so is to define entailment just through the logical order as it is done, e.g., in $FOUR_2$ by using $\leq_t$ (see Definition 3.2).

But in $SIXTEEN_3$—as it was noted above—we actually have *two* distinct logical orders (one for truth and one for falsity), and it would be hardly justifiable to prefer the truth order over the (non-)falsity order as "the most proper" representative of the notion of logical inference. This means that we get at least three

Y. Shramko and H. Wansing, *Truth and Falsehood*, Trends in Logic 36,
DOI: 10.1007/978-94-007-0907-2_4, © Springer Science+Business Media B.V. 2011

options: to consider the logic of the truth order (only), to deal with the logic of the falsity order (only), and to define a logic based on both orderings.

Recall that we consider languages $\mathscr{L}_t$, $\mathscr{L}_f$, and $\mathscr{L}_{tf}$ syntactically defined in Backus–Naur form as follows:

$$\mathscr{L}_t : A ::= p|{\sim_t}|\wedge_t|\vee_t$$
$$\mathscr{L}_f : A ::= p|{\sim_f}|\wedge_f|\vee_f$$
$$\mathscr{L}_{tf} : A ::= p|{\sim_t}|{\sim_f}|\wedge_t|\vee_t|\wedge_f|\vee_f$$

Belnap [4, p. 518] thinks of a logic as "rules for generating and evaluating inferences". That is, any logic has to provide means at least either for generating (valid) inferences (what is usually done by constructing a suitable syntactic logical system (calculus)), or for evaluating them (usually, by formulating a suitable class of semantical models). Let us first concentrate on the latter task. Typically a (semantic) definition of an entailment relation provides a method for checking the validity of (i.e., for evaluating) any inference. And clearly, we can use both the *truth order* and the *falsity order* of *SIXTEEN*$_3$ to obtain, in a straightforward way, the corresponding definitions (for arbitrary formulas $A, B \in \mathscr{L}_{tf}$). Consider first the ordering $\leq_t$.

Definition 4.1 $A \models_t^{16} B$ iff $\forall v^{16}(v^{16}(A) \leq_t v^{16}(B))$.[1]

This definition gives a precise semantic characterization of the logic that corresponds to the order $\leq_t$ in *SIXTEEN*$_3$. That is, the (semantically defined) logic $(\mathscr{L}_{tf}, \models_t^{16})$ is the set of all statements $A \models_t^{16} B$ with $A, B \in \mathscr{L}_{tf}$ such that for every 16-valuation v^{16}, $v^{16}(A) \leq_t v^{16}(B)$. Paraphrasing Belnap [4, p. 518] we might state that having an argument involving any inferential connections in the language $\mathscr{L}_{tf}$, we can now unambiguously decide whether "it is a good one" from the standpoint of the truth order.

The entailment relation in accordance with the falsity order can be naturally defined as follows.

Definition 4.2 $A \models_f^{16} B$ iff $\forall v^{16}(v^{16}(B) \leq_f v^{16}(A))$.

In this way the (semantically defined) logic $(\mathscr{L}_{tf}, \models_f^{16})$ is introduced as the set of all statements $A \models_f^{16} B$ with $A, B \in \mathscr{L}_{tf}$ such that for every 16-valuation v^{16}, $v^{16}(B) \leq_f v^{16}(A)$. It provides an apparatus for evaluating inferences from the standpoint of the falsity order.

It is not difficult to see that $\models_t^{16}$ and $\models_f^{16}$ present in fact different relations, since, e.g. $A \wedge_t B \models_t^{16} A$ holds, whereas $A \wedge_t B \models_f^{16} A$ does not. Thus, the trilattice *SIXTEEN*$_3$ comes with *two* natural definitions of *non-equivalent* entailment

[1] This definition introduces a relation between (two) formulas. It is not difficult to extend it to the case when we deal with a set of premises (see Definition 1.5), or to a relation between arbitrary sets of formulas (see, e.g., Definition 8.12 below).

relations reflecting an increase of truth and decrease of falsity. Therefore, it seems quite natural to conceive of the unified logic of *SIXTEEN*$_3$ as a *bi-consequence system* comprising two kinds of entailment relations.[2] This consideration leads us to the following definition:

Definition 4.3 The bi-consequence logic $(\mathscr{L}_{tf}, \models_t^{16}, \models_f^{16})$ is the set of all true statements $A \models_x^{16} B$, where $A, B \in \mathscr{L}_{tf}$, and $x = t$ or $x = f$.

The peaceful co-existence of two entailment relations in one and the same logic may be useful because it may well make a difference whether we reason along the truth order or the non-falsity order. The idea is that our reasoning may comprise more than just one kind of valid inferences, and we should be able to use the suitable argumentation pattern as applicable.

4.2 First-Degree Systems for *SIXTEEN*$_3$

Our next task is to obtain rules for *generating* (in a systematic way) valid inferences. To this effect we have to formalize the logics $(\mathscr{L}_{tf}, \models_t^{16})$ and $(\mathscr{L}_{tf}, \models_f^{16})$, i.e., to characterize them syntactically by means of suitable deductive systems. In this chapter we approach this task using the apparatus of first-degree consequence. We introduce complete (first-degree) systems for some important fragments of the language $\mathscr{L}_{tf}$, leaving the syntactic formalization of truth entailment and falsity entailment in extensions of $(\mathscr{L}_{tf}, \models_t^{16})$ and $(\mathscr{L}_{tf}, \models_f^{16})$ to the next chapter.

4.2.1 The Languages $\mathscr{L}_t$, $\mathscr{L}_f$ and Systems $\mathbf{FDE}_t^t$, $\mathbf{FDE}_f^f$

Let us first separately investigate logics that are generated by algebraic operations solely determined by the truth order and logics that are generated by algebraic operations solely determined by the falsity order.

We start with considering the language $\mathscr{L}_t$ with $\wedge_t$, $\vee_t$ and $\sim_t$ as propositional connectives. Having a 16-valuation v^{16}, we can standardly use Definition 3.7 (1)–(3) and Propositions 3.2 (1), (2) and 3.4 (1) for evaluating any formula of the language $\mathscr{L}_t$.

We prove now an important lemma, which will make our semantical considerations much easier:

[2] Cf. the formalism of "biconsequence relations" developed by A. Bochman in [38].

Lemma 4.1 *In SIXTEEN*$_3$ *the following clauses are all equivalent (for any* $A, B \in \mathscr{L}_t$):

$$\text{(a)}\ \forall v^{16}\big(\mathbf{F} \in v^{16}(B) \Rightarrow \mathbf{F} \in v^{16}(A)\big) \quad \text{(b)}\ \forall v^{16}\big(\mathbf{N} \in v^{16}(B) \Rightarrow \mathbf{N} \in v^{16}(A)\big)$$
$$\text{(c)}\ \forall v^{16}\big(\mathbf{T} \in v^{16}(A) \Rightarrow \mathbf{T} \in v^{16}(B)\big) \quad \text{(d)}\ \forall v^{16}\big(\mathbf{B} \in v^{16}(A) \Rightarrow \mathbf{B} \in v^{16}(B)\big).$$

Proof This is a generalization (and, in fact, an *extension*) of Dunn's analogous result for *FOUR*$_2$ (see, e.g., the proof of Proposition 4 in [80, p. 10]).

$(a) \Rightarrow (b)$: First, we define for any valuation v^{16} a *t-counterpart* valuation $v^{16\prime}$ as follows:

$$\mathbf{T} \in v^{16\prime}(p) \Leftrightarrow \mathbf{B} \in v^{16}(p); \quad \mathbf{B} \in v^{16\prime}(p) \Leftrightarrow \mathbf{T} \in v^{16}(p);$$
$$\mathbf{F} \in v^{16\prime}(p) \Leftrightarrow \mathbf{N} \in v^{16}(p); \quad \mathbf{N} \in v^{16\prime}(p) \Leftrightarrow \mathbf{F} \in v^{16}(p).$$

An easy induction extends $v^{16\prime}$ to any formula of $\mathscr{L}_t$.

Now, let $\forall v^{16}\big(\mathbf{F} \in v^{16}(B) \Rightarrow \mathbf{F} \in v^{16}(A)\big)$. Assume $\exists v^{16}(\mathbf{N} \in v^{16}(B)$ and $\mathbf{N} \notin v^{16}(A))$. Then $\mathbf{F} \in v^{16\prime}(B)$ and $\mathbf{F} \notin v^{16\prime}(A)$. A contradiction.

$(b) \Rightarrow (c)$: For any valuation v^{16} we define a *t-dual* valuation v^{16*} as follows:

$$\mathbf{T} \in v^{16*}(p) \Leftrightarrow \mathbf{N} \notin v^{16}(p); \quad \mathbf{B} \in v^{16*}(p) \Leftrightarrow \mathbf{F} \notin v^{16}(p);$$
$$\mathbf{F} \in v^{16*}(p) \Leftrightarrow \mathbf{B} \notin v^{16}(p); \quad \mathbf{N} \in v^{16*}(p) \Leftrightarrow \mathbf{T} \notin v^{16}(p);$$

and show by induction that it can be extended to any formula of $\mathscr{L}_t$.

Let $\forall v^{16}\big(\mathbf{N} \in v^{16}(B) \Rightarrow \mathbf{N} \in v^{16}(A)\big)$. Assume $\exists v^{16}(\mathbf{T} \in v^{16}(A)$ and $\mathbf{T} \notin v^{16}(B))$. Then $\mathbf{N} \notin v^{16*}(A)$ and $\mathbf{N} \in v^{16*}(B)$. A contradiction.

$(c) \Rightarrow (d)$: Let a *t*-counterpart valuation be defined as above, and let $\forall v^{16}(\mathbf{T} \in v^{16}(A) \Rightarrow \mathbf{T} \in v^{16}(B))$. Assume $\exists v^{16}\big(\mathbf{B} \in v^{16}(A)$ and $\mathbf{B} \notin v^{16}(B)\big)$. Then $\mathbf{T} \in v^{16\prime}(A)$ and $\mathbf{T} \notin v^{16\prime}(B)$. A contradiction.

$(d) \Rightarrow (a)$: Let a *t*-dual valuation be defined as above. Let $\forall v^{16}(\mathbf{B} \in v^{16}(A) \Rightarrow \mathbf{B} \in v^{16}(B))$. Assume $\exists v^{16}(\mathbf{F} \in v^{16}(B)$ and $\mathbf{F} \notin v^{16}(A))$. Then $\mathbf{B} \notin v^{16*}(B)$ and $\mathbf{B} \in v^{16*}(A)$. A contradiction. □

To what extent Lemma 4.1 simplifies our analysis should be clear from the following lemma:

Lemma 4.2 *For any* $A, B \in \mathscr{L}_t$: $A \models_t^{16} B$ *iff* $\forall v^{16}(\mathbf{T} \in v^{16}(A) \Rightarrow \mathbf{T} \in v^{16}(B))$.

Proof In view of the previous lemma, our claim is equivalent with

For any $A, B \in \mathscr{L}_t$: $A \models_t^{16} B$ iff

$$(*) \quad \forall v^{16}\forall y(((\exists x \in \mathbf{16})\ y \in x \,\&\, T \in y) \Rightarrow (y \in v^{16}(A) \Rightarrow y \in v^{16}(B)))$$
$$\& \ \forall v^{16}\forall y(((\exists x \in \mathbf{16})\ y \in x \,\&\, T \notin y) \Rightarrow (y \in v^{16}(B) \Rightarrow y \in v^{16}(A))).$$

$\Rightarrow$: If $A \models_t^{16} B$, by definition, $\forall v^{16}(v^{16}(A) \leq_t v^{16}(B))$, and thus (i) $v^{16}(A)^t \subseteq v^{16}(B)^t$ and (ii) $v^{16}(B)^{-t} \subseteq v^{16}(A)^{-t}$. By (i), if $T \in y \in v^{16}(A)$, then $y \in v^{16}(B)$. By (ii), if $T \notin y \in v^{16}(B)$, then $y \in v^{16}(A)$.

$\Leftarrow$: Suppose (*) holds. We must show that 1. $v^{16}(A)^t \subseteq v^{16}(B)^t$ and 2. $v^{16}(B)^{-t} \subseteq v^{16}(A)^{-t}$. Ad 1.: Let $y \in v^{16}(A)^t$. Then $T \in y$ and $y \in v^{16}(A)$. Suppose $y \notin v^{16}(B)^t$. Then $y \notin v^{16}(B)$ or $T \notin y$. But since $T \in y$, we have $y \notin v^{16}(B)$. Since $y \in v^{16}(A)$, by (*), $y \in v^{16}(B)$, a contradiction. Ad 2.: Let $y \in v^{16}(B)^{-t}$. Then $T \notin y$ and $y \in v^{16}(B)$. Suppose $y \notin v^{16}(A)^{-t}$. Then $y \notin v^{16}(A)$ or $T \in y$. But since $T \notin y$, we have $y \notin v^{16}(A)$. By (*), $y \notin v^{16}(B)$, a contradiction. □

Now we are in a good position to answer the question about the syntactic system that corresponds to the entailment relation introduced by Definition 4.1 when $A, B \in \mathscr{L}_t$. For formulas built up from $\wedge_t$, $\vee_t$ and $\sim_t$ this relation can be axiomatized by a (first-degree) consequence system which we call $\mathbf{FDE}^t_t$. The superscript indicates the type of language used, and the subscript explicates the kind of consequence. The system is thus a pair $(\mathscr{L}_t, \vdash_t)$, where $\vdash_t$ is a binary relation (consequence) on the language $\mathscr{L}_t$ satisfying the following postulates (axiom schemes and rules of inference):

$a_t1.\ A \wedge_t B \vdash_t A$
$a_t2.\ A \wedge_t B \vdash_t B$
$a_t3.\ A \vdash_t A \vee_t B$
$a_t4.\ B \vdash_t A \vee_t B$
$a_t5.\ A \wedge_t (B \vee_t C) \vdash_t (A \wedge_t B) \vee_t C$
$a_t6.\ A \vdash_t \sim_t\sim_t A$
$a_t7.\ \sim_t\sim_t A \vdash_t A$

$r_t1.\ A \vdash_t B, B \vdash_t C / A \vdash_t C$
$r_t2.\ A \vdash_t B, A \vdash_t C / A \vdash_t B \wedge_t C$
$r_t3.\ A \vdash_t C, B \vdash_t C / A \vee_t B \vdash_t C$
$r_t4.\ A \vdash_t B / \sim_t B \vdash_t \sim_t A.$

Note that the postulates of $\mathbf{FDE}^t_t$ are direct analogues of the postulates of the *first-degree entailment* system $\mathbf{E}_{fde}$ from [3, p. 158] (Dunn in [77] dubs it $\mathbf{R}_{fde}$, emphasizing thus the fact that it is simultaneously a fragment of both relevance logics **E** and **R**). In our view, this observation reinforces the claim that the logic of first-degree entailment is a significant logical system that can be arrived at using different well-motivated approaches.

First, we prove the consistency (soundness) of $\mathbf{FDE}^t_t$ relative to $\models^{16}_t$.

Theorem 4.1 *For any $A, B \in \mathscr{L}_t$: If $A \vdash_t B$, then $A \models^{16}_t B$.*

Proof Taking into account Lemma 4.2, it suffices to prove that (1) if $A \vdash_t B$ is an axiom of $\mathbf{FDE}^t_t$, then $\forall v^{16}(\mathbf{T} \in v^{16}(A) \Rightarrow \mathbf{T} \in v^{16}(B))$, and (2) all the rules of $\mathbf{FDE}^t_t$ preserve this property. This is mainly a routine check (employing Propositions 3.2 (1), (2) and 3.4 (1)) and can be safely left to the reader, except for r_t4 which, in addition, needs Lemma 4.1 for its justification. Indeed, assume $A \models^{16}_t B$, i.e., $\forall v^{16}(\mathbf{T} \in v^{16}(A) \Rightarrow \mathbf{T} \in v^{16}(B))$ (Lemma 4.2). Then, by Lemma 4.1, $\forall v^{16}(\mathbf{N} \in v^{16}(B) \Rightarrow \mathbf{N} \in v^{16}(A))$. Suppose $\sim_t B \not\models^{16}_t \sim_t A$, i.e., $\exists v^{16}(\mathbf{T} \in v^{16}(\sim_t B)$ and

$\mathbf{T} \notin v^{16}(\sim_t A))$. Then $\exists v^{16}(\mathbf{N} \in v^{16}(B)$ and $\mathbf{N} \notin v^{16}(A))$ (Proposition 3.4)—a contradiction. □

To prove completeness we have to construct a suitable canonical model. Let a *theory* be a set of sentences closed under $\vdash_t$ (i.e., for every theory α, if $A \in \alpha$ and $A \vdash_t B$, then $B \in \alpha$) and $\wedge_t$ (if $A \in \alpha$ and $B \in \alpha$, then $A \wedge_t B \in \alpha$). A theory α is *prime* iff the following holds: if $A \vee_t B \in \alpha$, then $A \in \alpha$ or $B \in \alpha$. The following fact about prime theories (Lindenbaum's lemma) is very well known; a proof is given, e.g., in [80, p. 13]:

Lemma 4.3 *For any A and* $B \in \mathscr{L}_t$, *if* $A \nvdash_t B$, *then there exists a prime theory* α *such that* $A \in \alpha$ *and* $B \notin \alpha$.

We next consider *ordered pairs* of prime theories. For any ordered pair of prime theories $\mathscr{T} = \langle \alpha_1, \alpha_2 \rangle$ we define the canonical 16-valuation $v^{16}_{\mathscr{T}}$ as follows:

$$\begin{array}{ll} \mathbf{N} \in v^{16}_{\mathscr{T}}(p) \text{ iff } \sim_t p \in \alpha_1; & \mathbf{F} \in v^{16}_{\mathscr{T}}(p) \text{ iff } \sim_t p \in \alpha_2; \\ \mathbf{T} \in v^{16}_{\mathscr{T}}(p) \text{ iff } p \in \alpha_1; & \mathbf{B} \in v^{16}_{\mathscr{T}}(p) \text{ iff } p \in \alpha_2. \end{array}$$

Now we can show that the canonical 16-valuation so defined is naturally extended to any formula of the language:

Lemma 4.4 *Let* $v^{16}_{\mathscr{T}}$ *be defined as above. Then for any formula* $A \in \mathscr{L}_t$:

$$\begin{array}{ll} \mathbf{N} \in v^{16}_{\mathscr{T}}(A) \text{ iff } \sim_t A \in \alpha_1; & \mathbf{F} \in v^{16}_{\mathscr{T}}(A) \text{ iff } \sim_t A \in \alpha_2; \\ \mathbf{T} \in v^{16}_{\mathscr{T}}(A) \text{ iff } A \in \alpha_1; & \mathbf{B} \in v^{16}_{\mathscr{T}}(A) \text{ iff } A \in \alpha_2. \end{array}$$

Proof This is a usual induction on the construction of formulas. We show only the case with negation while leaving other cases to the reader.

Let $A = \sim_t B$, and suppose the lemma holds for B. Then we have:

$\mathbf{N} \in v^{16}_{\mathscr{T}}(\sim_t B) \Leftrightarrow \mathbf{T} \in v^{16}_{\mathscr{T}}(B)$ (Proposition 3.4) $\Leftrightarrow B \in \alpha_1$ (inductive assumption) $\Leftrightarrow \sim_t\sim_t B \in \alpha_1$ (a_t6).

$\mathbf{F} \in v^{16}_{\mathscr{T}}(\sim_t B) \Leftrightarrow \mathbf{B} \in v^{16}_{\mathscr{T}}(B)$ (Proposition 3.4) $\Leftrightarrow B \in \alpha_2$ (inductive assumption) $\Leftrightarrow \sim_t\sim_t B \in \alpha_2$ (a_t6).

$\mathbf{T} \in v^{16}_{\mathscr{T}}(\sim_t B) \Leftrightarrow \mathbf{N} \in v^{16}_{\mathscr{T}}(B)$ (Proposition 3.4) $\Leftrightarrow \sim_t B \in \alpha_1$ (inductive assumption).

$\mathbf{B} \in v^{16}_{\mathscr{T}}(\sim_t B) \Leftrightarrow \mathbf{F} \in v^{16}_{\mathscr{T}}(B)$ (Proposition 3.4) $\Leftrightarrow \sim_t B \in \alpha_2$ (inductive assumption). □

And finally, the following theorem establishes completeness:

Theorem 4.2 *For any* $A, B \in \mathscr{L}_t$: *If* $A \models^{16}_t B$, *then* $A \vdash_t B$.

Proof Let $A \models^{16}_t B$. For the sake of contradiction, assume $A \nvdash_t B$. Then, by Lemma 4.3, there exists a prime theory α such that $A \in \alpha$ and $B \notin \alpha$. Consider a pair of prime theories $\mathscr{T} = \langle \beta_1, \beta_2 \rangle$ such that $\alpha = \beta_1$, and β_2 is arbitrary. Then we have $\mathbf{T} \in v^{16}_{\mathscr{T}}(A)$ and $\mathbf{T} \notin v^{16}_{\mathscr{T}}(B)$. A contradiction (by Lemma 4.2). □

It is quite natural to suppose that the logic determined by the falsity order in a given language should be perfectly *dual* to the logic of the truth order in this language. Namely, we should be able to obtain axiomatic presentations of these logics from one another simply by exchanging the '*t*' and '*f*' subscripts in the postulates of the corresponding systems.

Thus, we can consider the language $\mathscr{L}_f$ with propositional connectives $\wedge_f$, $\vee_f$, and $\sim_f$. A 16-valuation is extended to compound formulas of $\mathscr{L}_f$ by Definition 3.7 (4)–(6) and Propositions 3.2 (3), (4) and 3.4 (2). We obtain the system $\mathbf{FDE}^f_f = (\mathscr{L}_f, \vdash_f)$ just by replacing $\wedge_t$, $\vee_t$, $\sim_t$ and $\vdash_t$ in the axioms and rules of $\mathbf{FDE}^t_t$ by $\wedge_f$, $\vee_f$, $\sim_f$, and $\vdash_f$, respectively. The proofs of soundness and completeness of $\mathbf{FDE}^f_f$ with respect to the relation $\models^{16}_f$ (restricted to the language $\mathscr{L}_f$) introduced by Definition 4.2 are analogous *mutatis mutandis* to the corresponding proofs for $\mathbf{FDE}^t_t$.

4.2.2 The Language $\mathscr{L}_{tf}$ for $\leq_t$ and $\leq_f$

The systems $\mathbf{FDE}^t_t$ and $\mathbf{FDE}^f_f$ are quite standard. Each of these systems is formulated in a customary language with conjunction, disjunction and negation, and it is equipped with a suitable entailment relation corresponding to the truth ordering and the falsity ordering in *SIXTEEN*$_3$, respectively. But what kind of logical systems can we get by axiomatizing the relations $\models^{16}_t$ (introduced by Definition 4.1) and $\models^{16}_f$ (Definition 4.2) when they are extended to richer fragments of the language $\mathscr{L}_{tf}$ (or even up to the whole language)?

Let us first consider the relation $\models^{16}_t$. It turns out that in the presence of formulas with $\wedge_f$, $\vee_f$ and $\sim_f$, the previous Lemmas 4.1 and 4.2 require some essential modifications (restrictions).

Lemma 4.5 *For any* $A, B \in \mathscr{L}_{tf}$ *in SIXTEEN*$_3$*, the clause (a) from Lemma 4.1 is equivalent to the clause (c), and (b) is equivalent to (d).*

Proof We prove here only the case $(c) \Rightarrow (a)$ because the other cases are analogous.

For any 16-valuation v^{16} we define a *tf-dual* valuation $v^{16\#}$ as follows:

$$\mathbf{T} \in v^{16\#}(p) \Leftrightarrow \mathbf{F} \notin v^{16}(p); \quad \mathbf{B} \in v^{16\#}(p) \Leftrightarrow \mathbf{N} \notin v^{16}(p);$$
$$\mathbf{F} \in v^{16\#}(p) \Leftrightarrow \mathbf{T} \notin v^{16}(p); \quad \mathbf{N} \in v^{16\#}(p) \Leftrightarrow \mathbf{B} \notin v^{16}(p).$$

An induction on the construction of formulas shows that these conditions also hold for any formula $A \in \mathscr{L}_{tf}$. Let $\forall v^{16}(\mathbf{T} \in v^{16}(A) \Rightarrow \mathbf{T} \in v^{16}(B))$ and assume there exists a 16-valuation v^{16} such that $\mathbf{F} \in v^{16}(B)$ and $\mathbf{F} \notin v^{16}(A)$. Then $\mathbf{T} \notin v^{16\#}(B)$ and $\mathbf{T} \in v^{16\#}(A)$, which contradicts the assumption. □

Lemma 4.6 *For any* $A, B \in \mathscr{L}_{tf}$:

$$A \models_t^{16} B \text{ iff } (a)\ \forall v^{16}(\mathbf{T} \in v^{16}(A) \Rightarrow \mathbf{T} \in v^{16}(B)) \textit{ and}$$
$$(b)\ \forall v^{16}(\mathbf{B} \in v^{16}(A) \Rightarrow \mathbf{B} \in v^{16}(B)).$$

Proof Mutatis mutandis, as in Lemma 4.2, taking into account that (*) from Lemma 4.2 is also equivalent to the right hand side of the statement in the present lemma. □

The logical system that should correspond to the logic $(\mathscr{L}_{tf}, \models_t^{16})$ defined in the previous section can be denoted as $\mathbf{FDE}_t^{tf} = (\mathscr{L}_{tf}, \vdash_t)$.

As a first step toward axiomatizing $\mathbf{FDE}_t^{tf}$, we enrich the language $\mathscr{L}_t$ by another negation operator and consider the language $\mathscr{L}_{t+\sim_f} := \{\wedge_t, \vee_t, \sim_t, \sim_f\}$. Now we can introduce the system $\mathbf{FDE}_t^{t+\sim_f} = (\mathscr{L}_{t+\sim_f}, \vdash_t)$, where $\vdash_t$ is the binary relation on $\mathscr{L}_{t+\sim_f}$ satisfying the axioms and rules of inference $a_t1 - a_t7$, $r_1 - r_4$ stated above, as well as the following additional postulates for $\sim_f$:

$a_t8.\ A \vdash_t \sim_f\sim_f A$
$a_t9.\ \sim_f\sim_f A \vdash_t A$
$a_t10.\ \sim_f\sim_t A \vdash_t \sim_t\sim_f A$

$r_t5.\ A \vdash_t B / \sim_f A \vdash_t \sim_f B.$

It is not difficult to show (using Lemmas 4.5 and 4.6) that Theorem 4.1 holds for any formula of the language $\mathscr{L}_{t+\sim_f}$. That is, we have:

Theorem 4.3 *For any* $A, B \in \mathscr{L}_{t+\sim_f}$: *If* $A \vdash_t B$, *then* $A \models_t^{16} B$.

Note that the following statements are theorems of $\mathbf{FDE}_t^{t+\sim_f}$:

$t_t1.\ \sim_f(A \wedge_t B) \vdash_t \sim_f A \wedge_t \sim_f B$;
$t_t2.\ \sim_f A \wedge_t \sim_f B \vdash_t \sim_f(A \wedge_t B)$;
$t_t3.\ \sim_f A \vee_t \sim_f B \vdash_t \sim_f(A \vee_t B)$;
$t_t4.\ \sim_f(A \vee_t B) \vdash_t \sim_f A \vee_t \sim_f B$;
$t_t5.\ \sim_t\sim_f A \vdash_t \sim_f\sim_t A$.

To prove completeness, we continue to deal with prime theories closed under $\vdash_t$. But now for any theory α, we define the set of formulas

$$\alpha^* := \{A \mid \sim_f A \in \alpha\}.$$

Lemma 4.7 *Let* α *be a theory and let* α^* *be defined as above. Then:*

1. α^* *is a theory;*
2. $\sim_f A \in \alpha^*$ *iff* $A \in \alpha$;
3. α^* *is prime iff* α *is prime.*

Proof

1. Assume $A \vdash_t B$ and $A \in \alpha^*$. Then, by r_t5 $\sim_f A \vdash_t \sim_f B$ and by definition of $\alpha^*, \sim_f A \in \alpha$. Hence $\sim_f B \in \alpha$, and thus $B \in \alpha^*$. Next, assume $A \in \alpha^*$ and $B \in \alpha^*$. Then $\sim_f A \in \alpha$ and $\sim_f B \in \alpha$. Hence, $\sim_f A \wedge_t \sim_f B \in \alpha$, and by t_t2 $\sim_f(A \wedge_t B) \in \alpha$. By definition of $\alpha^*, A \wedge_t B \in \alpha^*$.
2. $\sim_f A \in \alpha^* \Leftrightarrow \sim_f\sim_f A \in \alpha$ (by definition) $\Leftrightarrow A \in \alpha$ (by a_t8, a_t9).
3. $\Rightarrow$: Assume α is not prime. Then there are A and B such that $A \vee_t B \in \alpha$, and $A \notin \alpha$, and $B \notin \alpha$. Then, by (2) above, $\sim_f(A \vee_t B) \in \alpha^*$, and $\sim_f A \notin \alpha^*$, and $\sim_f B \notin \alpha^*$. By t_t4, $\sim_f A \vee_t \sim_f B \in \alpha^*$, and hence α^* is not prime. $\Leftarrow$: Assume α^* is not prime. Then there are A and B such that $A \vee_t B \in \alpha^*$, and $A \notin \alpha^*$, and $B \notin \alpha^*$. By definition of α^*, we have: $\sim_f(A \vee_t B) \in \alpha$, and $\sim_f A \notin \alpha$, and $\sim_f B \notin \alpha$. Arguing as above we conclude that α is not prime. □

The definition of α^* and Lemma 4.7 immediately call to mind the famous "Routley star operator" used for defining a negation operator in the "Australian semantics" for relevance logic. In fact, the Routley star $*$ represents an algebraic operation known as *involution* (see, e.g., [1, 32, 138]). In view of this, $\sim_f$ can be naturally interpreted as an object language *involution connective* with respect to $\vdash_t$, whereas $\sim_t$ stands for a negation relative to $\vdash_t$. It is quite remarkable that *SIXTEEN*$_3$ allows us to deal with $\sim_f$ and $\sim_t$ simultaneously, thereby delivering interesting new evidence for a deep interrelation between the "Australian" and the "American" semantics (cf. [79, pp. 45–47]).

For any prime theory α we define the canonical 16-valuation v^{16}_α as follows[3]:

$$\mathbf{N} \in v^{16}_\alpha(p) \text{ iff } \sim_t p \in \alpha; \quad \mathbf{F} \in v^{16}_\alpha(p) \text{ iff } \sim_t p \in \alpha^*;$$
$$\mathbf{T} \in v^{16}_\alpha(p) \text{ iff } p \in \alpha; \quad \mathbf{B} \in v^{16}_\alpha(p) \text{ iff } p \in \alpha^*.$$

We can then prove a valuation lemma:

Lemma 4.8 *Let v^{16}_α be defined as above. Then for any formula $A \in \mathscr{L}_{t+\sim_f}$:*

$$\mathbf{N} \in v^{16}_\alpha(A) \textit{ iff } \sim_t A \in \alpha; \quad \mathbf{F} \in v^{16}_\alpha(A) \textit{ iff } \sim_t A \in \alpha^*;$$
$$\mathbf{T} \in v^{16}_\alpha(A) \textit{ iff } A \in \alpha; \quad \mathbf{B} \in v^{16}_\alpha(A) \textit{ iff } A \in \alpha^*.$$

Proof Again, as in the proof of Lemma 4.4, we consider only the case with negation. Let $A = \sim_f B$, and the lemma holds for B. Then we have:

$\mathbf{N} \in v^{16}_\alpha(\sim_f B) \Leftrightarrow \mathbf{F} \in v^{16}_\alpha(B)$ (Proposition 3.4) $\Leftrightarrow \sim_t B \in \alpha^*$ (inductive assumption) $\Leftrightarrow \sim_f\sim_t B \in \alpha$ (definition of α^*) $\Leftrightarrow \sim_t\sim_f B \in \alpha$ (a_t10).

[3] Incidentally, it turns out that it is possible to apply a similar construction in the completeness proof for $\mathbf{FDE}^t_t$, too. Namely, consider $\alpha^* := \{A \mid \sim_t A \notin \alpha\}$. Then, even for formulas of the "pure" language $\mathscr{L}_t$, we can simply define a canonical valuation v^{16}_α for any prime theory α as above (instead of dealing with arbitrary pairs of prime theories and the canonical valuation $v^{16}_{\mathscr{T}}$). However, in this case such a construction is not necessary, and one can content oneself with just pairs of theories which are totally independent of each other.

$\mathbf{F} \in v^{16}_\alpha(\sim_f(B)) \Leftrightarrow \mathbf{N} \in v^{16}_\alpha(B)$ (Proposition 3.4) $\Leftrightarrow \sim_t B \in \alpha$ (inductive assumption) $\Leftrightarrow \sim_f\sim_t B \in \alpha^*$ (Lemma 4.7.2) $\Leftrightarrow \sim_t\sim_f B \in \alpha^*$ (a_t10).

$\mathbf{T} \in v^{16}_\alpha(\sim_f(B)) \Leftrightarrow \mathbf{B} \in v^{16}_\alpha(B)$ (Proposition 3.4) $\Leftrightarrow B \in \alpha^*$ (inductive assumption) $\Leftrightarrow \sim_f(B) \in \alpha$ (definition of α^*).

$\mathbf{B} \in v^{16}_\alpha(\sim_f(B)) \Leftrightarrow \mathbf{T} \in v^{16}_\alpha(B)$ (Proposition 3.4) $\Leftrightarrow B \in \alpha$ (inductive assumption) $\Leftrightarrow \sim_f(B) \in \alpha^*$ (Lemma 4.7.2). □

This immediately leads us to a proof of the desired completeness theorem:

Theorem 4.4 *For any* $A, B \in \mathscr{L}_{t+\sim_f}$: *If* $A \models^{16}_t B$, *then* $A \vdash_t B$.

Proof Let $A \models^{16}_t B$. Assume $A \nvdash_t B$. Then, by Lemma 4.3 there exists a prime theory α such that $A \in \alpha$ and $B \notin \alpha$. Taking the canonical valuation v^{16}_α, we have $\mathbf{T} \in v^{16}_\alpha A$ and $\mathbf{T} \notin v^{16}_\alpha B$. A contradiction (by Lemma 4.6). □

It is interesting to observe that by setting $\sim A := \sim_f\sim_t A$, we obtain another unary connective, which appears to be another kind of negation. It is not difficult to show that the following statements are theorems of $\mathbf{FDE}^{t+\sim_f}_t$:

t_t6. $\sim(A \wedge_t B) \vdash_t \sim A \vee_t \sim B$;
t_t7. $\sim A \wedge_t \sim B \vdash_t \sim(A \vee_t B)$;
t_t8. $\sim A \vee_t \sim B \vdash_t \sim(A \wedge_t B)$;
t_t9. $\sim(A \vee_t B) \vdash_t \sim A \wedge_t \sim B$;
t_t10. $A \vdash_t \sim\sim A$;
t_t11. $\sim\sim A \vdash_t A$,

Also contraposition holds:

r_t6. $A \vdash_t B / \sim B \vdash_t \sim A$.

Semantically, $\sim$ is characterized by Proposition 3.4 (3) and the following definition:

Definition 4.4 For any A and B: $v^{16}(\sim A) = -_{tf}v^{16}(A)$.

The logic $(\mathscr{L}_{tf}, \models^{16}_f)$ and the system $\mathbf{FDE}^{tf}_f$ are defined in analogy to the definitions of $(\mathscr{L}_{tf}, \models^{16}_t)$ and $\mathbf{FDE}^{tf}_t$. On the way to axiomatizing $\mathbf{FDE}^{tf}_f$, we consider the language $\mathscr{L}_{f+\sim_t} := \{\wedge_f, \vee_f, \sim_f, \sim_t\}$. We can introduce the system $\mathbf{FDE}^{f+\sim_t}_f = (\mathscr{L}_{f+\sim_t}, \vdash_f)$, where $\vdash_f$ is the binary relation on $\mathscr{L}_{f+\sim_t}$ satisfying the dualized versions of the axioms and rules of $(\mathscr{L}_{t+\sim_f}, \vdash_t)$.

As to the connective $\sim$ defined above, it can be shown that in $\mathbf{FDE}^{f+\sim_t}_f$ the statements $t_f6 - t_f11$ and the rule r_f6 hold, which are obtained from $t_t6 - t_t11$ and r_t6 by uniformly replacing the subscript '*t*' with '*f*'. Thus, whereas $\sim_t$ is a negation connective relative to $\leq_t$ but an involution connective relative to $\leq_f$ and $\sim_f$ is a negation for $\leq_f$ but an involution for $\leq_t$, $\sim$ is a negation connective with respect to *both* logical orderings. Therefore, $\sim$ can be seen as a *generalized logical negation*.

The problem of constructing complete logical systems for the whole language $\mathscr{L}_{tf}$ will be considered in the next chapter.

4.3 First-Degree Everywhere

In the previous section we have, in particular, shown that the logics determined *separately* by the algebraic operations under the truth order and under the falsity order in *SIXTEEN*$_3$ coincide with the logic of *FOUR*$_2$, namely they turned out to be first-degree entailment. In the present section we extend this result to the infinite case and argue to the effect that Belnap's strategy of generalizing the set $\mathbf{2} = \{T, F\}$ of classical truth values is not only coherent but also stabilizing. At any stage, no matter how far it goes, the logic of the truth (falsity) order is again first-degree entailment.

Let X be a basic set of truth values, $\mathscr{P}^1(X) := \mathscr{P}(X)$ and $\mathscr{P}^n(X) := \mathscr{P}(\mathscr{P}^{n-1}(X))$ for $1 < n$, $n \in \mathbb{N}$. We obtain an infinite collection of sets of generalized truth values by considering the sets $\mathscr{P}^n(\mathbf{4})$. As the starting point of our construction we choose $\mathscr{P}^n(\mathbf{4})$ and not $\mathscr{P}^n(\mathbf{2})$ because the truth order in *FOUR*$_2$ is not just a truth order but rather a truth-and-falsity order. The information ordering $\leq_i$ on any set of generalized truth values is just the subset relation. In order to define a truth ordering $\leq_t$ on $\mathscr{P}^n(\mathbf{4})$, we define for every $x \in \mathscr{P}^n(\mathbf{4})$ the set x^t of its "truth-containing" elements and the set x^{-t} of its "truthless" elements:

$$x^t := \{y_0 \in x \mid (\exists y_1 \in y_0)\,(\exists y_2 \in y_1) \ldots (\exists y_{n-1} \in y_{n-2})\; T \in y_{n-1}\};$$
$$x^{-t} := \{y_0 \in x \mid \neg(\exists y_1 \in y_0)\,(\exists y_2 \in y_1) \ldots (\exists y_{n-1} \in y_{n-2})\; T \in y_{n-1}\}.$$

To define a falsity ordering $\leq_f$ on $\mathscr{P}^n(\mathbf{4})$, we define for every $x \in \mathscr{P}^n(\mathbf{4})$ the set x^f of its "falsity-containing" elements and the set x^{-f} of its "falsityless" elements analogously:

$$x^f := \{y_0 \in x \mid (\exists y_1 \in y_0)\,(\exists y_2 \in y_1) \ldots (\exists y_{n-1} \in y_{n-2})\; F \in y_{n-1}\};$$
$$x^{-f} := \{y_0 \in x \mid \neg(\exists y_1 \in y_0)\,(\exists y_2 \in y_1) \ldots (\exists y_{n-1} \in y_{n-2})\; F \in y_{n-1}\}.$$

Thus, $x^{-t} = x \setminus x^t$ and $x^{-f} = x \setminus x^f$.[4] We call x t-positive (t-negative, f-positive, f-negative) iff x^t (x^{-t}, x^f, x^{-f}) is non-empty. We denote by $\mathscr{P}^n(\mathbf{4})^t$ ($\mathscr{P}^n(\mathbf{4})^{-t}$, $\mathscr{P}^n(\mathbf{4})^f$, $\mathscr{P}^n(\mathbf{4})^{-f}$) the set of all t-positive (t-negative, f-positive, f-negative) elements of $\mathscr{P}^n(\mathbf{4})$.

Now all three partial orders can be introduced for any $\mathscr{P}^n(\mathbf{4})$ by means of Definition 3.5 suitably modified (i.e., with x^t, x^{-t}, x^f, and x^{-f} redefined as above). We next define an important class of trilattices which we call *Belnap trilattices*:

[4] The definitions of x^t, x^{-t}, x^f and x^{-f} generalize the corresponding definitions from Sect. 3.5.

Definition 4.5 A *Belnap trilattice* is a structure

$$\mathcal{M}_3^n := (\mathcal{P}^n(\mathbf{4}), \sqcap_i, \sqcup_i, \sqcap_t, \sqcup_t, \sqcap_f, \sqcup_f),$$

where $\sqcap_i$ ($\sqcap_t$, $\sqcap_f$) is the lattice meet and $\sqcup_i$ ($\sqcup_t$, $\sqcup_f$) is the lattice join with respect to the ordering $\leq_i$ ($\leq_t$, $\leq_f$) on $\mathcal{P}^n(\mathbf{4})$.

Thus, *SIXTEEN*$_3$ ($=\mathcal{M}_3^1$) is the smallest Belnap trilattice. Furthermore, if we have a Belnap trilattice, we may wish to equip it with appropriate operations of inversion, as it is done, e.g., for *SIXTEEN*$_3$ in accordance with Definition 3.6. Let us generalize the approach to t-, f-, and i-inversions developed in Sect. 3.5. Recall that in a multilattice with several partial orders an operation of this kind should not only invert the corresponding ordering, but also preserve all the other orders.

Definition 4.6 Let $\mathcal{M}_n = (S, \leq_1, \ldots, \leq_n)$ be a multilattice and $1 \leq j \leq n$. Then a unary operation $-_j$ on S is said to be a (pure) j-inversion iff the following conditions are satisfied:

$$\begin{array}{ll} (iso) & x \leq_1 y \Rightarrow -_j x \leq_1 -_j y; \\ & \vdots \\ (anti) & x \leq_j y \Rightarrow -_j y \leq_j -_j x; \\ & \vdots \\ (iso) & x \leq_n y \Rightarrow -_j x \leq_n -_j y; \\ (per2) & -_j -_j x = x. \end{array}$$

Moreover, for certain orderings it could be useful to consider combined inversion operations so that, e.g., 23-inversion would invert simultaneously both $\leq_2$ and $\leq_3$, leaving the other partial orders untouched.

Condition (*anti*) from Definition 4.6 means that $-_j$ is antitone with respect to $\leq_j$, whereas the last condition is just the period two property. Both conditions determine on a given multilattice a *dual automorphism* for $\leq_j$, i.e., the property $x \leq_j y$ iff $-_j y \leq_j -_j x$. Birkhoff [34] calls a dual automorphism of period 2 *involution*. However, nowadays it is more customary to call an involution just any operator of period 2. Dunn [3, p. 193] calls an operation satisfying the conditions (*anti*) and (*per*2) from Definition 4.6 *intensional complementation* and remarks that it is not generally a Boolean complementation. In [81, Sect. 3.13] various kinds of *non-classical complementation* are considered by introducing various conditions for these operations. In a multilattice-framework, though, it is important that any inversion defined relative to some partial order is not only antitone with respect to this very order but also *isotone* with respect to all the remaining orderings.

Note that Definition 4.6 is a sort of a "frame-definition" which only sets up the basic conditions for some operation, *provided this operation in fact exists*. By itself this definition does not guarantee the existence of the corresponding operation in every case (for a given structure). Table 3.1 from Sect. 3.5 defines t-inversion and f-inversion for *SIXTEEN*$_3$. The following theorem guarantees the existence of such operations for any Belnap trilattice.

Theorem 4.5 *For any Belnap trilattice* $\mathcal{M}_3^n$ *there exist t-inversions and f-inversions on* $\mathcal{P}^n(\mathbf{4})$.

Proof For any $\mathcal{M}_3^n$ we can define an operation of t-inversion in a canonical way as follows. Let $x \in \mathcal{P}^n(\mathbf{4})$. If x is empty, we define $-_t x = x$. If $x \neq \varnothing$, we define $-_t x$ by considering the elements $y \in x$. Every $y \in x$ contains, at some depth of nesting, elements from $\mathbf{4}$, i.e., $\varnothing$, $\{T\}$, $\{F\}$, or $\{T,F\}$. We replace these elements according to the following instruction:

$\varnothing$ is replaced by $\{T\}$
$\{T\}$ is replaced by $\varnothing$
$\{F\}$ is replaced by $\{F,T\}$
$\{F,T\}$ is replaced by $\{F\}$.

Example If x is the value $\{\{\varnothing,\{F\},\{F,T\}\},\{\{T\},\{F,T\}\}\}$, then $-_t x = \{\{\{T\},\{F,T\},\{F\}\},\{\varnothing,\{F\}\}\}$.

In other words, for every element of $\mathbf{4}$ in $\mathcal{P}^n(\mathbf{4})$, $-_t$ introduces the classical T, where it is absent and excludes T from where it is present. Obviously, this definition of $-_t x$ preserves the information order $\leq_i$ since x and $-_t x$ have the same cardinality. The falsity ordering $\leq_f$ is preserved, too, because the inclusion or exclusion of T has no effect on the presence or absence of F. And clearly, the truth ordering $\leq_t$ is inverted by definition and $-_t -_t x = x$.

The canonical definition of an f-inversion is analogous. Let again $x \in \mathcal{P}^n(\mathbf{4})$. If x is empty, we define $-_f x = x$. If $x \neq \varnothing$, every $y \in x$ contains at some depth of nesting elements from $\mathbf{4}$. We replace these elements according to the following rule:

$\varnothing$ is replaced by $\{F\}$
$\{T\}$ is replaced by $\{F,T\}$
$\{F\}$ is replaced by $\varnothing$
$\{F,T\}$ is replaced by $\{T\}$,

and observe that $-_f$ so defined satisfies the conditions required by Definition 4.6. □

Thus, in what follows we can, without loss of generality, consider Belnap trilattices with t-inversions and f-inversions.

Clearly, the lattice top $\mathbf{1}_t$ of $\mathcal{M}_3^n$ with respect to the truth order $\leq_t$ (the lattice top $\mathbf{1}_f$ with respect to the falsity order $\leq_f$) is $\mathcal{P}^{n-1}(\mathbf{4})^t$ ($\mathcal{P}^{n-1}(\mathbf{4})^f$), and the lattice bottom $\mathbf{0}_t$ with respect to $\leq_t$ ($\mathbf{0}_f$ with respect to $\leq_f$) is $\mathcal{P}^{n-1}(\mathbf{4})^{-t}$ ($\mathcal{P}^{n-1}(\mathbf{4})^{-f}$). Since $\mathbf{0}_t \leq_t -_t\mathbf{1}_t$ and since $\mathbf{0}_t \leq_t -_t\mathbf{1}_t$ iff $\mathbf{1}_t \leq_t -_t\mathbf{0}_t$, $\mathbf{1}_t = -_t\mathbf{0}_t$. Moreover, $\mathbf{0}_t = -_t\mathbf{1}_t$, $\mathbf{1}_f = -_f\mathbf{0}_f$, and $\mathbf{0}_f = -_f\mathbf{1}_f$. Note that any operation $-_t$ ($-_f$) satisfying the conditions of Definition 4.6 satisfies the De Morgan laws with respect to $\sqcap_t$ and $\sqcup_t$ ($\sqcap_f$ and $\sqcup_f$).

Lemma 4.9

(a) *Suppose, x is t-positive. Then*

1. $x \in (y \sqcap_t z)$ *iff* $(x \in y$ *and* $x \in z)$ *and*
2. $x \in (y \sqcup_t z)$ *iff* $(x \in y$ *or* $x \in z)$.

(b) *Suppose, x is t-negative. Then*

1. $x \in (y \sqcap_t z)$ *iff* $(x \in y$ *or* $x \in z)$ *and*
2. $x \in (y \sqcup_t z)$ *iff* $(x \in y$ *and* $x \in z)$.

Proof (a) 1. $\Rightarrow$: Obvious from the definition of meets and $\leq_t$. $\Leftarrow$: Suppose $x \in y$, $x \in z$, but $x \notin (y \sqcap_t z)$. Let $X := y^{-t} \cup z^{-t}$. Then $\{x\} \cup X \leq_t y$ and $\{x\} \cup X \leq_t z$, but $\{x\} \cup X \nleq_t (y \sqcap_t z)$, implying that $y \sqcap_t z$ is not the greatest lower bound. (a) 2. $\Leftarrow$: Obvious from the definition of joins and $\leq_t$. $\Rightarrow$: Suppose $x \in (y \sqcup_t z)$ and $x \notin y$, $x \notin z$. Then $y \leq_t (y \cup z)^t$ and $z \leq_t (y \cup z)^t$, but $(y \sqcup_t z) \nleq_t (y \cup z)^t$, implying that $(y \sqcup_t z)$ is not the least upper bound. (b) 1. $\Leftarrow$: Obvious from the definition of meets and $\leq_t$. $\Rightarrow$: Suppose $x \in (y \sqcap_t z)$ and $x \notin y$, $x \notin z$. Then $(y \cup z)^{-t} \leq_t y$ and $(y \cup z)^{-t} \leq_t z$, but $(y \cup z)^{-t} \nleq (y \sqcap_t z)$ (since $(y \sqcap_t z)^{-t} \nsubseteq (y \cup z)^{-t-t} = (y \cup z)^{-t}$), implying that $(y \sqcap_t z)$ is not the greatest lower bound. (b) 2. $\Rightarrow$: Obvious from the definition of joins and $\leq_t$. $\Leftarrow$: Suppose $x \in y$, $x \in z$, but $x \notin (y \sqcup_t z)$. Define now $X := y^t \cup z^t$. Then $y \leq_t \{x\} \cup X$, $z \leq_t \{x\} \cup X$, and $(y \sqcup_t z) \nleq_t \{x\} \cup X$, from which it follows that $(y \sqcup_t z)$ is not the least upper bound. □

Lemma 4.10 *For any* $\mathscr{P}^n(\mathbf{4})$:

1. $x^{-t} \neq \varnothing$ *iff* $(-_t x)^t \neq \varnothing$.
2. $x^t \neq \varnothing$ *iff* $(-_t x)^{-t} \neq \varnothing$.

Proof First, we observe that $-_t\varnothing = \varnothing$. This holds, because $\varnothing \subseteq -_t\varnothing$ iff $\varnothing \leq_i -_t \varnothing$ iff $-_t\varnothing \leq_i -_t -_t\varnothing$ (from left to right directly by condition (*iso*) Definition 4.6; from right to left by (*iso*) and (*per*2) the same definition) iff $-_t\varnothing \leq_i \varnothing$ iff $-_t\varnothing \subseteq \varnothing$. (1) $\Rightarrow$: Suppose $y \in x^{-t}$ but $(-_t x)^t = \varnothing$. Then $-_t x \leq_t (-_t x)^t$. Since $(-_t x)^t = \varnothing = -_t\varnothing$, we have $-_t x \leq_t -_t \varnothing$ and thus $\varnothing \leq_t x$, a contradiction with $y \in x^{-t}$. $\Leftarrow$: Suppose $y \in (-_t x)^t$ but $x^{-t} = \varnothing$. Then $\varnothing \leq_t x$ and hence $-_t x \leq_t \varnothing$, a contradiction with $y \in (-_t x)^t$. (2) $\Rightarrow$: Suppose $y \in x^t$ but $(-_t x)^{-t} = \varnothing$. Then $(-_t x)^{-t} \leq_t -_t x$. Therefore $x \leq_t \varnothing$, a contradiction with $y \in x^t$. $\Leftarrow$: Suppose $y \in (-_t x)^{-t}$ but $x^t = \varnothing$. Then $x \leq_t -_t \varnothing$ and hence $\varnothing \leq_t -_t x$, a contradiction with $y \in (-_t x)^{-t}$. □

Consider again the languages $\mathscr{L}_t$, $\mathscr{L}_f$, and $\mathscr{L}_{tf}$ introduced in Sect. 4.1. An n-valuation is a function v^n from the set of propositional variables into $\mathscr{P}^n(\mathbf{4})$. We may employ Definition 3.7 (with the obvious replacement of v^{16} by v^n) to extend any n-valuation to an interpretation in $\mathscr{P}^n(\mathbf{4})$ for arbitrary formulas of the corresponding languages.

Now we can generalize Definitions 4.1 and 4.2 and define the notions of t-entailment and f-entailment for any n (for arbitrary formulas A, B from $\mathscr{L}_{tf}$):

Definition 4.7 $A \models_t^n B$ iff $\forall v^n\, (v^n(A) \leq_t v^n(B))$.

Definition 4.8 $A \models_f^n B$ iff $\forall v^n\, (v^n(B) \leq_f v^n(A))$.

Our next task will be to axiomatize these notions for the languages $\mathscr{L}_t$ and $\mathscr{L}_f$ separately.

We shall use the following lemma to show that the syntactic first-degree entailment system is complete with respect to the semantic entailment relation $\models_t^n$, provided this relation is restricted to $\mathscr{L}_t$-formulas.

Lemma 4.11 *The following two statements are equivalent for all* $A, B \in \mathscr{L}_t$:

1. $\forall v^n \forall y\, (y \in v^n(A)^t \Rightarrow y \in v^n(B)^t)$;
2. $\forall v^n \forall y\, (y \in v^n(B)^{-t} \Rightarrow y \in v^n(A)^{-t})$.

Proof To prove the lemma, we shall construct "dual" valuations from a given valuation. For every $y \in \mathscr{P}^{n-1}(\mathbf{4})^{-t}$, let g_y be a map from $\mathscr{P}^{n-1}(\mathbf{4})^{-t}$ into $\mathscr{P}^{n-1}(\mathbf{4})^t$, and for every y in $\mathscr{P}^{n-1}(\mathbf{4})^t$, let $f_y\colon \mathscr{P}^{n-1}(\mathbf{4})^t \longrightarrow \mathscr{P}^{n-1}(\mathbf{4})^{-t}$). In view of Lemma 4.10, we may choose g_y and f_y such that $g_y(y) \in x^t$ iff $y \in -_t x$ and $f_y(y) \in x^{-t}$ iff $y \in -_t x$. Given a valuation v^n and y in $\mathscr{P}^{n-1}(\mathbf{4})$, we stipulate:

$$
\begin{aligned}
&g_y(y) \in v_y^{n*}(p) \text{ iff } y \notin v^n(p)\\
&\qquad x \in v_y^{n*}(p) \text{ iff } f_x(x) \in v_y^{n*}(p) \text{ if } x \in \mathscr{P}^{n-1}(\mathbf{4})^t \text{ and } x \neq g_y(y)\\
&f_y(y) \in v_y^{n*}(p) \text{ iff } y \notin v^n(p)\\
&\qquad x \in v_y^{n*}(p) \text{ iff } g_x(x) \in v_y^{n*}(p) \text{ if } x \in \mathscr{P}^{n-1}(\mathbf{4})^{-t} \text{ and } x \neq f_y(y)
\end{aligned}
$$

We must show that v^{n*} can be extended to a valuation function for any $A \in \mathscr{L}_t$. For $x \neq g_y(y)$ and $x \neq f_y(y)$, the claim follows from the remaining cases.

Case 1 $A = (\sim_t B)$.

$$
\begin{aligned}
g_y(y) \in v_y^{n*}(\sim_t B) &\Leftrightarrow g_y(y) \in -_t v_y^{n*}(B)\\
&\Leftrightarrow f_{g_y(y)}(g_y(y)) \in v_y^{n*}(B)\\
&\Leftrightarrow g_y(y) \notin v^n(B)\\
&\Leftrightarrow y \notin -_t v^n(B)\\
&\Leftrightarrow y \notin v^n(\sim_t B).
\end{aligned}
$$

The case for $f_y(y)$ is analogous.

Case 2 $A = (B \wedge_t C)$. We must show that (a) $g_y(y) \in v_y^{n*}(B \wedge_t C)$ iff $y \notin v^n(B \wedge_t C)$ and (b) $f_y(y) \in v_y^{n*}(B \wedge_t C)$ iff $y \notin v^n(B \wedge_t C)$. (a) $\Leftarrow$: Suppose $g_y(y) \notin v_y^{n*}(B \wedge_t C)$

and $y \notin v^n(B \wedge_t C)$. Since $v^n(B)^{-t} \subseteq (v^n(B) \sqcap_t v^n(C))^{-t}$ and $v^n(C)^{-t} \subseteq (v^n(B) \sqcap_t v^n(C))^{-t}$, $y \notin v^n(B)$ and $y \notin v^n(C)$. From Lemma 4.9 and $g_y(y) \notin (v_y^{n*}(B) \sqcap_t v_y^{n*}(C))$ it follows that $g_y(y) \notin v_y^{n*}(B)$ or $g_y(y) \notin v_y^{n*}(C)$. Thus $y \in v^n(B)$ or $y \in v^n(C)$, a contradiction. $\Rightarrow$: Use the induction hypothesis and Lemma 4.9. (b): analogous.

Case 3 $A = (B \vee_t C)$. The proof is similar to the previous case.

Now we can complete the proof of the lemma. (i) $\Rightarrow$ (ii): Suppose (i) holds and there exists a t-negative element y and a valuation v^n such that (a) $y \in v^n(B)$ but (b) $y \notin v^n(A)$. By (a) and the definition of v_y^{n*}, $g_y(y) \notin v_y^{*n}(B)$ and (by (i) and (b)) $g_y(y) \in v_y^{*n}(B)$, a contradiction. (ii) $\Rightarrow$ (i): analogous. □

Corollary 4.1 *For any* $A, B \in \mathscr{L}_t$:

$$A \models_t^n B \text{ iff } \forall v^n (x \in v^n(A)^t \Rightarrow x \in v^n(B)^t).$$

Proof In view of the previous lemma, the claim is equivalent with: for any $A, B \in \mathscr{L}_t$, $A \models_t^n B$ iff conditions (i) and (ii) of Lemma 4.1 hold. We have:

$$\begin{aligned} A \models_t^n B \ &\text{iff } \forall v^n \ (v^n(A) \leq_t v^n(B)) \\ &\text{iff } \forall v^n \ (v^n(A)^t \subseteq v(B)^t \ \& \ v^n(B)^{-t} \subseteq v^n(A)^{-t}) \\ &\text{iff (i) and (ii)} \end{aligned}$$

□

Consider again the proof system $\mathbf{FDE}_t^t = (\mathscr{L}_t, \vdash_t)$ introduced in Sect. 4.2.1.

Theorem 4.6 *(Soundness) For all* A, $B \in \mathscr{L}_t$, $A \vdash_t B$ *implies* $A \models_t^n B$.

Proof We show that for every axiom $A \vdash_t B$, we have $\forall v^n \ (v^n(A) \leq_t v^n(B))$, and that this property is preserved by the rules. This is a simple exercise; consider for example rule r_t4: Suppose $v^n(A) \leq_t v^n(B)$ but $v^n(\sim_t B) \nleq_t v^n(\sim_t A)$. Then $v^n(\sim_t B) \nleq_t v^n(\sim_t A) \Leftrightarrow -_t v^n(\sim_t A) \nleq_t -_t v^n(\sim_t B) \Leftrightarrow v^n(\sim_t \sim_t A) \nleq_t v^n(\sim_t \sim_t B) \Leftrightarrow v^n(A) \nleq_t v^n(B)$. □

To prove completeness, we construct a simple canonical model employing the notion of a (prime) theory as defined in Sect. 4.2.1. Clearly, Lemma 4.3 (Lindenbaum) continues to hold.

Let α be a prime theory. We define the canonical n-valuations $v_y^{n\tau}$ as follows, using the functions g_y and f_y from Lemma 4.11:

$$g_y(y) \in v_y^{n\tau}(p) \text{ iff } p \in \alpha$$
$$x \in v_y^{n\tau}(p) \text{ iff } f_x(x) \in v_y^{n\tau}(p) \text{ if } x \in \mathscr{P}^{n-1}(\mathbf{4})^t \text{ and } x \neq g_y(y)$$
$$f_y(y) \in v_y^{n\tau}(p) \text{ iff } \sim_t p \in \alpha$$
$$x \in v_y^{n\tau}(p) \text{ iff } g_x(x) \in v_y^{n\tau}(p) \text{ if } x \in \mathscr{P}^{n-1}(\mathbf{4})^{-t} \text{ and } x \neq f_y(y)$$

Lemma 4.12 *The canonical n-valuations $v^{n\tau}$ can be extended to a valuation function for any $A \in \mathscr{L}_t$.*

Proof For $x \neq g_y(y)$ and $x \neq f_y(y)$, the claim follows from the remaining cases.

Case 1 $A = (\sim_t B)$.

$$\begin{aligned} f_y(y) \in v_y^{n\tau}(\sim_t B) &\Leftrightarrow f_y(y) \in -_t v_y^{n\tau}(B) \\ &\Leftrightarrow g_{f_y(y)}(f_y(y)) \in v_y^{n\tau}(B) \\ &\Leftrightarrow B \in \alpha \\ &\Leftrightarrow \sim_t \sim_t B \in \alpha. \end{aligned}$$

The case for $g_y(y)$ is analogous.

Case 2 $A = (B \wedge_t C)$. We must show that (a) $g_y(y) \in v_y^{n\tau}(B \wedge_t C)$ iff $(B \wedge_t C) \in \alpha$ and (b) $f_y(y) \in v_y^{n\tau}(B \wedge_t C)$ iff $\sim_t(B \wedge_t C) \in \alpha$. (b) Assume $\sim_t(B \wedge_t C) \in \alpha$. By the De Morgan laws, primeness, and closure under $\vdash_t$, this is the case iff $\sim_t B \in \alpha$ or $\sim_t C \in \alpha$. By the induction hypothesis, the latter holds iff $f_y(y) \in v^{n\tau}B$ or $f_y(y) \in v^{n\tau}C$, which, by Lemma 4.9, holds iff $f_y(y) \in (v^{n\tau}B \sqcap_t v^{n\tau}C)$ iff $f_y(y) \in v^{n\tau}(B \wedge_t C)$. (a): analogous.

Case 3 $A = (B \vee_t C)$. The proof is similar to the previous case. □

Theorem 4.7 *(Completeness) For all A, $B \in \mathscr{L}_t$, $A \models_t^n B$ implies $A \vdash_t B$.*

Proof Suppose $A \models_t^n B$ but $A \nvdash_t B$. Then there exists a prime theory α such that $A \in \alpha$ but $B \notin \alpha$. Then $g_y(y) \in v^{n\tau}(A)$ but $g_y(y) \notin v^{n\tau}(B)$ and thus there exits a valuation v^n such that $v^n(A)^t \not\subseteq v^n(B)^t$. □

By making all the necessary adjustments—namely by replacing uniformly the subscript t with f in all proofs and formulations—we also obtain the following theorem:

Theorem 4.8 *For any $A, B \in \mathscr{L}_f$: $A \models_f^n B$ iff $A \vdash_f B$.*

4.4 Hyper-Contradictions and Generalizations of Priest's Logic

In [197], Graham Priest argues in favor of generalizing the semantics of his *Logic of Paradox* presented in [196]. According to Priest [197, p. 237], "[t]here is growing evidence that the logical paradoxes (and perhaps some other kinds of assertions) are both true and false", and since he claims that "a sentence must have *some* value at least", Priest's preferred set of truth values is $\mathbf{3} = \{\mathbf{F}, \mathbf{T}, \mathbf{B}\}$. **F** means "is false *only*" and **T** means "is true *only*", whereas **B** is to be understood as "is *both* true and false". The elements from **3** again can be represented as certain *non-empty* subsets of the set of classical truth values T (Truth) and F (Falsehood). Thus, again, $\mathbf{T} = \{T\}$, $\mathbf{F} = \{F\}$ and $\mathbf{B} = \{F, T\}$.

Priest suggests considering "higher-order" combinations of truth values from **3** and beyond. The motivation for this is a "revenge Liar" argument, leading to so-called "impossible values" or *hyper-contradictions* [197, p. 239]. Here we present a slightly modified version:

> Consider the sentence: (*) *This sentence is false only*. Against the background of **3**, (*) is either (i) true only, (ii) false only, or (iii) both true and false. (i): If (*) is true only, what (*) says is true and hence the sentence is true only *and* false only. In other words, (*) takes the impossible value $\boldsymbol{B}= \{\{F\}, \{T\}\}$ not available in **3**. (ii): If (*) is false only, what (*) says is not true, and thus the sentence is either true only or both true and false. Hence (*) is either false only and true only, or it is false only and both true and false. That is, (*) takes the impossible value $\boldsymbol{B}$ or the impossible value $\{\{F\}, \mathbf{B}\}$. (iii): Suppose (*) is both true and false. Then in particular it is true and thus takes an impossible value $\{\{F\}, \mathbf{B}\}$.

Incidentally, it is well-known that another version of **3**, namely the set $\mathbf{3}' = \{\mathbf{F}, \mathbf{T}, \mathbf{N}\}$ (the set of truth values of Kleene's three-valued logics), where **N** stands for "unknown" or "is *neither* true nor false" ($=\varnothing$, a "truth-value gap") also gives rise to a revenge Liar:

> Consider the sentence: (**) *This sentence is false*[5] *or neither true nor false*. Against the background of $\mathbf{3}'$, (**) is either (i) true, (ii) false, or (iii) neither true nor false. (i): If (**) is true, we have to consider two cases. If (**) is false, (**) takes the impossible value $\boldsymbol{B}$; if (**) is neither true nor false, it takes the impossible value $\{\mathbf{N}, \{T\}\}$. (ii): If (**) is false, what (**) says is not the case. Hence the sentence is true and takes the impossible value $\boldsymbol{B}$. (iii): Suppose (**) is neither true nor false. Then in particular it is not true, and hence (**) takes the impossible value $\{\mathbf{N}, \{T\}\}$.

According to Priest, a sentence always takes at least some value, and the paradoxes reveal that some sentences are both true and false; by contrast according to Keith Simmons [240, p. 119], "[t]he claim that Liar sentences are gappy seems natural enough—after all, the assumption that they are true or false leads to a contradiction."

In any case, both (*) and (**) show that admittedly the only way to escape the revenge Liar is to introduce higher-order truth values such as $\{\{F\}, \{T\}\}$, $\{\{T\}, \{F, T\}\}$ and so on. To do so, Priest defines for any non-empty set of truth values S_n the corresponding higher-order set S_{n+1} as follows: $S_{n+1} = \mathscr{P}(S_n)\backslash\{\varnothing\}$ for all $n \in \mathbb{N}$, where S_0 is just the set **2** of classical truth values ($= \{F, T\}$). Then he introduces the following definition for evaluating compound formulas on each level:

Definition 4.9 Given the classical truth value functions $\wedge_0, \vee_0, \sim_0$ on S_0:

1. $x \wedge_{n+1} y = \{z : \exists x' \in x \exists y' \in y(z = x' \wedge_n y')\}$;
2. $x \vee_{n+1} y = \{z : \exists x' \in x \exists y' \in y(z = x' \vee_n y')\}$;
3. $\sim_{n+1} x = \{z : \exists x' \in x(z = \sim_n x')\}$.

---

[5] In the absence of **B** we can well refer to the values $\{F\}$ and $\{T\}$ just as "false" and "true", respectively.

Next, Priest defines the map $\sigma(x) = \{x\}$ and shows that σ is an isomorphism between any S_n and $\sigma[S_n]$ ($= \{\sigma(x) | x \in S_n\}$). By virtue of this fact, Priest identifies S_n with $\sigma[S_n]$, $\wedge_n$ with $\wedge_{n+1}$ restricted to $\sigma[S_n]$, etc. He then defines the set $S = \bigcup_n S_n$ and introduces on S generalized logical operators $\wedge, \vee$ and $\neg$ in an analogous way (so that, e.g., $\wedge = \bigcup_n \wedge_n$, etc.). Finally he singles out the set of designated values $\mathcal{D}$ so that a value is designated just if it contains T at some depth of membership. This allows him to define a relation of logical consequence in the usual way, i.e., by introducing valuation functions v of the sentences of the language under consideration into S.

Definition 4.10 $\Sigma \models A$ iff $\forall v : \exists B \in \Sigma\ v(B) \notin \mathcal{D}$, or $v(A) \in \mathcal{D}$.

The main result of [197] is that $\models$ in fact coincides with the consequence relation of Priest's Logic of Paradox from [196], i.e., $\models = \models_1$.[6] That is, Priest tells us, "hyper-contradictions make no difference: the first contradiction $\{1,0\}$ of S_1 changes the consequence relation... Subsequent contradictions have no effect" [197, p. 241].

Pragati Jain [133] extends the result by Priest in that she does not collect the S_n together to form the set S, but keeps each S_n distinct and defines semantic consequence relations $\models_n$ for any n accordingly. Then she shows that if we define the sets $\mathcal{D}_n$ of designated values following Priest's definition (a truth value is designated just in case T occurs in it somewhere), the following holds: for each n, $\models_n = \models_1$. However, as Jain points out, this result (as well as the result by Priest) holds only relative to the given choice of designated elements, and a different choice would have produced different results. Thus, the coincidence with the Logic of Paradox does not necessarily mean that there is something "special" about S_1. Rather it indicates that there is something special about the choice of designated values in each case, see also Chap. 9.

But what should be the criteria for choosing the set of designated truth values? Is it a completely arbitrary (subjective) procedure, or can we provide some theoretical framework which would determine the choice? And, more generally, what are the relationships between (changes of) the set of designated truth values and the relation of logical entailment? It is well-known, for example, that the difference between Kleene's three-valued logic and Priest's Logic of Paradox can be described as a different choice of designated values. Whereas in Kleene's logic **T** is designated, in Priest's logic the values **T** and **B** are designated, see, for instance, [198, pp. 122, 124].

Instead of specifying a set of designated values and defining entailment as the preservation of possessing a designated value, entailment may also be defined from a given partial order $\leq$ on the (non-empty) set of truth values by setting: $A \models B$ iff for every valuation v, $v(A) \leq v(B)$ (cf. Definitions 4.1 and 4.2). Given a set $\mathcal{D}$ of designated truth values, a partial order $\leq_{\mathcal{D}}$ may be defined by requiring that

[6] One obtains the relation $\models_1$ if one applies Definition 4.10 to the set S_1.

$x \leq_{\mathscr{D}} y$ iff $(x \in \mathscr{D} \Rightarrow y \in \mathscr{D})$. It seems that for a given partial order $\leq$, there is no canonical way of defining a set $\mathscr{D}_{\leq}$ of designated values such that the entailment relations coming with $\leq$ and $\mathscr{D}_{\leq}$ coincide. The approach to entailment based on a partial order appears to be more general than the approach based on designated values.

Our informal motivation for introducing **16** in Sect. 3.3 was to extend Belnap's "computerized" interpretation from *one* computer to bundles of interconnected computers (computer networks). When it comes, as in Priest's case, to motivating generalized truth values by pointing to impossible truth values or hyper-contradictions,[7] one may notice that we are not forced to choose between truth value gaps and truth value gluts. In both cases, revenge Liar sentences seem to speak in favor of generalized truth values.

Consider again Definition 4.9. Unfortunately, Priest does not supply it with a theoretical justification except for remarking briefly that this way of defining propositional connectives is "obvious" [197, p. 237]. However, its apparent obviousness notwithstanding, Definition 4.9 still has some vulnerable points. In particular, it cannot be naturally extended to a construction that would allow the empty set to enter at every stage. More specifically, if we let $S^g_{n+1} = \mathscr{P}(S_n)$, then, as Priest himself mentions, Definition 4.9 gives the extension of any truth functor according to the rule "gap-in, gap-out" [197, p. 242], e.g., for S^g_1 so defined, $\varnothing \wedge_1 x = \varnothing \vee_1 x = \sim_1 \varnothing = \varnothing$. But such an extension of S_1 would not be identical (as one could expect) with *FOUR*$_2$, where, e.g., $\mathbf{N} \wedge \mathbf{F}$ amounts to **F** and not to **N**. As Jain [133, § 4] remarks, this situation is caused by the fact that Definition 4.9 treats truth functions in terms of the members of each argument, but $\varnothing$ has no members. It means that the approach proposed by Priest cannot naturally be extended to the sets **4**, **16**, etc., nor can it be applied to the set $S'_1 = \mathbf{3}'(= \{\mathbf{F}, \mathbf{T}, \mathbf{N}\})$ taken as the set of truth values of Kleene's *strong* three-valued logic and its possible generalizations.[8]

This seems to be a rather unwanted outcome of Definition 4.9, for usually $S_1 = \mathbf{3}$ and $S'_1 = \mathbf{3}'$ are considered just substructures of $S^g_1 = \mathbf{4}$ (the algebraic and logical properties of which are determined by *FOUR*$_2$).

---

[7] It is interesting to note that if we talk about "impossible truth values", all the values from **16** except for **N** are impossible from the point of view of **4**, and at the level of **2**, all the values of **4** are impossible—not only **N** and **B**—because no element of **4** belongs to **2**. If a sentence is said to be hyper-contradictory iff it necessarily takes impossible truth values, and if a sentence is called contradictory iff it necessarily takes all the available truth values, then a hyper-contradictory sentence need not be contradictory. In the classical setting, the impossible value the Liar takes is the set of all values available at the level of **2**. The sentence (*), however, although necessarily taking a value not available in **3**, not necessarily takes the value **FTB**. It might just take the value **FT**.

[8] However, this definition apparently reflects the spirit of Kleene's *weak* three-valued logic which assigns the value **N** to any compound formula in which some part has been assigned **N** [92].

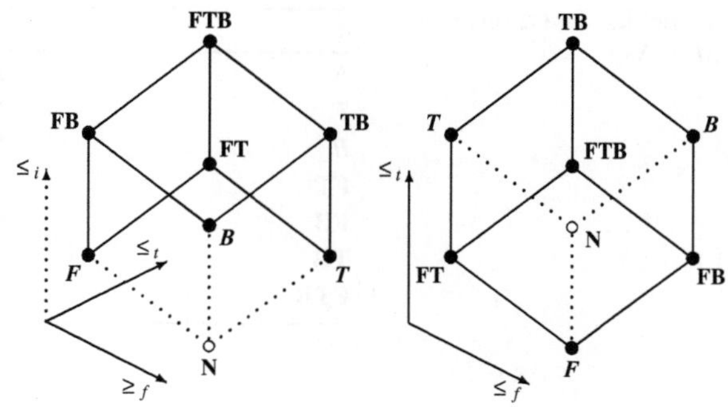

Fig. 4.1 Bi-and-a-half-lattice $SEVEN_{2.5}$ and trilattice $EIGHT_3$

Let us dwell in more detail upon Priest's S_2, which is in effect the set **7** = {***F***, ***T***, ***B***, **FT**, **FB**, **TB**, **FTB**}.[9] By applying to this set the "multilattice approach" developed in the present book, it appears that—interestingly enough—its algebraic structure constitutes what can be called a *bi-and-a-half*-lattice $SEVEN_{2.5}$ (cf. [231, p. 783]). We represent this lattice in Fig. 4.1 in two slightly different projections.

One can clearly observe here the complete lattices under $\leq_t$ and $\leq_f$, but the information order fails to form a lattice. Under $\leq_i$ we merely have a *semi*lattice with **FTB** as a top but with no bottom. However, $SEVEN_{2.5}$ can be directly extended to a trilattice $EIGHT_3$ by adding **N** as a bottom element for $\leq_i$. The dotted lines in Fig. 4.1 present the result of such an extension. Note that $EIGHT_3$ is *not* a Belnap trilattice. However, both $SEVEN_{2.5}$ and $EIGHT_3$ are sublattices of $SIXTEEN_3$, and in this respect the relationship between these multilattices perfectly corresponds to the relationship between their bases.

We again label by means of $\sqcap_t$ and $\sqcup_t$ meet and join under $\leq_t$ and by $\sqcap_f$ and $\sqcup_f$ the corresponding lattice-operations under $\leq_f$ in $SEVEN_{2.5}$. If we now examine Priest's Definition 4.9 with regard to its conformity with the algebraic operations in $SEVEN_{2.5}$, we can state the slightly "eclectic" nature of the operations introduced by this definition. Namely, it turns out that, e.g., Priest's $\wedge_2$ behaves like $\sqcap_t$ for some elements of **7** and like $\sqcup_f$ for the others. In particular, we have $T \wedge_2 B = B = T \sqcup_f B$, but $F \wedge_2 B = F = F \sqcap_t B$. Thus, Definition 4.9 does not allow one to discriminate between the truth order and the falsity order. However, we believe that logical connectives generated separately under the truth order and the falsity order in $SEVEN_{2.5}$ deserve special investigation.

However, first we need to explore the possibility of introducing a negation operator. It seems rather natural to do this by means of a suitable inversion operation as introduced by Definition 4.6. In Sect. 3.5 we have formulated concrete definitions for *t*-, *f*- and *i*-inversions (Table 3.1) and have thus secured the existence of these operations in $SIXTEEN_3$. Moreover, Theorem 4.5 guarantees the

[9] Observe the close resemblance of **7** to Jaina seven-valued logic, where ***B*** is usually interpreted as "non-assertible" (see, e.g., [113]).

Table 4.1 Inversions in *SEVEN*$_{2.5}$

a	$-_t a$	$-_f a$
F	**TB**	*F*
T	*T*	**FB**
B	*F*	*T*
FT	**TB**	**FB**
FB	**FB**	**T**
TB	**F**	**TB**
FTB	**FTB**	**FTB**

existence of t- and f-inversions for *any* Belnap trilattice. Unfortunately, *SEVEN*$_{2.5}$ appears to be not so perfect.

Proposition 4.1 *It is impossible in SEVEN*$_{2.5}$ *to define a pure t-inversion.*

Proof Suppose there exists a pure t-inversion satisfying all the conditions from Definition 4.6. Then **T** and **FB** should be fixed points of such an operation because they are bottom and top relative to $\leq_f$, and thus any change in their position would result in changing the falsity order, too.

Now, having **T** as a fixed point, consider elements **F** and **FT**. We have both $\mathbf{F} \leq_t \mathbf{T}$ and $\mathbf{FT} \leq_t \mathbf{T}$. It is not difficult to see that it is impossible to define $-_t\mathbf{F}$ and $-_t\mathbf{FT}$ to guarantee both $-_t\mathbf{T} \leq_t -_t\mathbf{F}$ and $-_t\mathbf{T} \leq_t -_t\mathbf{FT}$. Indeed, there is only one candidate left for such value: **TB**. But then we would be forced to take simultaneously $-_t\mathbf{TB} = \mathbf{F}$ and $-_t\mathbf{TB} = \mathbf{FT}$ (to secure the "period two" property), which would be a violation of functionality. □

Analogously it is not difficult to show that there exists no pure f-inversion in *SEVEN*$_{2.5}$. This result may seem rather disappointing as it means a considerable complication for introducing negation operators suitable for *SEVEN*$_{2.5}$. More specifically, we cannot apply Definition 4.6 in full generality, and thus have to find some other way of defining the notion of negation in our seven-valued logic.[10]

An obvious way out of this situation is to try to weaken Definition 4.6 by weakening (or even giving up) some of its conditions. In the spirit of [81, p. 89], we call a unary operation $-_j$ a *subminimal* j-inversion iff it only satisfies conditions (*anti*) and (*iso*) of Definition 4.6. That is, a subminimal inversion, although it reverses the corresponding partial order, is not an involution. Then we could just vary several conditions in the vicinity of (*per*2), thus obtaining various kinds of (non-classical) inversions. E.g., adding to (*anti*) and (*iso*) the condition $x \leq_j -_j -_j x$ would lead us to a so-called *quasiminimal* j-inversion etc. (cf. [81, pp. 88–92] and also [82, 229]). It turns out that a subminimal t-inversion and f-inversion can be defined in *SEVEN*$_{2.5}$ as presented in Table 4.1. This gives to our disposal an operation for governing an object language negation, even if this

---

[10] Note, incidentally, that Priest's Definition 4.9 is of little use in the context of *SEVEN*$_{2.5}$, for according to this definition $\sim_2\mathbf{FTB} = \mathbf{FTB}$ and $\sim_2\mathbf{FT} = \mathbf{FT}$. Then, having $\mathbf{FT} <_t \mathbf{FTB}$, we get also $\sim_2\mathbf{FT} <_t \sim_2\mathbf{FTB}$ which surely is unwanted.

operation appears to be quite weak. However, it is surely better to have a weak negation than not to have any negation at all.

Now, having lattice operations of meet, join, and subminimal inversion (for both $\leq_t$ and $\leq_t$), consider again language $\mathscr{L}_{tf}$ and its fragments. Valuations v^7 and their extension to compound formulas are introduced in a standard way (see *mutatis mutandis* Definition 3.7), as are corresponding entailment relations: $\models_t^7$ and $\models_f^7$ (see Definitions 4.7 and 4.8 suitably modified).[11] As a result we get logics $(\mathscr{L}_{tf}, \models_t^7)$ and $(\mathscr{L}_{tf}, \models_f^7)$ (semantically defined) as well as their fragments (e.g., the logic $(\mathscr{L}_t, \models_t^7)$ etc.).

Thus, the multilattice-approach has led us to several seven-valued logics, each of which differs essentially from Priest's Logic of Paradox. We leave the problem of finding suitable axiomatizations of these new logics for future work.

4.5 An Approach to a Generalization of Kleene's Logic: A Tetralattice

In [293] Dmitry Zaitsev has proposed taking, as a starting point for a generalization procedure, neither the set of classical truth values **2** nor the set of Belnap's truth values **4** but three truth values which resemble very much the truth values of Kleene's three-valued logic which form the set $\mathbf{3}' = \mathbf{4} \setminus \{\mathbf{B}\}$. More specifically, he considers the set of "reported values" $\{\mathbf{a}, \mathbf{d}, \mathbf{u}\}$ as an initial base. The elements of this set represent possible "reports" regarding the question of whether a certain sentence is asserted. Then value **a** means "*yes*, the sentence is *asserted*", value **d** means "*no*, the sentence is *denied*", and value **u** means "I don't know, I am *uncertain*".

We will reconstruct Zaitsev's approach by applying it directly to the set $\mathbf{3}'$. Let us have some information source that may operate with the following initial truth values: **T**—*truth* (positive certainty, positive information); **F**—*falsehood* (negative certainty, negative information); **N**—*neither truth nor falsehood* (uncertainty, absence of information). Now, if we have several such sources that—in the sense of Belnap—supply information to a "higher order" computer, then we naturally arrive at the set **8**, the powerset of $\mathbf{3}'$, as the set of truth values which are at the disposal of this computer:

1. $\mathbf{N} = \varnothing$	5. $\mathbf{NF} = \{\mathbf{N}, \mathbf{F}\}$
2. $\boldsymbol{N} = \{\mathbf{N}\}$	6. $\mathbf{NT} = \{\mathbf{N}, \mathbf{T}\}$
3. $\boldsymbol{F} = \{\mathbf{F}\}$	7. $\mathbf{FT} = \{\mathbf{F}, \mathbf{T}\}$
4. $\boldsymbol{T} = \{\mathbf{T}\}$	8. $\mathbf{NFT} = \{\mathbf{N}, \mathbf{F}, \mathbf{T}\}$.

[11] Note once again that, unlike Priest, we do not explicitly make use of any set of designated truth values but canonically define entailment through a logical order.

One has to pay attention to the difference between the values *N* and **N** already emphasized on page 55. Whereas **N** symbolizes the absence of any information (at a "zero level"), *N* means that exactly this information about the absence of information has been transferred "one level higher", i.e., our computer has received the specific information that a source is *uncertain* about whether some sentence is classically true or classically false. By interpreting this difference, Zaitsev remarks (notation adjusted): "**N** means that the computer does not know, whereas *N* means that the source does not know, which turns to be [a] more informative answer (in the later case the computer *knows* that the source does not know)" [293, p. 269].

This differentiation between information about uncertainty and pure non-informativeness can find its expression in the corresponding orderings of the above eight truth values. Namely, we may differentiate between the usual information order (standardly defined as the subset-relation) and a new "uncertainty order", which traces the transfer of the specific element **N** from one generalized truth value to another.

Definition 4.11 Let

$$x^t = \{y \in x \mid \mathbf{T} = y\}, \quad x^n = \{y \in x \mid \mathbf{N} = y\}, \quad x^f = \{y \in x \mid \mathbf{F} = y\},$$
$$x^{-t} = \{y \in x \mid \mathbf{T} \neq y\}, \quad x^{-n} = \{y \in x \mid \mathbf{N} \neq y\}, \quad x^{-f} = \{y \in x \mid \mathbf{F} \neq y\}.$$

Then

$$\begin{array}{l} x \leq_i y \text{ iff } x \subseteq y, \\ x \leq_n y \text{ iff } x^n \subseteq y^n \text{ and } y^{-n} \subseteq x^{-n}, \\ x \leq_f y \text{ iff } x^f \subseteq y^f \text{ and } y^{-f} \subseteq x^{-f}, \\ x \leq_t y \text{ iff } x^t \subseteq y^t \text{ and } y^{-t} \subseteq x^{-t}. \end{array}$$

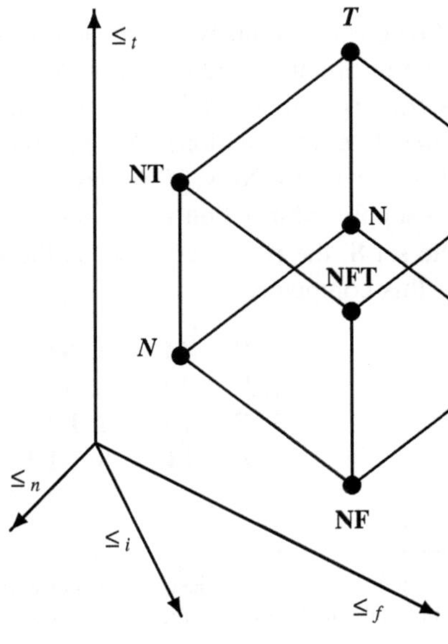

Fig. 4.2 The tetralattice *EIGHT*$_4$

In this way we obtain the tetralattice $EIGHT_4$ as an algebraic structure that results from the combination of four complete lattices $\langle \mathbf{8}, \leq_i \rangle$, $\langle \mathbf{8}, \leq_n \rangle$, $\langle \mathbf{8}, \leq_f \rangle$, $\langle \mathbf{8}, \leq_t \rangle$ presented by a Hasse diagram in Fig. 4.2.

Note that in this diagram the axes for $\leq_i$ and $\leq_n$ are only approximately traced. Clearly, tetralattice $EIGHT_4$ is a sublattice of the trilattice $SIXTEEN_3$. Again, both $\leq_t$ and $\leq_f$ represent logical orders, and hence, the operations of meet and join associated with these orders can be naturally seen as determining the corresponding object language connectives of conjunctions and disjunctions. Important algebraic properties of these operations are represented in the following proposition (cf. Proposition 3.2):

Proposition 4.2

(1) $\mathbf{T} \in x \sqcap_t y \Leftrightarrow \mathbf{T} \in x$ *and* $\mathbf{T} \in y$

$\mathbf{F} \in x \sqcap_t y \Leftrightarrow \mathbf{F} \in x$ *or* $\mathbf{F} \in y$

$\mathbf{N} \in x \sqcap_t y \Leftrightarrow \mathbf{N} \in x$ *or* $\mathbf{N} \in y$

(2) $\mathbf{T} \in x \sqcup_t y \Leftrightarrow \mathbf{T} \in x$ *or* $\mathbf{T} \in y$

$\mathbf{F} \in x \sqcup_t y \Leftrightarrow \mathbf{F} \in x$ *and* $\mathbf{F} \in y$

$\mathbf{N} \in x \sqcup_t y \Leftrightarrow \mathbf{N} \in x$ *and* $\mathbf{N} \in y$

(3) $\mathbf{T} \in x \sqcap_f y \Leftrightarrow \mathbf{T} \in x$ *or* $\mathbf{T} \in y$

$\mathbf{F} \in x \sqcap_t y \Leftrightarrow \mathbf{F} \in x$ *and* $\mathbf{F} \in y$

$\mathbf{N} \in x \sqcap_t y \Leftrightarrow \mathbf{N} \in x$ *or* $\mathbf{N} \in y$

(4) $\mathbf{T} \in x \sqcup_f y \Leftrightarrow \mathbf{T} \in x$ *and* $\mathbf{T} \in y$

$\mathbf{F} \in x \sqcup_f y \Leftrightarrow \mathbf{F} \in x$ *or* $\mathbf{F} \in y$

$\mathbf{N} \in x \sqcup_f y \Leftrightarrow \mathbf{N} \in x$ *and* $\mathbf{N} \in y$

Table 4.2 defines operations of *t*-inversion, *f*-inversion, *n*-inversion and *i*-inversion for $EIGHT_4$ in accordance with Definition 4.6.

Zaitsev [293, p. 272] observes that, first, the operations $-_i$ and $-_n$ are coincident, and, second, the operation $-_n$ can be defined through $-_t$ and $-_f$ as follows: $-_t -_f -_t x = -_f -_t -_f x = -_n x$. Hence, it is enough to consider just two inversion operations ($-_t$ and $-_f$) and the corresponding object language negations.

It is not difficult to see that for inversions in $EIGHT_4$ the following proposition holds (cf. Proposition 3.4):

Table 4.2 Inversions in $EIGHT_4$

a	$-_t a$	$-_f a$	$-_n a$	$-_i a$
N	**NFT**	**NFT**	**NFT**	**NFT**
N	**NT**	**NF**	**FT**	**FT**
F	**FT**	**NT**	**NF**	**NF**
T	**NF**	**FT**	**NT**	**NT**
NF	*T*	*N*	*F*	*F*
NT	*N*	*F*	*T*	*T*
FT	*F*	*T*	*N*	*N*
NFT	**N**	**N**	**N**	**N**

Proposition 4.3

$$\begin{array}{llll} (1) & \mathbf{T} \in -_t x \Leftrightarrow \mathbf{T} \notin x & (2) & \mathbf{T} \in -_f x \Leftrightarrow \mathbf{N} \notin x \\ & \mathbf{F} \in -_t x \Leftrightarrow \mathbf{N} \notin x & & \mathbf{F} \in -_f x \Leftrightarrow \mathbf{F} \notin x \\ & \mathbf{N} \in -_t x \Leftrightarrow \mathbf{F} \notin x & & \mathbf{N} \in -_f x \Leftrightarrow \mathbf{T} \notin x. \end{array}$$

We can again consider the language $\mathscr{L}_{tf}$ with connectives $\wedge_t, \vee_t, \sim_t, \wedge_f, \vee_f, \sim_f$, bearing in mind the possibility of separating it into two distinct sublanguages $\mathscr{L}_t$ and $\mathscr{L}_f$ with corresponding connectives. A valuation function v^8 (an 8-valuation) can standardly be defined as a map from the set of propositional variables into **8**, extended to compound formulas by the following conditions (cf. Definition 3.7):

Definition 4.12 For any $\mathscr{L}_{tf}$-formulas A and B:

$$\begin{array}{ll} 1.\, v^8(A \wedge_t B) = v^8(A) \sqcap_t v^8(B); & 4.\, v^8(A \wedge_f B) = v^8(A) \sqcup_f v^8(B); \\ 2.\, v^8(A \vee_t B) = v^8(A) \sqcup_t v^8(B); & 5.\, v^8(A \vee_f B) = v^8(A) \sqcap_f v^8(B); \\ 3.\, v^8(\sim_t A) = -_t v^8(A); & 6.\, v^8(\sim_f A) = -_f v^8(A). \end{array}$$

By means of the truth order ($\leq_t$) and the falsity order ($\leq_f$), entailment relations for *EIGHT*$_4$ can be introduced in analogy to Definitions 4.1 and 4.2.

Definition 4.13 For arbitrary formulas A and B of $\mathscr{L}_{tf}$,

$$A \models_t^8 B \text{ iff } \forall v^8 (v^8(A) \leq_t v^8(B)).$$

Definition 4.14 For arbitrary formulas A and B of $\mathscr{L}_{tf}$,

$$A \models_f^8 B \text{ iff } \forall v^8 (v^8(B) \leq_f v^8(A)).$$

Notice that even if we disregard the order $\leq_n$, the resulting trilattice *EIGHT*$_3$ is *not* a Belnap trilattice. Nevertheless, as Zaitsev observes [293, p. 273], the relations $\models_t^8$ and $\models_f^8$, restricted correspondingly to the languages $\mathscr{L}_t$ and $\mathscr{L}_f$, are axiomatized exactly by the logic of first-degree entailment (cf. the result presented in Sect. 4.3). Indeed, $\langle \mathbf{8}, \leq_t \rangle$ and $\langle \mathbf{8}, \leq_f \rangle$ are just variations of the well-known lattice M_0 (see [3, p. 198]), and it is shown in [3, §18.8] that M_0 is a characteristic matrix for the system $\mathbf{E}_{fde}$ of "tautological entailments".

Zaitsev aims to construct a logical system which combines the logical connectives generated by *both* orderings $\leq_t$ and $\leq_f$ within a joint framework. For this purpose he extends the language $\mathscr{L}_{tf}$ with a propositional constant **t**, considering thus the language $\mathscr{L}_{tf+\mathbf{t}}$. The constant **t** plays a technical role as it is needed for a completeness proof; moreover, as Zaitsev remarks, its introduction "is justified by the idea that some generally invalid derivations can be quite legitimate under the assumption of reasoning within a complete and consistent theory" [293, p. 273].

The valuation function v^8 can be extended to the constant **t** as follows:

Definition 4.15 $v^8(\mathbf{t}) = \mathbf{NFT}$.

Then Zaitsev formulates a (first-degree) consequence system, which we mark here by $\mathbf{FDE}_t^{tf+\mathbf{t}}$ as a pair $(\mathscr{L}_{tf+\mathbf{t}}, \vdash_t)$, where $\vdash_t$ is a consequence relation for which the following deductive postulates hold:

A1. $A \wedge_t B \vdash_t A$
A2. $A \wedge_t B \vdash_t B$
A3. $A \vdash_t A \vee_t B$
A4. $B \vdash_t A \vee_t B$
A5. $A \wedge_t (B \vee_t C) \vdash_t (A \wedge_t B) \vee_t C$
A6. $A \wedge_f (B \vee_f C) \vdash_t (A \wedge_f B) \vee_f C$
A7. $A \wedge_t B \vdash_t A \wedge_f B$
A8. $A \vee_f B \vdash_t A \vee_t B$
A9. $\sim_t\sim_t A \vdash_t A$
A10. $A \vdash_t \sim_t\sim_t A$
A11. $\sim_f\sim_f A \vdash_t A$
A12. $A \vdash_t \sim_f\sim_f A$
A13. $\mathbf{t} \vdash_t \sim_t A \vee_t A$
A14. $\sim_f\sim_t\sim_f A \vdash_t \sim_t\sim_f\sim_t A$
A15. $\sim_t\sim_f\sim_t A \vdash_t \sim_f\sim_t\sim_f A$
A16. $\sim_f(A \vee_t B) \vdash_t \sim_f A \vee_t \sim_f B$
A17. $\sim_f A \wedge_t \sim_f B \vdash_t \sim_f(A \wedge_t B)$

R1. $A \vdash_t B, B \vdash_t C / A \vdash_t C$
R2. $A \vdash_t B, A \vdash_t C / A \vdash_t B \wedge_t C$
R3. $A \vdash_t C, B \vdash_t C / A \vee_t B \vdash_t C$
R4. $A \vdash_t B / \sim_t B \vdash_t \sim_t A$
R5. $A \vdash_t B, \mathbf{t} \vdash_t \sim_f A / \mathbf{t} \vdash_t \sim_f B$
R6. $\mathbf{t} \vdash_t A \wedge_f B / \mathbf{t} \vdash_t A \wedge_t B$
R7. $\mathbf{t} \vdash_t A \vee_t B / \mathbf{t} \vdash_t A \vee_f B$

In [293] a proof is offered stating that the system $\mathbf{FDE}_t^{tf+\mathbf{t}}$ is a sound and complete axiomatization of the relation $\models_t^8$ introduced by Definition 4.13. A dual result also demonstrates that the system $\mathbf{FDE}_f^{tf+\mathbf{t}}$, which results from $\mathbf{FDE}_t^{tf+\mathbf{t}}$ by interchanging all the t-subscripts and f-subscripts, is an adequate formalization of the relation $\models_f^8$ introduced by Definition 4.14.

4.6 Uncertainty Versus Lack of Information

Note that Zaitsev interprets his reported value **u** in epistemic terms. Moreover, in [293, p. 267] he claims that Belnap's value **N** is ambiguous. This view of uncertainty as a truth value has been criticized in [273, p. 267]. Whereas Belnap's value

N from **4** is a "told neither True nor False" sign, Zaitsev's reported value **u** is introduced as a "reported don't know" sign. If a proposition p is marked by a value containing **u**, then at least one source of information has supplied the information that it is ignorant concerning p. In contrast to this, if p is marked by Belnap's **N**, then no information has been given concerning p. If the proposition p is marked by a value from **16** containing **N**, then at least one informant has supplied the information that no information has been given concerning p. Contrary to what Zaitsev [293] says, there is no "ambiguity of value **N**".

For each of his three basic values, Zaitsev defines a partial order on $\mathscr{P}(\{\mathbf{a},\mathbf{d},\mathbf{u}\}) = \mathbf{8}$ and thereby obtains the tetralattice *EIGHT*$_4$. Using his notation, the four relations on **8** are defined as follows:

$$
\begin{aligned}
x \leq_{\mathbf{a}} y &\text{ iff } x^{\mathbf{a}} \subseteq y^{\mathbf{a}} \text{ and } y^{-\mathbf{a}} \subseteq x^{-\mathbf{a}},\\
x \leq_{\mathbf{d}} y &\text{ iff } x^{\mathbf{d}} \subseteq y^{\mathbf{d}} \text{ and } y^{-\mathbf{d}} \subseteq x^{-\mathbf{d}},\\
x \leq_{\mathbf{u}} y &\text{ iff } x^{\mathbf{u}} \subseteq y^{\mathbf{u}} \text{ and } y^{-\mathbf{u}} \subseteq x^{-\mathbf{u}},\\
x \leq_{i} y &\text{ iff } x \subseteq y,
\end{aligned}
$$

where $x^{\mathbf{a}} := \{y \in x \mid a = y\}, x^{-\mathbf{a}} := \{y \in x \mid a \neq y\}, x^{\mathbf{u}} := \{y \in x \mid u = y\}, x^{-\mathbf{u}} := \{y \in x \mid u \neq y\}, x^{\mathbf{d}} := \{y \in x \mid d = y\}$, and $x^{-\mathbf{d}} := \{y \in x \mid d \neq y\}$.

Since there are three basic values, the relations $\leq_{\mathbf{a}}$ and $\leq_{\mathbf{d}}$ are not inverses of each other. Zaitsev refers to the partial orders $\leq_{\mathbf{a}}$, $\leq_{\mathbf{d}}$, and $\leq_{\mathbf{u}}$ as truth, falsity, and uncertainty orderings, respectively. This terminology may be questioned. Consider, for example, Zaitsev's truth (alias assertion) ordering $\leq_{\mathbf{a}}$ depicted in Fig. 4.3. The fact that $\{\mathbf{a},\mathbf{u}\} <_{\mathbf{a}} \{\mathbf{a}\}$, $\{\mathbf{a},\mathbf{d}\} <_{\mathbf{a}} \{\mathbf{a}\}$, and $\{\mathbf{a},\mathbf{u},\mathbf{d}\} <_{\mathbf{a}} \{\mathbf{a},\mathbf{u}\}$ may be seen as an indication that $\leq_{\mathbf{a}}$ is a true-only ordering and not just a truth ordering since **a** is an element of all the three values.

Analogously, one can define the pentalattice *SIXTEEN*$_5$ on **16** by defining the following five partial orders on this set of generalized truth values:

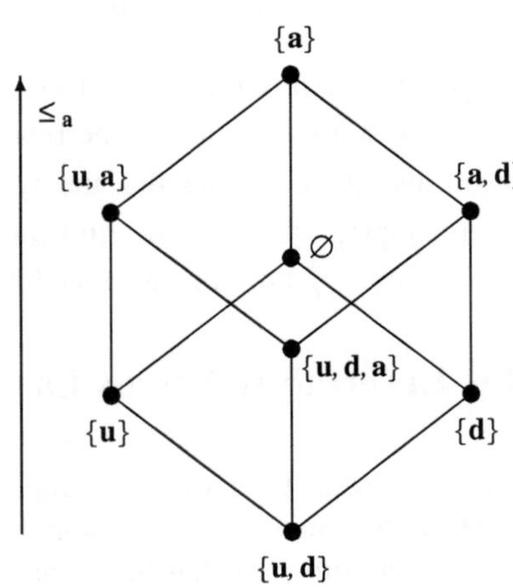

Fig. 4.3 Zaitsev's assertion order on **8**

$$
\begin{aligned}
&x \leq_{\mathbf{N}} y \text{ iff } x^{\mathbf{N}} \subseteq y^{\mathbf{N}} \text{ and } y^{-\mathbf{N}} \subseteq x^{-\mathbf{N}},\\
&x \leq_{\mathbf{T}} y \text{ iff } x^{\mathbf{T}} \subseteq y^{\mathbf{T}} \text{ and } y^{-\mathbf{T}} \subseteq x^{-\mathbf{T}},\\
&x \leq_{\mathbf{F}} y \text{ iff } x^{\mathbf{F}} \subseteq y^{\mathbf{F}} \text{ and } y^{-\mathbf{F}} \subseteq x^{-\mathbf{F}},\\
&x \leq_{\mathbf{B}} y \text{ iff } x^{\mathbf{B}} \subseteq y^{\mathbf{B}} \text{ and } y^{-\mathbf{B}} \subseteq x^{-\mathbf{B}},\\
&x \leq_{i} y \text{ iff } x \subseteq y,
\end{aligned}
$$

where $x^{\sharp} := \{y \in x \mid \sharp = y\}$ and $x^{-\sharp} := \{y \in x \mid \sharp \neq y\}$, for $\sharp \in \{\mathbf{T}, \mathbf{F}, \mathbf{N}, \mathbf{B}\}$.

Obviously, $\{\{T\}\}$ is the top-element of $\leq_{\mathbf{T}}$ on **16**, but for reasons analogous to the ones just stated for $\leq_{\mathbf{a}}$, $\leq_{\mathbf{T}}$ seems to be a "told True-only" ordering rather than a "told True" ordering. A convincing "told True" ordering $\preceq_{\mathbf{T}}$ is definable on the powerset of **16**. In this definition the value **T** plays a role corresponding to the role of the classical value T in the definition of $\leq_t$ in *SIXTEEN*$_3$.

Definition For every x, y in $\mathscr{P}(\mathbf{16})$, $x \preceq_{\mathbf{T}} y$ iff $x^{\mathbf{T}} \subseteq y^{\mathbf{T}}$ and $y^{-\mathbf{T}} \subseteq x^{-\mathbf{T}}$, where $x^{\mathbf{T}}$ and $x^{-\mathbf{T}}$ are now defined as follows: $x^{\mathbf{T}} := \{y \in x | \mathbf{T} \in y\}$ and $x^{-\mathbf{T}} := \{y \in x | \mathbf{T} \notin y\}$.

Orderings $x \preceq_{\mathbf{N}} y$, $x \preceq_{\mathbf{F}} y$, and $x \preceq_{\mathbf{B}} y$ can be defined analogously.

Note also that information and uncertainty in *EIGHT*$_4$ do not interact in a way that might be expected. Zaitsev's uncertainty order is not the inverse of the information order; an increase (decrease) of information fails to imply a decrease (increase) of uncertainty.

Ignorance in the context of first-degree entailment is to be understood as lack of information, and, clearly, information overflow must be taken into account as well. If ignorance is understood as lack of *belief* instead of lack of knowledge and *if* an information source can diagnose itself as "uncertain," then it would seem that one should allow it the ability to diagnose itself as "conflicted" (**c**) as well. This does not amount to claiming that an agent may have the ability to be aware of inconsistent beliefs, because inconsistent beliefs must be distinguished from conflicting, antagonistic beliefs. Statements of conflicting beliefs are satisfiable in neighborhood models (also called Scott-Montague models or minimal models), while ascriptions of inconsistent beliefs are not, cf. [259]. In other words, if $B_i p$ stands for "agent i believes that p", then $B_i p \wedge B_i \neg p$ is satisfiable in neighborhood semantics, whereas $B_i(p \wedge \neg p)$ is not. As reported values, **u** and **c** then seem to stand and fall together, and, obviously, the set of reported values $\{\mathbf{a}, \mathbf{d}, \mathbf{u}, \mathbf{c}\}$ will give us 16 marked values instead of eight.

Chapter 5
Axiom Systems for Trilattice Logics

Abstract This chapter is devoted to introducing and discussing the axiom systems for trilattice logics obtained by Odintsov (Studia Logica 93:407–428, 2009). First-degree proof systems for logics related to $SIXTEEN_3$ in the language $\mathcal{L}_{tf}$ are introduced. Moreover, the language $\mathcal{L}_{tf}$ is extended by a truth implication $\to_t$, a falsity implication $\to_f$, or both. Adding implications to the truth and falsity vocabulary of $\mathcal{L}_{tf}$ is of independent interest. We start with considering two ways of defining a many-valued logic: the method of valuation systems (matrices) and the lattice approach as exemplified by the bilattice $FOUR_2$ and the trilattice $SIXTEEN_3$. We conclude that the lattice approach has the advantage of admitting a general and uniform way of defining implications. An implication connective is required for Odintsov's Hilbert-style proof systems with *modus ponens* as the only rule of inference. These Hilbert-style systems axiomatize truth entailment and falsity entailment in the languages comprising at least one implication. The axiom systems are instructive insofar as they reveal that falsity conjunction and falsity disjunction have an indeterministic interpretation with respect to truth entailment in the trilattice $SIXTEEN_3$, whereas truth conjunction and truth disjunction have an indeterministic interpretation with respect to falsity entailment in $SIXTEEN_3$. Moreover, the matrix presentation of the algebraic operations of $SIXTEEN_3$ will be used in Chap. 6 to obtain cut-free sequent calculi for truth entailment and falsity entailment in the languages with or without truth or falsity implication.

5.1 Truth Value Lattices and the Implication Connective

In Sect. 1.6 we specified two main kinds of logical structures, viz. structures of truth values: valuation systems (matrices) and truth-value lattices. These two kinds of logical structures may be employed for presenting (multiple-valued) logics in two different ways. The first way consists in singling out a non-empty set of designated values and defining entailment as the preservation of membership in this set of designated values from the premises to the conclusions.

Y. Shramko and H. Wansing, *Truth and Falsehood*, Trends in Logic 36,
DOI: 10.1007/978-94-007-0907-2_5, © Springer Science+Business Media B.V. 2011

The second way makes use of the fact that the truth values of certain many-valued logics constitute lattices, see, for example, [210]. According to Arieli and Avron [7, p. 25] it is even the case that "[w]hen using multiple-valued logics, it is usual to order the truth values in a lattice structure". Then one can (i) define a lattice order $\leq$ on the set $\mathscr{V}$ of truth values, (ii) interpret logical operations as operations on the lattice $(\mathscr{V}, \leq)$, and (iii) stipulate that a formula A entails a formula B iff for every homomorphic valuation function v, $v(A) \leq v(B)$.

In Sect. 3.4 we introduced the notion of a generalized truth value as a subset of some basic set and also the idea of a multilattice by which the generalized truth values can be organized. As a paradigmatic case of a multilattice, we have considered the trilattice $SIXTEEN_3$.

An advantage of the lattice approach is that within this approach, there exits a uniform way of interpreting an implication connective by a lattice operation, namely interpreting implication as the residuum of the lattice meet. Dually, one may interpret a connective of co-implication (cf. [271] and the references given there) as the residuum of the lattice join. Introducing an implication connective interpreted as the residuum of the lattice meet is a central component in Odintsov's definition of Hilbert-style proof systems for truth and falsity entailment in $SIXTEEN_3$ to be considered in this chapter. The introduction of implications in trilattice logics is thus not only of interest in general, but it can be motivated by a comparison with the matrix approach and the definition of axiom systems with *modus ponens* as the only rule of inference.

More specifically, if $(\mathscr{V}, \leq)$ is a complete lattice with lattice meet $\sqcap$ and lattice top $\top$, an implication connective $\rightarrow$ can be canonically defined by the following evaluation clause, cf. [130, p. 29]:

$$v(A \rightarrow B) = \bigsqcup\{x \mid x \sqcap v(A) \leq v(B)\}.$$

The lattice operation corresponding to $\rightarrow$ is called the *residuum* of $\sqcap$ and it ensures that $\rightarrow$ satisfies the Deduction Theorem (in the following semantical shape): For every valuation v, $\top \leq v(A \rightarrow B)$ iff $v(A) \leq v(B)$. If v is a valuation function, then $v(A) \leq v(B)$ iff $\top \sqcap v(A) \leq v(B)$ iff $\top \leq \bigsqcup\{x \mid x \sqcap v(A) \leq v(B)\}$ iff $\top \leq v(A \rightarrow B)$ (iff $\top = v(A \rightarrow B)$).

As the Deduction Theorem may be seen as a defining characteristic of an implication connective, this way of introducing an implication gives rise to defining two implication connectives for logics of $SIXTEEN_3$, where we are interested in the decrease of falsity and hence in the inverse of $\leq_f$:

$$v(A \rightarrow_t B) = \sqcup_t\{x \mid x \sqcap_t v(A) \leq_t v(B)\}$$
$$v(A \rightarrow_f B) = \sqcap_f\{x \mid x \sqcup_f v(A) \geq_f v(B)\}.$$

The language $\mathscr{L}_{tf}$ may thus be extended by these implication connectives.

Definition 5.1 The languages $\mathscr{L}_{tf}^{\rightarrow_t}$, $\mathscr{L}_{tf}^{\rightarrow_f}$, and $\mathscr{L}_{tf}^{*}$ are defined as $\mathscr{L}_{tf} \cup \{\rightarrow_t\}$, $\mathscr{L}_{tf} \cup \{\rightarrow_f\}$, and $\mathscr{L}_{tf} \cup \{\rightarrow_t, \rightarrow_f\}$, respectively.

5.2 From First-Degree Proof Systems to Proof Systems with *Modus Ponens*

A proof system is a non-empty set of inference rules. In *axiomatic* proof systems, one usually aims at keeping to a minimum the number of genuine inference rules, that is, rules with a non-empty set of premises. Instead, the emphasis of the syntactic presentation is on axiom schemes, i.e., schematic formulas that are assumed to be self-evident insofar as they are taken to be provable from the empty set of premises.[1] In axiom systems for propositional logics, often only a single genuine inference rule is assumed, namely *modus ponens*: From A and $(A \to B)$ infer B, where $\to$ is usually the unique implication connective in the object language. Let $\triangleq$ stand for syntactic identity of two expressions. A derivation of a formula A from a finite set Δ of formulas is then a sequence of formulas $A_1, \ldots, A_n$ $(n \geq 1)$ such that $A_n \triangleq A$, and each of the formulas A_i $(1 \leq i \leq n)$ is either an instance of one of the schematic axioms or there is j and k with $j < i$, $k < i$ and A_i is derivable from A_j and A_k by applying *modus ponens*. A formula A is derivable from a finite set Δ of formulas $(\Delta \vdash A)$ just in case there exists a derivation of A from Γ. If Δ and Γ are finite sets of formulas, then Γ is derivable from Δ iff there exists an $A \in \Gamma$ with $\Delta \vdash A$. In implication-free languages, *modus ponens* is not available, of course, so a derivability relation is defined differently, for example, by means of a first-degree proof system using single-antecedent and single-conclusion derivability statements (sequents) of the form $A \vdash B$ as inputs and outputs of inference rules. First-degree proof systems for certain trilattice logics were dealt with in Chap. 4.

Odintsov [183] has presented axiomatizations of truth entailment and falsity entailment in $SIXTEEN_3$ in the enriched languages $\mathscr{L}_{tf}^{\to_t}$, $\mathscr{L}_{tf}^{\to_f}$, and $\mathscr{L}_{tf}^{*}$. A central idea for defining these axiomatizations is to present the trilattice $SIXTEEN_3$ as a four-component twist-structure over the two-element Boolean algebra.[2] This

---

[1] Alternatively, one may work with concrete axioms and use a rule of uniform substitution of arbitrary formulas for atomic formulas.

[2] The term 'twist-structure' was introduced by Kracht [147]. Given an algebra $\mathfrak{A} = (\mathsf{A}, \{f_{\mathfrak{A}} \mid f \in \mathscr{F}\})$, for each n-ary function symbol $f \in \mathscr{F}$, the n-place operation $f_{\mathfrak{A}}$ is defined on the direct product of a finite number of copies of the universe A. The operations are not defined componentwise but in a "twisted" way. Twist structures for logics with strong negation have been introduced in [84, 258], see also [182]. Truth entailment in $FOUR_2$, for example, can be represented as a twist-structure over the two-element lattice. An element $x \in \mathbf{4}$ is represented as a pair (i, j) of values of characteristic functions from x into $\{0, 1\}$ such that $i = 1$ iff $T \in x$ and $j = 1$ iff $F \in x$. The lattice operations $\sqcap_t$, $\sqcup_t$, and $-_t$ are then defined as the following "twisted" operations:

$$(x, y) \sqcup_t (x', y') := (x \vee x', y \wedge y'),$$
$$(x, y) \sqcap_t (x', y') := (x \wedge y, x' \vee y'),$$
$$-_t(x, y) := (y, x),$$

matrix presentation of the algebraic operations of $SIXTEEN_3$ will give rise to the definition of cut-free sequent calculi for truth and falsity entailment in Chap. 6.

5.3 Odintsov's Axiomatization of Truth Entailment and Falsity Entailment in $SIXTEEN_3$

It follows from results in universal algebra that the relations of truth entailment and falsity entailment in the language $\mathscr{L}_{tf}$ can be finitely axiomatized.[3] An algebra is congruence-distributive iff its lattice of congruence relations is a distributive lattice. A class of algebras is congruence-distributive iff each of its members is congruence distributive. Since the trilattice $SIXTEEN_3$ lies within a congruence distributive variety, Baker's famous finite basis theorem applies, see [15]. Baker's theorem says that every congruence-distributive variety which is generated by a finite number of finite algebras and has a finite language is finitely axiomatizable. Knowing that a finite axiomatization exists is one thing. However, finding such an axiomatization is another story.

In order to define axiom systems for truth entailment and falsity entailment in $SIXTEEN_3$, Odintsov [183] starts with an auxiliary, first-degree axiomatic proof system $\vdash_{\text{base}}$ for the language $\mathscr{L}_{tf}$. In a second step, he defines two extensions $\vdash_T$ and $\vdash_B$ of the basic system $\vdash_{\text{base}}$ and shows that $\models_t = \vdash_T \cap \vdash_B$. Similarly, Odintsov defines first-degree proof systems $\vdash^f_{\text{base}}$, $\vdash^f_T$, and $\vdash^f_B$ for the language $\mathscr{L}_{tf}$, such that $\models_f = \vdash^f_T \cap \vdash^f_B$. Since the proof systems $\vdash_T$ and $\vdash_B$, and respectively, $\vdash^f_T$ and $\vdash^f_B$ do not have *modus ponens* as their only rule of inference, it is not clear how to axiomatize their intersections.

Miura's Theorem [174] (Theorem 4.5 in [13]) on the axiomatization of intersections of superintuitionistic logics presented as Hilbert-style proof systems uses the Deduction Theorem for intuitionistic implication and applies to propositional logics with axiomatizations comprising *modus ponens* as the only inference rule. Let **IPL** denote intuitionistic propositional logic in the language $\{\wedge, \vee, \rightarrow, \neg\}$, let Δ and Γ be sets of formulas (additional axioms) in the language of **IPL**, and let **IPL**+Δ (**IPL**+Γ) be the logic axiomatized by adding Δ (Γ) to the axioms of **IPL**. We write $A(p_1, \ldots, p_m)$ to indicate that all propositional variables of A belong to $\{p_1, \ldots, p_m\}$.

(Footnote 2 continued)
where $\wedge$ and $\vee$ are the classical truth functions interpreting the conjunction and the disjunction connective, respectively. If $\wedge$, $\vee$, and $\rightarrow$ represent the operations of an implicative lattice, and the binary operation $\sqsupset_t$ is defined by

$$(a,b) \sqsupset_t (c,d) = (a \rightarrow c, a \wedge d),$$

the result is a semantics for Nelson's constructive paraconsistent logic **N4**, see [181, 182].

[3] This has been pointed out to us by Anatol Reibold.

The *repeatedless disjunction* of formulas $A(p_1,\ldots,p_n)$ and $B(p_1,\ldots,p_m)$ in the language of **IPL** is the formula $A\underline{\vee}B := A(p_1,\ldots,p_n)\vee B(p_{n+1},\ldots,p_{n+m})$. The disjuncts of $A\underline{\vee}B$ have no propositional variables in common.

Theorem 5.1 (*Miura* [174]) *If* $L_1 = \mathbf{IPL} + \{A_i \mid i \in I\}$ *and* $L_2 = \mathbf{IPL} + \{B_j \mid i \in J\}$ *for some index sets I and J, then* $L_1 \cap L_2 = \mathbf{IPL} + \{A_i\underline{\vee}B_j \mid i \in I, j \in J\}$.

To apply Miura's Theorem, Odintsov considers the implicational languages $\mathscr{L}_{tf}^{\rightarrow_t}$, $\mathscr{L}_{tf}^{\rightarrow_f}$, and $\mathscr{L}_{tf}^{*}$. Moreover, Odintsov introduces a simple and elegant matrix presentation of the algebraic operations in *SIXTEEN*$_3$. The trilattice *SIXTEEN*$_3$ is thereby presented as a "twist structure" because its operations "are defined componentwise on several components and are "twisted" in some way on the others" [183]. Every element x of **16** is a subset of $\mathscr{P}(\mathbf{2})$; therefore it can be represented as a 2×2-matrix of values of characteristic functions:

$$\begin{vmatrix} n & f \\ t & b \end{vmatrix}$$

where each element of the matrix is an element of $\{0, 1\}$ (understood as the set of classical truth values) and the following equivalences hold:

$$n = 1 \text{ iff } \mathbf{N} \in x; \quad f = 1 \text{ iff } \mathbf{F} \in x; \quad t = 1 \text{ iff } \mathbf{T} \in x; \quad b = 1 \text{ iff } \mathbf{B} \in x.$$

In order to emphasize that falsity is usually taken to be preserved from the conclusions of a valid inference to the premises (in the sense that if all the conclusions are false, then at least one of the premises is), in Chap. 3 we interpreted falsity conjunction $\wedge_f$ by the lattice *join* with respect to the falsity order $\leq_f$ of *SIXTEEN*$_3$ and falsity disjunction $\vee_f$ with respect to the lattice *meet* of $\leq_f$. We stipulated that $v(A \wedge_f B) := v(A) \sqcup_f v(B)$ and $v(A \vee_f B) := v(A) \sqcap_f v(B)$.[4] Odintsov [183] deviates from this definition by setting: $v(A \wedge_f B) := v(A) \sqcap_f v(B)$ and $v(A \vee_f B) := v(A) \sqcup_f v(B)$. In this case, however, not only truth but also falsity is preserved from the premise to the conclusion of a valid (single conclusion and single premise) inference.

Definition 5.2 Let A and B be $\mathscr{L}_{tf}$-formulas. Then

$$A \models_t B \text{ iff } \forall v\ (v(A) \leq_t v(B)), \qquad A \models_f B \text{ iff } \forall v\ (v(A) \leq_f v(B)).$$

To allow a better comparison with [183], in what follows, we will adopt the interpretation of $\wedge_f$ and $\vee_f$ by $\sqcap_f$ and $\sqcup_f$, respectively, and Definition 5.2. Also for the sake of comparison, in the rest of this chapter, we follow Odintsov in denoting

[4] In this chapter we will omit the superscript '16' when dealing with a 16-valuation v^{16}.

by $\leq_f$ the inverse of the falsity order (as defined in Chap. 3) and in denoting by $\sqcap_f$ ($\sqcup_f$) the lattice meet (join) of the inverted falsity order.

Proposition 5.1 (*Odintsov* [183]) *Let $\wedge$ and $\vee$ be the classical truth functions of conjunction and disjunction, and let the matrices*

$$\begin{vmatrix} n & f \\ t & b \end{vmatrix} \quad \textit{and} \quad \begin{vmatrix} n' & f' \\ t' & b' \end{vmatrix}$$

represent elements of **16**. *Then the following equations hold*:

$$\begin{vmatrix} n & f \\ t & b \end{vmatrix} \sqcap_t \begin{vmatrix} n' & f' \\ t' & b' \end{vmatrix} = \begin{vmatrix} n \vee n' & f \vee f' \\ t \wedge t' & b \wedge b' \end{vmatrix}$$

$$\begin{vmatrix} n & f \\ t & b \end{vmatrix} \sqcup_t \begin{vmatrix} n' & f' \\ t' & b' \end{vmatrix} = \begin{vmatrix} n \wedge n' & f \wedge f' \\ t \vee t' & b \vee b' \end{vmatrix}$$

$$-_t \begin{vmatrix} n & f \\ t & b \end{vmatrix} = \begin{vmatrix} t & b \\ n & f \end{vmatrix}$$

$$\begin{vmatrix} n & f \\ t & b \end{vmatrix} \sqcap_f \begin{vmatrix} n' & f' \\ t' & b' \end{vmatrix} = \begin{vmatrix} n \wedge n' & f \vee f' \\ t \wedge t' & b \vee b' \end{vmatrix}$$

$$\begin{vmatrix} n & f \\ t & b \end{vmatrix} \sqcup_f \begin{vmatrix} n' & f' \\ t' & b' \end{vmatrix} = \begin{vmatrix} n \vee n' & f \wedge f' \\ t \vee t' & b \wedge b' \end{vmatrix}$$

$$-_f \begin{vmatrix} n & f \\ t & b \end{vmatrix} = \begin{vmatrix} f & n \\ b & t \end{vmatrix}.$$

Proof These equations follow immediately from Propositions 3.2 and 3.4 (taking into account that there we have interpreted $\wedge_f$ by $\sqcup_f$ and $\vee_f$ by $\sqcap_f$). □

Definition 5.3 The operations $\sqsupset_t$ and $\sqsupset_f$ on **16** are defined as follows:

$$\begin{vmatrix} n & f \\ t & b \end{vmatrix} \sqsupset_t \begin{vmatrix} n' & f' \\ t' & b' \end{vmatrix} := \begin{vmatrix} \neg n \wedge n' & \neg f \wedge f' \\ t \rightarrow t' & b \rightarrow b' \end{vmatrix}$$

$$\begin{vmatrix} n & f \\ t & b \end{vmatrix} \sqsupset_f \begin{vmatrix} n' & f' \\ t' & b' \end{vmatrix} := \begin{vmatrix} n \rightarrow n' & \neg f \wedge f' \\ t \rightarrow t' & \neg b \wedge b' \end{vmatrix}$$

where $\wedge$, $\rightarrow$ and $\neg$ are the truth functions of Boolean conjunction, implication, and negation, respectively.

The definition of a valuation function v is now extended by the clauses:

$$v(A \rightarrow_t B) = v(A) \sqsupset_t v(B), \quad v(A \rightarrow_f B) = v(A) \sqsupset_f v(B).$$

The following proposition justifies the above definition of $\sqsupset_t$ and $\sqsupset_f$. It shows that the definition is correct in the sense that it in fact results in a semantical version of the Deduction Theorem.

Proposition 5.2 (*Odintsov* [183]) *For every x, y, $z \in \mathbf{16}$,*

1. $x \leq_t y \sqsupset_t z$ *iff* $x \sqcap_t y \leq_t z$, *and*
2. $x \leq_f y \sqsupset_f z$ *iff* $x \sqcap_f y \leq_f z$.

Corollary 5.1

1. *For every $x, y \in \mathbf{16}$, $x \leq_t y$ iff $x \sqsupset_t y = \mathbf{TB}$.*
2. *For every formula $A, B \in \mathscr{L}_{tf}^{\rightarrow_t}$, $A \models_t B$ iff $\forall v(v(A \rightarrow_t B) = \mathbf{TB})$.*
3. *For every $x, y \in \mathbf{16}$, $x \leq_f y$ iff $x \sqsupset_f y = \mathbf{NT}$.*
4. *For every formula $A, B \in \mathscr{L}_{tf}^{\rightarrow_f}$, $A \models_f B$ iff $\forall v(v(A \rightarrow_f B) = \mathbf{NT})$.*

The two biconditionals $A \leftrightarrow_t B$ and $A \leftrightarrow_f B$ are defined as $(A \rightarrow_t B) \wedge_t (B \rightarrow_t A)$ and $(A \rightarrow_f B) \wedge_f (B \rightarrow_f A)$, respectively, and the classical negation $\neg A$ of A can be defined in $\mathscr{L}_{tf}^{\rightarrow_t}$ as $A \rightarrow_t \sim_t(p \rightarrow_t p)$ and in $\mathscr{L}_{tf}^{\rightarrow_f}$ as $A \rightarrow_f \sim_f(p \rightarrow_f p)$, where p is some fixed propositional variable.

Odintsov axiomatizes the semantically presented logic (set of formulas) $L^t = \{A \mid \forall v(v(A) = \mathbf{TB}), A \in \mathscr{L}_{tf}^{\rightarrow_t}\}$ as the intersection of two logics L_T and L_B presented as Hilbert-style proof systems with *modus ponens* as the only rule of inference. Similarly, he axiomatizes the logic $L^{t*} = \{A \mid \forall v(v(A) = \mathbf{TB}), A \in \mathscr{L}_{tf}^{*}\}$ as the intersection of two logics L_T^* and L_B^* presented as axiom systems with *modus ponens* as the only inference rule. Axiomatizations for the semantically defined logics $L^f = \{A \mid \forall v(v(A) = \mathbf{NT}), A \in \mathscr{L}_{tf}^{\rightarrow_f}\}$ and $L^{f*} = \{A \mid \forall v(v(A) = \mathbf{NT}), A \in \mathscr{L}_{tf}^{*}\}$ can be obtained analogously. We will here focus on the systems explicitly treated by Odintsov, that is, L^t and L^{t*}. The other axiom systems surveyed in Table 5.1 (except for those under (1) and (2)) are not explicitly listed in [183]. We are now confronted with 24 axiom systems related to the trilattice $SIXTEEN_3$. The survey in Table 5.1 systematizes the overall picture.

5.3.1 First-Degree Calculi

The first-degree proof systems $\vdash_{\text{base}}$, $\vdash_T$, and $\vdash_B$ for the implication-free language $\mathscr{L}_{tf}$ provide the starting point of the further development. A canonical model construction in the style of the canonical model construction from Chap. 4 is used to show that $(\mathscr{L}_{tf}, \models_t)$ is sound and complete with respect to $\vdash_T$ and that $(\mathscr{L}_{tf}, \models_f)$ is sound and complete with respect to $\vdash_B$. We use the notation $A \dashv\vdash B$ to state both $A \vdash B$ and $B \vdash A$. If a sequent $A \vdash B$ is provable in $\vdash_\sharp$ ($\sharp \in \{\text{base}, T, B\}$, we will also write $A \vdash_\sharp B$.

Definition 5.4 The first-degree proof system $\vdash_{\text{base}}$ consists of the following sequents and inference rules:

1. $A \wedge_t B \vdash A, \quad A \wedge_t B \vdash B$
2. $A \vdash A \vee_t B, \quad B \vdash A \vee_t B$

Table 5.1 Axiom systems related to $SIXTEEN_3$.

Axiom system	Language	Relation to $\models_t$ or $\models_f$	Inference rules
(1) $\vdash_{base}$	$\mathscr{L}_{tf}$	$\vdash_{base} \subseteq \models_t$	Rules for $\vdash$
$\vdash_T$	$\mathscr{L}_{tf}$	$\models_t = \vdash_T \cap \vdash_B$	Rules for $\vdash$
$\vdash_B$	$\mathscr{L}_{tf}$	$\models_t = \vdash_T \cap \vdash_B$	Rules for $\vdash$
(2) $\vdash^f_{base}$	$\mathscr{L}_{tf}$	$\vdash^f_{base} \subseteq \models_f$	Rules for $\vdash$
$\vdash^f_T$	$\mathscr{L}_{tf}$	$\models_f = \vdash^f_T \cap \vdash^f_B$	Rules for $\vdash$
$\vdash^f_B$	$\mathscr{L}_{tf}$	$\models_f = \vdash^f_T \cap \vdash^f_B$	Rules for $\vdash$
(3) L_{base}	$\mathscr{L}^{\to_t}_{tf}$	$L_{base} \subseteq \models_t$	Just m. p. for $\to_t$
L_T	$\mathscr{L}^{\to_t}_{tf}$	$L^t = L_T \cap L_B$	Just m. p. for $\to_t$
L_B	$\mathscr{L}^{\to_t}_{tf}$	$L^t = L_T \cap L_B$	Just m. p. for $\to_t$
(4) L'_{base}	$\mathscr{L}^{\to_t}_{tf}$	$L'_{base} \subseteq \models_f$	Just m. p. for $\to_t$
L'_T	$\mathscr{L}^{\to_t}_{tf}$	$L' = L'_T \cap L'_B$	Just m. p. for $\to_t$
L'_B	$\mathscr{L}^{\to_t}_{tf}$	$L' = L'_T \cap L'_B$	Just m. p. for $\to_t$
(5) L^f_{base}	$\mathscr{L}^{\to_f}_{tf}$	$L^f_{base} \subseteq \models_f$	Just m. p. for $\to_f$
L^f_T	$\mathscr{L}^{\to_f}_{tf}$	$L^f = L^f_T \cap L^f_B$	Just m. p. for $\to_f$
L^f_B	$\mathscr{L}^{\to_f}_{tf}$	$L^f = L^f_T \cap L^f_B$	Just m. p. for $\to_f$
(6) $L^{f'}_{base}$	$\mathscr{L}^{\to_f}_{tf}$	$L^{f'}_{base} \subseteq \models_t$	Just m. p. for $\to_f$
$L^{f'}_T$	$\mathscr{L}^{\to_f}_{tf}$	$L^{f'} = L^{f'}_T \cap L^{f'}_B$	Just m. p. for $\to_f$
$L^{f'}_B$	$\mathscr{L}^{\to_f}_{tf}$	$L^{f'} = L^{f'}_T \cap L^{f'}_B$	Just m. p. for $\to_f$
(7) L^{t*}_{base}	$\mathscr{L}^*_{tf}$	$L^{t*}_{base} \subseteq \models_t$	Just m. p. for $\to_t$ and $\to_f$
L^{t*}_T	$\mathscr{L}^*_{tf}$	$L^{t*} = L^{t*}_T \cap L^{t*}_B$	Just m. p. for $\to_t$ and $\to_f$
L^{t*}_B	$\mathscr{L}^*_{tf}$	$L^{t*} = L^{t*}_T \cap L^{t*}_B$	Just m. p. for $\to_t$ and $\to_f$
(8) L^{f*}_{base}	$\mathscr{L}^*_{tf}$	$L^{f*}_{base} \subseteq \models_f$	Just m. p. for $\to_t$ and $\to_f$
L^{f*}_T	$\mathscr{L}^*_{tf}$	$L^{f*} = L^{f*}_T \cap L^{f*}_B$	Just m. p. for $\to_t$ and $\to_f$
L^{*f}_B	$\mathscr{L}^*_{tf}$	$L^{*f} = L^{f*}_T \cap L^{f*}_B$	Just m. p. for $\to_t$ and $\to_f$

3. $A \wedge_t (B \vee_t C) \dashv\vdash (A \wedge_t B) \vee_t (A \wedge_t C)$
4. $A \vee_t (B \wedge_t C) \dashv\vdash (A \vee_t B) \wedge_t (A \vee_t C)$
5. $A \dashv\vdash \sim_t \sim_t A$
6. $A \dashv\vdash \sim_f \sim_f A$
7. $\sim_t \sim_f A \dashv\vdash \sim_f \sim_t A$
8. $\sim_f A \dashv\vdash \sim_t \sim_f \sim_t A$
9. $\sim_t A \dashv\vdash \sim_f \sim_t \sim_f A$
10. $\sim_t (A \wedge_t B) \dashv\vdash \sim_t A \vee_t \sim_t B$
11. $\sim_t (A \vee_t B) \dashv\vdash \sim_t A \wedge_t \sim_t B$
12. $\sim_t (A \wedge_f B) \dashv\vdash \sim_t A \wedge_f \sim_t B$
13. $\sim_t (A \vee_f B) \dashv\vdash \sim_t A \vee_f \sim_t B$
14. $\sim_f (A \wedge_f B) \dashv\vdash \sim_f A \vee_f \sim_f B$
15. $\sim_f (A \vee_f B) \dashv\vdash \sim_f A \wedge_f \sim_f B$
16. $\sim_f (A \wedge_t B) \dashv\vdash \sim_f A \wedge_t \sim_f B$

17. $\sim_f (A \vee_t B) \dashv\vdash \sim_f A \vee_t \sim_f B$
18. $\sim_f\sim_t (A \wedge_t B) \dashv\vdash \sim_f\sim_t A \vee_t \sim_f\sim_t B$
19. $\sim_f\sim_t (A \vee_t B) \dashv\vdash \sim_f\sim_t A \wedge_t \sim_f\sim_t B$
20. $\sim_f\sim_t (A \wedge_f B) \dashv\vdash \sim_f\sim_t A \vee_f \sim_f\sim_t B$
21. $\sim_f\sim_t (A \vee_f B) \dashv\vdash \sim_f\sim_t A \wedge_f \sim_f\sim_t B$

$$\frac{A \vdash B \quad B \vdash C}{A \vdash C} \qquad \frac{A \vdash B \quad A \vdash C}{A \vdash B \wedge_t C} \qquad \frac{A \vdash C \quad B \vdash C}{A \vee_t B \vdash C}$$

Definition 5.5 The proof system $\vdash_T$ is obtained from $\vdash_{\text{base}}$ by adding the axioms:

$$1^T \; A \wedge_f B \dashv\vdash A \wedge_t B \quad \text{and} \quad 2^T \; A \vee_f B \dashv\vdash A \vee_t B.$$

The proof system $\vdash_B$ is obtained from $\vdash_{\text{base}}$ by adding the axioms:

$$1^B \; A \wedge_f B \dashv\vdash A \vee_t B \quad \text{and} \quad 2^B \; A \vee_f B \dashv\vdash A \wedge_t B.$$

As Odintsov points out, $\vdash_T$ and $\vdash_B$ are not definitional extensions of $\vdash_{\text{base}}$. A definiendum may be replaced in any syntactic context by its definiens with unharmed mutual interderivability (and unharmed provable equivalence). Axiom 1^T notwithstanding, in $\vdash_T$ the falsity conjunction $A \wedge_f B$ is not *defined* by the truth conjunction $A \wedge_t B$ because $A \wedge_f B$ is not replaceable with unharmed interderivability by $A \wedge_t B$. Replacement is harmed in the context of $\sim_t$ and $\sim_f$. The breakdown of replacement (interchangeability) in the scope of a negation operator is well-known from constructive logics with strong negation, see, for example, [2, 182, 231, 264, 267, 270, 271]. These logics are conservative extensions of positive intuitionistic logic in which the double-negation laws and the De Morgan laws hold for conjunction $\wedge$, disjunction $\vee$ and strong negation $\sim$. The strong negation $\sim(A \rightarrow B)$ of an intuitionistic implication is interderivable with $(A \wedge \sim B)$. However, whereas $\sim\sim(A \rightarrow B)$ *is* interderivable with $(A \rightarrow B)$, $(A \rightarrow B)$ *is not* interderivable with $\sim(A \wedge \sim B)$ in these logics. In $\vdash_T$, for example, $A \wedge_f B$ and $A \wedge_t B$ are interderivable by Axiom 1^T. By Axiom 14 of $\vdash_{\text{base}}$, $\sim_f(A \wedge_f B)$ and $\sim_f A \vee_f \sim_f B$ are interderivable. By Axiom 16, $\sim_f(A \wedge_t B)$ is interderivable with $\sim_f A \wedge_f \sim_f B$. However, $\sim_f A \vee_f \sim_f B$ and $\sim_f A \wedge_f \sim_f B$ fail to be interderivable in $\vdash_T$.

Odintsov observes that if one deletes from $\vdash_{\text{base}}$ all axioms containing $\vee_f$ and $\wedge_f$ and adds to 1^T and 2^T the following axioms for $\wedge_f$ and $\vee_f$:

$$\sim_f(A \wedge_f B) \dashv\vdash \sim_f(A \vee_t B),\ \sim_t(A \wedge_f B) \dashv\vdash \sim_t(A \vee_t B),$$
$$\sim_f\sim_t(A \wedge_f B) \dashv\vdash \sim_f\sim_t(A \wedge_t B),$$
$$\sim_f(A \vee_f B) \dashv\vdash \sim_f(A \wedge_t B),\ \sim_t(A \vee_f B) \dashv\vdash \sim_t(A \wedge_t B),$$
$$\sim_f\sim_t(A \vee_f B) \dashv\vdash \sim_f\sim_t(A \vee_t B),$$

one obtains another axiomatization of $\vdash_T$. In this case, a definition of $A \wedge_f B$ and $A \vee_f B$ is given in the sense that conditions for unharmed interderivability *in different syntactic contexts* are stated. Analogously, one can obtain another axiomatization of $\vdash_B$.

As already mentioned, it can be shown that truth entailment in $\mathscr{L}_{tf}$ coincides with the intersection of $\vdash_T$ and $\vdash_B$. From the axioms 1^T, 2^T, 1^B, and 2^B it becomes clear that $\vee_f$ and $\wedge_f$ have a non-deterministic interpretation with respect to truth entailment. In order to establish this characterization of $\models_t$, Odintsov considers for every valuation v its co-ordinate valuations v_t, v_f, v_n, and v_b. The co-ordinate valuations are classical valuations and they are defined as follows:

$$v(p) = \begin{vmatrix} v_n(p) & v_f(p) \\ v_t(p) & v_b(p) \end{vmatrix}$$

The co-ordinate valuations can be used to define co-ordinate entailment relations. We consider two sets of such entailment relations, one associated with $\models_t$ and the other with $\models_f$.

Definition 5.6 Let $A, B \in \mathscr{L}_{tf}$ and let $\geq$ and $\leq$ stand for the usual ordering on $\mathbb{N}$.

$$\begin{array}{ll} A \models_N B \text{ iff } \forall v(v_n(A) \geq v_n(B)), & A \models_F B \text{ iff } \forall v(v_f(A) \geq v_f(B)); \\ A \models_T B \text{ iff } \forall v(v_t(A) \leq v_t(B)); & A \models_B B \text{ iff } \forall v(v_b(A) \leq v_b(B)); \\ A \models^f_N B \text{ iff } \forall v(v_n(A) \leq v_n(B)); & A \models^f_F B \text{ iff } \forall v(v_f(A) \geq v_f(B)); \\ A \models^f_T B \text{ iff } \forall v(v_t(A) \leq v_t(B)); & A \models^f_B B \text{ iff } \forall v(v_b(A) \geq v_b(B)). \end{array}$$

Proposition 5.3 $\models_t = \models_T \cap \models_B \cap \models_F \cap \models_N$ *and* $\models_f = \models^f_T \cap \models^f_B \cap \models^f_F \cap \models^f_N$.

Proof It is enough to note that for $\begin{vmatrix} n & f \\ t & b \end{vmatrix}$ and $\begin{vmatrix} n' & f' \\ t' & b' \end{vmatrix} \in \mathbf{16}$, the following equivalences hold:

$$\begin{vmatrix} n & f \\ t & b \end{vmatrix} \leq_t \begin{vmatrix} n' & f' \\ t' & b' \end{vmatrix} \text{ iff } \begin{matrix} n \geq n' & f \geq f' \\ t \leq t' & b \leq b' \end{matrix}$$

$$\begin{vmatrix} n & f \\ t & b \end{vmatrix} \leq_f \begin{vmatrix} n' & f' \\ t' & b' \end{vmatrix} \text{ iff } \begin{matrix} n \leq n' & f \geq f' \\ t \leq t' & b \geq b' \end{matrix}$$

□

Soundness theorems for the logics $\vdash_T$ and $\vdash_B$ follow by induction on the construction of proofs.

Theorem 5.2 *(Soundness) For all $\mathscr{L}_{tf}$-formulas A and B:*

1. $A \vdash_T B$ *implies* $A \models_T B$,
2. $A \vdash_T B$ *implies* $A \models_F B$,
3. $A \vdash_B B$ *implies* $A \models_B B$, *and*
4. $A \vdash_B B$ *implies* $A \models_N B$.

Completeness theorems for the logics $\vdash_T$ and $\vdash_B$ can be proved by applying the canonical model method. In other words, semantical evaluation is characterized by membership in certain sets of formulas. In this way, falsifying models can be defined for non-derivable sequents without making any ontological suppositions beyond the assumption that there exist the formulas of the language under consideration and that there exist sets of formulas. The sets of formulas required for the proof are, again, certain prime theories.

Definition 5.7 Let $\natural \in \{T, B\}$. A set of formulas α is closed under $\vdash_\natural$ iff it holds that (i) $A \in \alpha$ and $A \vdash_\natural B$ implies $B \in \alpha$ and (ii) $A \in \alpha$ and $B \in \alpha$ implies $A \wedge_t B \in \alpha$. A set of sentences α is a $\natural$-theory iff α is closed under $\vdash_\natural$. A $\natural$-theory α is prime iff it holds that if $A \vee_t B \in \alpha$, then $A \in \alpha$ or $B \in \alpha$.

The appropriate version of Lindenbaum's extension lemma can be proved.

Lemma 5.1 *(Extension Lemma) For $\natural \in \{T, B\}$ and $\mathscr{L}_{tf}$-formulas A and B, if $A \vdash_\natural B$ does not hold, then there exists a prime $\natural$-theory α such that $A \in \alpha$ and $B \notin \alpha$.*

The prime theories have all the properties needed to emulate in a "Truth Lemma" the evaluation conditions for co-ordinate valuation functions in terms of set membership. These co-ordinate valuations have to be defined in four different ways in order to capture the entailment relations $\models T$ and $\models F$ in the case of prime T-theories and the entailment relations $\models B$, and $\models N$ in the case of prime B-theories.

Lemma 5.2 *Let $\natural \in \{T, B\}$, α be a prime $\natural$-theory, and $A, B \in \mathscr{L}_{tf}$. Then*

- $A \wedge_t B \in \alpha$ *iff* $A \in \alpha$ *and* $B \in \alpha$;
- $A \vee_t B \in \alpha$ *iff* $A \in \alpha$ *or* $B \in \alpha$;
- $\sim_t\sim_t A \in \alpha$ *iff* $A \in \alpha$;
- $\sim_f\sim_f A \in \alpha$ *iff* $A \in \alpha$;
- $\sim_t\sim_f A \in \alpha$ *iff* $\sim_f\sim_t A \in \alpha$;
- $\sim_t\sim_f\sim_t A \in \alpha$ *iff* $\sim_f A \in \alpha$;
- $\sim_f\sim_t\sim_f A \in \alpha$ *iff* $\sim_t A \in \alpha$;
- $\sim_t (A \wedge_t B) \in \alpha$ *iff* $\sim_t A \in \alpha$ *or* $\sim_t B \in \alpha$;
- $\sim_t (A \vee_t B) \in \alpha$ *iff* $\sim_t A \in \alpha$ *and* $\sim_t B \in \alpha$;
- $\sim_f (A \wedge_t B) \in \alpha$ *iff* $\sim_f A \in \alpha$ *and* $\sim_f B \in \alpha$;
- $\sim_f (A \vee_t B) \in \alpha$ *iff* $\sim_f A \in \alpha$ *or* $\sim_f B \in \alpha$;
- $\sim_f\sim_t (A \wedge_t B) \in \alpha$ *iff* $\sim_f\sim_t A \in \alpha$ *or* $\sim_f\sim_t B \in \alpha$;
- $\sim_f\sim_t (A \vee_t B) \in \alpha$ *iff* $\sim_f\sim_t A \in \alpha$ *and* $\sim_f\sim_t B \in \alpha$;
- *if* $\natural = T$, *then* ($A \vee_f B \in \alpha$ *iff* $A \vee_t B \in \alpha$) *and* ($A \wedge_f B \in \alpha$ *iff* $A \wedge_t B \in \alpha$);
- *if* $\natural = B$, *then* ($A \vee_f B \in \alpha$ *iff* $A \wedge_t B \in \alpha$) *and* ($A \wedge_f B \in \alpha$ *iff* $A \vee_t B \in \alpha$).

Proof All these equivalences follow straightforwardly from the definition of the axioms of $\vdash_{\text{base}}$, $\vdash_T$, and $\vdash_B$ and the assumption that α is prime. □

Valuations for the four positions n, f, t, and b and prime T-theories are defined in two ways. The first definition is suitable for proving the completeness of $\vdash_T$ with respect to $\models_T$.

Definition 5.8 For every T-theory α, the valuation function $v^{\alpha,T}$ from atomic formulas into **16** is defined as follows:

$$v_n^{\alpha,T}(p) = 1 \Leftrightarrow \sim_t p \in \alpha \quad v_f^{\alpha,T}(p) = 1 \Leftrightarrow \sim_f\sim_t p \in \alpha$$
$$v_t^{\alpha,T}(p) = 1 \Leftrightarrow \quad p \in \alpha \quad v_b^{\alpha,T}(p) = 1 \Leftrightarrow \quad \sim_f p \in \alpha$$

Lemma 5.3 (*Truth Lemma, Odintsov* [183]) *Let* $A \in \mathscr{L}_{tf}$. *Then*

$$v_n^{\alpha,T}(A) = 1 \Leftrightarrow \sim_t A \in \alpha \quad v_f^{\alpha,T}(A) = 1 \Leftrightarrow \sim_f\sim_t A \in \alpha$$
$$v_t^{\alpha,T}(A) = 1 \Leftrightarrow \quad A \in \alpha \quad v_b^{\alpha,T}(A) = 1 \Leftrightarrow \quad \sim_f A \in \alpha$$

Proof The four claims are proved conjointly by simultaneous induction on the structure of A. The induction base holds by the definition of $v^{\alpha,T}$. We here consider just one exemplary case for the co-ordinate valuation $v_f^{\alpha,T}$. The other co-ordinate valuations can be treated similarly.

($\wedge_f$): A has the shape $C \wedge_f B$. Then $v_f^{\alpha,T}(C \wedge_f B) = 1$ iff $v_f^{\alpha,T}(C) = 1$ or $v_f^{\alpha,T}(B) = 1$. By the induction hypothesis, this disjunction holds iff $\sim_f\sim_t C \in \alpha$ or $\sim_f\sim_t B \in \alpha$.

By Lemma 5.2, this holds iff $\sim_f\sim_t C \vee_t \sim_f\sim_t B \in \alpha$. Since α is a T-theory, the latter is equivalent to $\sim_f\sim_t C \vee_f \sim_f\sim_t B \in \alpha$ and to $\sim_f\sim_t (C \wedge_f B) \in \alpha$, by deductive closure and Axiom 20. □

Theorem 5.3 *Let* $A, B \in \mathscr{L}_{tf}$. *Then* $A \models_T B$ *implies* $A \vdash_T B$.

Proof If $A \nvdash_T B$, then, by the Extension Lemma, there exists a prime T-theory α such that $A \in \alpha$ and $B \notin \alpha$. By the Truth Lemma 5.3, $v_t^{\alpha,F}(A) = 1$ and $v_t^{\alpha,F}(B) = 0$. Hence $v_t^{\alpha,F}(A) > v_t^{\alpha,F}(B)$, which means that $A \not\models_T B$. Therefore, $\models_T \subseteq \vdash_T$. □

In order to prove the completeness of $\vdash_T$ with respect to $\models_F$, Odintsov defined another set of co-ordinate valuations relative to a prime T-theory.

Definition 5.9 Let α be a prime T-theory. The valuation function $v^{\alpha,F}$ from atomic formulas into **16** is defined by the following equivalences:

$$v_n^{\alpha,F}(p) = 0 \Leftrightarrow \quad \sim_f p \in \alpha \quad v_f^{\alpha,F}(p) = 0 \Leftrightarrow \quad p \in \alpha$$
$$v_t^{\alpha,F}(p) = 0 \Leftrightarrow \sim_f\sim_t p \in \alpha \quad v_b^{\alpha,F}(p) = 0 \Leftrightarrow \sim_t p \in \alpha$$

Lemma 5.4 *Let* $A \in \mathscr{L}_{tf}$. *Then*

$$v_n^{\alpha,F}(A) = 0 \Leftrightarrow \quad \sim_f A \in \alpha \quad v_f^{\alpha,F}(A) = 0 \Leftrightarrow \quad A \in \alpha$$
$$v_t^{\alpha,F}(A) = 0 \Leftrightarrow \sim_f\sim_t A \in \alpha \quad v_b^{\alpha,F}(A) = 0 \Leftrightarrow \sim_t A \in \alpha$$

Lemma 5.4 is proved similarly to Lemma 5.3. Moreover, the proof of the following completeness theorem is analogous to the proof of Theorem 5.3.

Theorem 5.4 *Let* $A, B \in \mathscr{L}_{tf}$. *Then* $A \models_F B$ *implies* $A \vdash_F B$.

The completeness of $\vdash_B$ with respect to $\models_B$ and $\models_N$ can again be shown by defining suitable co-ordinate valuations.

Theorem 5.5 *Let* $A, B \in \mathscr{L}_{tf}$. *Then 1.* $A \models_B B$ *implies* $A \vdash_B B$ *and* 2. $A \models_N B$ *implies* $A \vdash_B B$.

Proof To prove the first claim, for every prime B-theory α, co-ordinate valuations $v_n^{\alpha,B}$, $v_f^{\alpha,B}$, $v_t^{\alpha,B}$, $v_b^{\alpha,B}$ are defined as follows:

$$\begin{array}{ll} v_n^{\alpha,B}(p) = 1 \Leftrightarrow \sim_f\sim_t p \in \alpha & v_f^{\alpha,B}(p) = 1 \Leftrightarrow \sim_t p \in \alpha \\ v_t^{\alpha,B}(p) = 1 \Leftrightarrow \quad \sim_f p \in \alpha & v_b^{\alpha,B}(p) = 1 \Leftrightarrow \quad p \in \alpha \end{array}$$

To prove the second claim, for every prime B-theory α, co-ordinate valuations $v_n^{\alpha,N}$, $v_f^{\alpha,N}$, $v_t^{\alpha,N}$, $v_b^{\alpha,N}$ are defined as follows:

$$\begin{array}{ll} v_n^{\alpha,N}(p) = 0 \Leftrightarrow \quad p \in \alpha & v_f^{\alpha,N}(p) = 0 \Leftrightarrow \quad \sim_f p \in \alpha \\ v_t^{\alpha,N}(p) = 0 \Leftrightarrow \sim_t p \in \alpha & v_b^{\alpha,N}(p) = 0 \Leftrightarrow \sim_f\sim_t p \in \alpha \end{array}$$

In both cases, it is shown by simultaneous induction on the structure of A that all the equivalences can be extended to an arbitrary $\mathscr{L}_{tf}$-formula A. Completeness then follows by the familiar contrapositive argument. □

We may now combine the Soundness Theorem 5.2, the Completeness Theorems 5.3, 5.4, 5.5, and Proposition 5.3 to observe that the entailment relation $\models_T$ is the intersection of two syntactically defined inference relations.

Theorem 5.6 (*Odintsov* [183]) $\models_t = \vdash_T \cap \vdash_B$.

From this characterization of $\models_t$, it is clear that the following semantical facts hold.[5]

Proposition 5.4 *Let* $A, B \in \mathscr{L}_{tf}$, *Then*

1. $A \wedge_t B \models_t A \wedge_f B$, $\quad A \wedge_t B \models_t A \vee_f B$; $\quad$ 2. $(A \wedge_f B) \wedge_t (A \vee_f B) \models_t A \wedge_t B$;
3. $A \wedge_f B \models_t A \vee_t B$, $\quad A \vee_f B \models_t A \vee_t B$; $\quad$ 4. $A \vee_t B \models_t (A \wedge_f B) \vee_t (A \vee_f B)$.

A completely parallel development can be given by a characterization of $\models_f$ as the intersection of two syntactic consequence relations defined as first-degree proof systems. The first step of this development is the definition of the consequence relation $\vdash^f_{\text{base}}$.

[5] Odintsov raises the question whether the addition of the corresponding sequents to $\vdash_{\text{base}}$ might result in an axiomatization of $\models_t$. This problem might be addressed in light of the results in [293].

Definition 5.10 The first-degree proof system $\vdash^f_{\text{base}}$ consists of the Axioms 5–21 of $\vdash_{\text{base}}$ and the following sequents and inference rules:

$1^f A \wedge_f B \vdash A, \quad A \wedge_f B \vdash B$
$2^f A \vdash A \vee_f B, \quad B \vdash A \vee_f B$
$3^f A \wedge_f (B \vee_f C) \dashv\vdash (A \wedge_f B) \vee_f (A \wedge_f C)$
$4^f A \vee_f (B \wedge_f C) \dashv\vdash (A \vee_f B) \wedge_f (A \vee_f C)$

$$\frac{A \vdash B \quad B \vdash C}{A \vdash C} \quad \frac{A \vdash B \quad A \vdash C}{A \vdash B \wedge_f C} \quad \frac{A \vdash C \quad B \vdash C}{A \vee_f B \vdash C}$$

Definition 5.11 The proof system $\vdash^f_T$ is obtained from $\vdash^f_{\text{base}}$ by adding the axioms 1^T and 2^T. The proof system $\vdash^f_B$ is obtained from $\vdash_{\text{base}}$ by adding the axioms 1^B and 2^B.

Proposition 5.3 states that $\models_f$ is the intersection of the entailment relations $\models^f_T$, $\models^f_B$, $\models^f_F$, and $\models^f_N$ from Definition 5.6. One can show that each of these co-ordinate entailment relations coincides either with $\vdash^f_T$ or with $\vdash^f_B$.

Theorem 5.7 (*Odintsov* [183]) *Let* $A, B \in \mathcal{L}_{tf}$. *Then*

$$A \vdash^f_T B \text{ iff } A \models^f_T B \text{ iff } A \models^f_F B; \qquad A \vdash^f_B B \text{ iff } A \models^f_B B \text{ iff } A \models^f_N B.$$

By Theorem 5.7 together with Proposition 5.3, it follows that $\models_f = \vdash^f_T \cap \vdash^f_B$.

5.3.2 *Systems with* Modus Ponens *as the Sole Rule of Inference*

As explained already, a Hilbert-style axiomatization with *modus ponens* as the only inference rule can be obtained for $\models_t$ and $\models_f$ if at least one of the implications $\rightarrow_t$ and $\rightarrow_f$ is taken into consideration. We consider $\mathcal{L}^{\rightarrow_t}_{tf}$.

Definition 5.12 The logic L_{base} is the smallest set of formulas in the language $\mathcal{L}^{\rightarrow_t}_{tf}$ closed under *modus ponens* for $\rightarrow_t$

$$\frac{A, \ A \rightarrow_t B}{B}$$

and containing the following axioms:

1. $A \rightarrow_t (B \rightarrow_t A)$
2. $(A \rightarrow_t (B \rightarrow_t C)) \rightarrow_t ((A \rightarrow_t B) \rightarrow_t (A \rightarrow_t C))$
3. $(A \wedge_t B) \rightarrow_t A$
4. $(A \wedge_t B) \rightarrow_t B$
5. $(A \rightarrow_t B) \rightarrow_t ((A \rightarrow_t C) \rightarrow_t (A \rightarrow_t (B \wedge_t C)))$

6. $A \to_t (A \vee_t B)$
7. $B \to_t (A \vee_t B)$
8. $(A \to_t C) \to_t ((B \to_t C) \to_t ((A \vee_t B) \to_t C))$
9. $A \vee_t (A \to_t B)$
10. $A \leftrightarrow_t \sim_t\sim_t A$
11. $A \leftrightarrow_t \sim_f\sim_f A$
12. $\sim_t\sim_f A \leftrightarrow_t \sim_f\sim_t A$
13. $\neg\sim_f A \leftrightarrow_{t\sim_f\neg A}$
14. $\neg\sim_t A \leftrightarrow_t \sim_t \neg A$
15. $\neg\sim_f\sim_t A \leftrightarrow_t \sim_f\sim_t \neg A$
16. $\sim_f A \leftrightarrow_t \sim_t\sim_f\sim_t A$
17. $\sim_t A \leftrightarrow_t \sim_f\sim_t\sim_f A$
18. $\sim_t (A \wedge_t B) \leftrightarrow_t (\sim_t A \vee_t \sim_t B)$
19. $\sim_t (A \vee_t B) \leftrightarrow_t (\sim_t A \wedge_t \sim_t B)$
20. $\sim_t (A \wedge_f B) \leftrightarrow_t (\sim_t A \wedge_f \sim_t B)$
21. $\sim_t (A \vee_f B) \leftrightarrow_t (\sim_t A \vee_f \sim_t B)$
22. $\sim_f (A \wedge_f B) \leftrightarrow_t (\sim_f A \vee_f \sim_f B)$
23. $\sim_f (A \vee_f B) \leftrightarrow_t (\sim_f A \wedge_f \sim_f B)$
24. $\sim_f (A \wedge_t B) \leftrightarrow_t (\sim_f A \wedge_t \sim_f B)$
25. $\sim_f (A \vee_t B) \leftrightarrow_t (\sim_f A \vee_t \sim_f B)$
26. $\sim_f\sim_t (A \wedge_t B) \leftrightarrow_t (\sim_f\sim_t A \vee_t \sim_f\sim_t B)$
27. $\sim_f\sim_t (A \vee_t B) \leftrightarrow_t (\sim_f\sim_t A \wedge_t \sim_f\sim_t B)$
28. $\sim_f\sim_t (A \wedge_f B) \leftrightarrow_t (\sim_f\sim_t A \vee_f \sim_f\sim_t B)$
29. $\sim_f\sim_t (A \vee_f B) \leftrightarrow_t (\sim_f\sim_t A \wedge_f \sim_f\sim_t B)$
30. $(A \to_t B) \leftrightarrow_t (\neg A \vee_t B)$
31. $\sim_t (A \to_t B) \leftrightarrow_t (\sim_t \neg A \wedge_t \sim_t B)$
32. $\sim_f (A \to_t B) \leftrightarrow_t (\sim_f A \to_t \sim_f B)$
33. $\sim_f\sim_t (A \to_t B) \leftrightarrow_t (\sim_f\sim_t \neg A \wedge_t \sim_f\sim_t B)$

Definition 5.13 The axiom systems L_T and L_B are defined as follows:

$$L_T := L_{\text{base}} \cup \{(A \wedge_f B) \leftrightarrow_t (A \wedge_t B), (A \vee_f B) \leftrightarrow_t (A \vee_t B)\},$$

$$L_B := L_{\text{base}} \cup \{(A \wedge_f B) \leftrightarrow_t (A \vee_t B), (A \vee_f B) \leftrightarrow_t (A \wedge_t B)\}.$$

Let Δ and Γ be sets of formulas (additional axioms) in the language $\mathscr{L}_{tf}^{\to_t}$ of L_{base}, and let $L_{\text{base}} + \Delta$ ($L_{\text{base}} + \Gamma$) be the logic axiomatized by adding Δ (Γ) to the axioms of L_{base}.

Theorem 5.8 *If* $L_1 = L_{\text{base}} + \{A_i \mid i \in I\}$ *and* $L_2 = L_{\text{base}} + \{B_j \mid i \in J\}$, *then* $L_1 \cap L_2 = L_{\text{base}} + \{A_i \underline{\vee_t} B_j \mid i \in I, j \in J\}$.

Proof Since L_{base} contains the standard axioms of positive intuitionistic logic in the connectives $\wedge_t$, $\vee_t$, and $\to_t$ (Axioms 1–8) and since L_{base} contains *modus ponens* as the only rule of inference, L_{base} satisfies the Deduction Theorem

$$\Theta \cup \{D\} \vdash E \text{ iff } \Theta \vdash D \rightarrow_t E.$$

All resources needed for the proof of Miura's Theorem (cf. [50, p. 111]) are available in L_{base}. Suppose that $C \in L_1 \cap L_2$. By the Deduction Theorem and properties of $\wedge_t$, it follows that there exist finite index sets I' and J' such that the formulas

$$\bigwedge_{t\, i \in I'} A'_i \rightarrow_t C, \qquad \bigwedge_{t\, j \in J'} B'_j \rightarrow_t C$$

belong to L_{base}, where every A'_i and every B'_j with $i \in I'$ and $j \in J'$ is a substitution instance of some A_k and B_l, for $k \in I$ and $l \in J$. By Axiom 8 and the distribution of truth disjunction over truth conjunction, it follows that

$$\bigwedge_{t\, i \in I',\, j \in J'} (A'_i \vee_t B'_j) \rightarrow_t C$$

belongs to L_{base}. Since $(A'_i \vee_t B'_j)$ is a substitution instance of $(A_i \underline{\vee_t} B_j)$, it follows that $C \in L_{\text{base}} + \{A_i \underline{\vee_t} B_j \mid i \in I, j \in J\}$.

If, on the other hand, $C \in L_{\text{base}} + \{A_i \underline{\vee_t} B_j \mid i \in I, j \in J\}$, then there is a finite set of substitution instances $(A'_i \vee_t B'_j)$ of $(A_i \underline{\vee_t} B_j)$ such that C is derivable from this set. With the help of Axioms 6 and 7, formula C is derivable both from the set of A'_i's and from the set of B'_j's. Therefore, $C \in L_1 \cap L_2$. □

Definition 5.14 The axiom system HL^t consists of L_{base} together with the following disjunction:

$$(((A \wedge_t B) \leftrightarrow_t (A \wedge_f B)) \wedge_t ((A \vee_t B) \leftrightarrow_t (A \vee_f B))) \vee_t$$
$$(((C \wedge_t D) \leftrightarrow_t (C \vee_f D)) \wedge_t ((C \vee_t D) \leftrightarrow_t (C \wedge_f D))).$$

Definiton 5.15 The logic L^*_{base} is the smallest set of formulas in the language $\mathscr{L}^*_{tf}$ closed under *modus ponens* for $\rightarrow_t$ and for $\rightarrow_f$, comprising the axioms of L_{base}, and containing the following axioms for $\rightarrow_f$:

34. $(A \rightarrow_f B) \leftrightarrow_t (\neg A \vee_f B)$
35. $\sim_t(A \rightarrow_f B) \leftrightarrow_t (\sim_t A \rightarrow_f \sim_t B)$
36. $\sim_f(A \rightarrow_f B) \leftrightarrow_t (\sim_f \neg A \wedge_f \sim_f B)$
37. $\sim_f\sim_t(A \rightarrow_f B) \leftrightarrow_t (\sim_f\sim_t \neg A \wedge_f \sim_f\sim_t B)$.

Definition 5.16 The axiom systems L^*_T and L^*_B are defined as follows:

$$L^*_T := L^*_{\text{base}} \cup \{(A \wedge_t B) \leftrightarrow_t (A \wedge_f B), (A \vee_t B) \leftrightarrow_t (A \vee_f B), (A \rightarrow_f B) \leftrightarrow_t (A \rightarrow_t B)\},$$
$$L^*_B := L^*_{\text{base}} \cup \{(A \wedge_t B) \leftrightarrow_t (A \vee_f B), (A \vee_t B) \leftrightarrow_t (A \wedge_f B), (A \rightarrow_f B) \leftrightarrow_t (\neg A \wedge_t B)\}.$$

Definition 5.17 The axiom system HL^{t*} consists of L^*_{base} and the following disjunction:

38. $(((A \wedge_t B) \leftrightarrow_t (A \wedge_f B)) \wedge_t ((A \vee_t B) \leftrightarrow_t (A \vee_f B)) \wedge_t ((A \rightarrow_f B) \leftrightarrow_t (A \rightarrow_t B))) \vee_t$
$(((C \wedge_t D) \leftrightarrow_t (C \vee_f D)) \wedge_t ((C \vee_t D) \leftrightarrow_t (C \wedge_f D)) \wedge_t ((C \rightarrow_f D) \leftrightarrow_t (\neg C \wedge_t D)))$.

Theorem 5.9 (*Odintsov* [183]) *Let A be a formula of $\mathscr{L}_{tf}^{\rightarrow_t}$. Then $A \in L^t$ iff $HL^t \vdash A$. Let A be a formula of $\mathscr{L}_{tf}^{*}$. Then $A \in L^{t*}$ iff $HL^{t*} \vdash A$.*

Proof The claims are shown by combining (a) soundness results for L_T and L_B obtained by induction on the complexity of proofs, (b) the canonical model method to get completeness results for L_T and L_B and (b) Miura's Theorem in the shape of Theorem 5.8.

Step (a) gives:

$$L_T \subseteq \{A \mid \forall v(v_t(A) = 1)\}, \quad L_T \subseteq \{A \mid \forall v(v_f(A) = 0)\};$$
$$L_B \subseteq \{A \mid \forall v(v_b(A) = 1)\}, \quad L_B \subseteq \{A \mid \forall v(v_n(A) = 0)\}.$$

Consider step (b) in the characterization of HL^t. Odintsov shows that

$$L_T \supseteq \{A \mid \forall v(v_t(A) = 1)\}, \quad L_T \supseteq \{A \mid \forall v(v_f(A) = 0)\}$$
$$L_B \supseteq \{A \mid \forall v(v_b(A) = 1)\}, \quad L_B \supseteq \{A \mid \forall v(v_n(A) = 0)\}.$$

Definition 5.18 Let $\sharp \in \{T, B\}$. A set of formulas α is closed under *modus ponens* for $\rightarrow_t$ iff from $A \in \alpha$ and $A \rightarrow_t B \in \alpha$ it follows that $B \in \alpha$. A set of sentences α is an $L_\sharp$-theory iff $L_\sharp \subseteq \alpha$ and α is closed under *modus ponens* for $\rightarrow_t$. A $\sharp$-theory α is prime iff it holds that if $A \vee_t B \in \alpha$, then $A \in \alpha$ or $B \in \alpha$.

The language $\mathscr{L}_{tf}^{\rightarrow_t}$ contains $\rightarrow_t$ (classical implication) as a primitive connective and $\neg$ (classical negation) as a defined connective. Prime $L_\sharp$-theories enjoy properties related to $\rightarrow_t$ and $\neg$ that enable the emulation of evaluation conditions for co-ordinate valuation functions in the language $\mathscr{L}_{tf}^{\rightarrow_t}$. Namely, in addition to all equivalences listed in Lemma 5.2 and due to Axioms 13–15 and 30–33 of L_{base}, the following equivalences hold:

- $A \rightarrow_t B \in \alpha$ iff $A \notin \alpha$ or $B \in \alpha$;
- $\neg A \in \alpha$ iff $A \notin \alpha$;
- $\neg \sim_t A \in \alpha$ iff $\sim_t \neg A \in \alpha$;
- $\neg \sim_f A \in \alpha$ iff $\sim_f \neg A \in \alpha$;
- $\neg \sim_f \sim_t A \in \alpha$ iff $\sim_f \sim_t \neg A \in \alpha$;
- $\sim_t (A \rightarrow_t B) \in \alpha$ iff $\sim_t \neg A \wedge_t \sim_t B \in \alpha$;
- $\sim_f (A \rightarrow_t B) \in \alpha$ iff $\sim_f A \rightarrow_t \sim_f B \in \alpha$;
- $\sim_f \sim_t (A \rightarrow_t B) \in \alpha$ iff $\sim_f \sim_t \neg A \wedge_t \sim_f \sim_t B \in \alpha$.

Since we are now characterizing sets of formulas instead of sets of sequents of the form $A \vdash B$, the Extension Lemma has to be adjusted.

Lemma 5.5 *Let $\sharp \in \{T, B\}$ and $A \in \mathscr{L}_{tf}^{\rightarrow_t}$. Then $A \notin L_\sharp$ implies that there exists a prime $L_\sharp$-theory α with $A \notin \alpha$.*

To obtain a first completeness result for L_T, for every L_T-theory α, four co-ordinate valuations $v_n^{\alpha,T}$, $v_f^{\alpha,T}$, $v_t^{\alpha,T}$, and $v_b^{\alpha,T}$ are assigned as in Definition 5.8.

Lemma 5.6 (*Truth Lemma, Odintsov* [183]) *Let $A \in \mathscr{L}_{tf}^{\rightarrow}$. Then*

$$v_n^{\alpha,T}(A) = 1 \Leftrightarrow \sim_t A \in \alpha \qquad v_f^{\alpha,T}(A) = 1 \Leftrightarrow \sim_f \sim_t A \in \alpha$$
$$v_t^{\alpha,T}(A) = 1 \Leftrightarrow \quad A \in \alpha \qquad v_b^{\alpha,T}(A) = 1 \Leftrightarrow \quad \sim_f A \in \alpha$$

Proof The proof is again by simultaneous induction on the structure of A. Consider by way of example the co-ordinate valuation $v_n^{\alpha,T}$ and the subcase $A \triangleq (B \rightarrow_t C)$. By the definition of $\sqsupset_t$, $v_n^{\alpha,T}(B \rightarrow_t C) = 1$ iff $v_n^{\alpha,T}(B) = 0$ and $v_n^{\alpha,T}(C) = 1$. By the induction hypothesis this means that $\sim_t B \notin \alpha$ and $\sim_t C \in \alpha$. Since α is an L_T-theory, $\sim_t B \notin \alpha$ iff $\neg \sim_t B \in \alpha$ iff $\sim_t \neg B \in \alpha$. Thus $v_n^{\alpha,T}(B \rightarrow_t C) = 1$ iff both $\sim_t \neg B \in \alpha$ and $\sim_t C \in \alpha$. Again by the properties of L_T-theories, the latter holds just in case $\sim_t (B \rightarrow_t C) \in \alpha$.

Completeness follows by using contraposition. If $A \notin L_T$, there exists a prime L_T-theory α with $t\,A \notin \alpha$. Hence $v_t^{\alpha,T}(A) = 0$, and therefore $\{A \mid \forall v(v_t(A) = 1)\} \subseteq L_T$. □

Moreover, to show that $L_T = \{A \mid \forall v(v_f(A) = 0)\}$, the co-ordinate valuations from Definition 5.9 can be used. To show that $L_B \supseteq \{A \mid \forall v(v_b(A) = 1)\}$ and $L_B \supseteq \{A \mid \forall v(v_n(A) = 0)\}$, one may use the co-ordinate valuations defined in the proof of Theorem 5.5.

Since L^t is the set of formulas that receive the value **TB** under any valuation, and since

$$\mathbf{TB} = \begin{vmatrix} 0 & 0 \\ 1 & 1 \end{vmatrix}$$

it follows that $L^t = L_T \cap L_B$. Similarly, it can be shown that $L^{t*} = L_T^* \cap L_B^*$. Applying Theorem 5.8 twice, the claims of Theorem 5.9 follow. □

5.4 Discussion

The axiom systems for trilattice logics enjoy all the advantages that Hilbert-style proof systems may have. In the first place, they are succinct syntactic presentations of logics related to *SIXTEEN*$_3$. In addition, they give insight about truth entailment and falsity entailment in *SIXTEEN*$_3$. The basic system L_{base} reveals that in the presence of $\rightarrow_t$, positive intuitionistic logic (Axioms 1–8) and positive classical propositional logic (Axioms 1–9) are subsystems of truth entailment. Moreover, Axioms 10–11 present double-negation laws, and Axioms 12–17 provide information about the interaction between classical negation, truth negation, and falsity negation. Axioms 18–29, which include certain De Morgan laws, may be seen as "rewrite rules" for negated conjunctions and disjunctions. Axiom 30 presents truth implication as a Boolean implication, and Axioms 31–33 may be regarded as rewrite instructions for negated truth implications. It seems fair to say that L_{base} is a systematic, transparent, and informative axiom system, and so is L^{*}_{base}, which subjoins to L_{base} an axiom presenting falsity implication as a Boolean conditional and axioms for negated falsity implications.

A special merit of the axiom systems HL^{t} and HL^{t*}, however, comes with the additional disjunctive axioms that allow one to capture the intersection of L_{T} and L_{B} and the intersection of L^{*}_{T} and L^{*}_{B}. In particular, Axiom 38 shows that falsity conjunction, falsity disjunction, and falsity implication have an indeterministic interpretation with respect to truth entailment in $SIXTEEN_3$. A falsity conjunction $(A \wedge_f B)$, disjunction $(A \vee_f B)$, and implication $(A \rightarrow_f B)$ is either to be understood as the truth conjunction $(A \wedge_t B)$, truth disjunction $(A \vee_t B)$, and truth implication $(A \rightarrow_t B)$, respectively, or as the truth disjunction $(A \vee_t B)$, truth conjunction $(A \wedge_t B)$, and $(\neg A \wedge_t B)$, respectively.

Analogously, HL^{f*} reveals that truth conjunction, truth disjunction, and truth implication have an indeterministic interpretation with respect to falsity entailment in $SIXTEEN_3$.[6]

Axiom systems such as HL^{t} and HL^{t*} display, however, not only the clear and significant conceptual merits axiom systems may have, but they also share certain limitations with other Hilbert-style proof systems. In particular they are, first and foremost, not especially user-friendly and, second, they are not suitable for manual or automatic proof search. Moreover, from the more philosophical point of view of anti-realistic approaches to meaning, the axioms of HL^{t} and HL^{t*} cannot be seen as isolating the proof-theoretical meaning of the connectives in $\mathscr{L}^{\rightarrow_t}_{tf}$ and $\mathscr{L}^{*}_{tf}$, respectively. Seen as a proof-theoretical meaning-assignment, Axiom 38 is massively holistic, because it lays down *simultaneously* the inferential meaning of all the binary connectives of $\mathscr{L}^{*}_{tf}$ and the defined connective $\neg$.

In the just-mentioned respects, Gentzen-style proof systems [117, 179, 252] and tableau calculi [59] are superior to Hilbert-style systems. In many cases, the inference rules are separated and exhibit only one logical operation, so that they can be viewed as non-holistic meaning assignments. These proof systems often are such that they not only provide a heuristics for proof-search but in the case of decidable logics also give rise to a decision procedure. In comparison with natural deduction and sequent-style proof systems, tableau calculi for modal and certain non-classical logics, however, are sometimes criticized for representing in an excessively obvious way semantical features in the inference rules. In addition to ordinary Gentzen-style natural deduction and sequent calculi, there are many

---

6 In view of this indeterministic interpretation of certain connectives, one might think of investigating relations to Avron's non-deterministic matrices, see [13, 14].

generalizations of the familiar framework, for example, higher-arity sequent systems, hypersequent calculi, tree-hypersequent calculi, and display calculi.[7]

In Chap. 6 we will define and investigate ordinary and higher-arity sequent systems for logics related to $SIXTEEN_3$. In particular, we will present sequent calculi for HL^t and HL^{t*} and other trilattice logics in $\mathscr{L}_{tf}^{\rightarrow_t}$, $\mathscr{L}_{tf}^{\rightarrow_f}$, and $\mathscr{L}_{tf}^{*}$.

[7] A discussion of various kinds of Gentzen-style proof systems from a methodological point of view can be found, for example, in [192, 265].

Chapter 6
Sequent Systems for Trilattice Logics

Abstract In the present chapter, we will define various standard and non-standard sequent calculi for logics related to the trilattice *SIXTEEN*$_3$. In a *first* step, we will introduce three sequent calculi G_B, F_B, and Q_B for Odintsov's first-degree proof system $\vdash_B$ presented in the previous chapter. The system G_B is a standard Gentzen-type sequent calculus, F_B is a four-place (horizontal) matrix sequent calculus, and Q_B is a quadruple (vertical) matrix sequent calculus. In contrast with G_B, the calculus F_B satisfies the subformula property, and the calculus Q_B reflects Odintsov's co-ordinate valuations associated with valuations in *SIXTEEN*$_3$. The mutual equivalence between G_B, F_B, and Q_B, the cut-elimination theorems for these calculi, and the decidability of $\vdash_B$ are proved. In addition, it is shown how the sequent systems for $\vdash_B$ can be extended to cut-free sequent calculi for Odintsov's L_B, which is an extension of $\vdash_B$ by adding classical implication and negation connectives. The axiom systems $L_T, L_B, L^f_T, L^f_B, L'_T, L'_B, L^{f'}_T$, and $L^{f'}_B$ from Chap. 5 are all auxiliary calculi in the sense that they do not capture truth entailment $\models_t$ or falsity entailment $\models_f$ in the respective languages. The chief axiom systems are those capturing the semantically defined logics $L^t, L^f, L', L^{f'}, L^{t*}$ and L^{f*}. In a *second* step, we will present cut-free sound and complete higher-arity sequent calculi for all these axiom systems. The semantical foundation of these calculi is provided by the co-ordinate valuations. A sequent calculus GL^* for truth entailment in *SIXTEEN*$_3$ in the full language $\mathcal{L}^*_{tf}$ is introduced and shown to be sound and complete and admitting of cut-elimination. This sequent calculus directly takes up Odintsov's presentation of *SIXTEEN*$_3$ as a twist-structure over the two-element Boolean algebra and enjoys all the nice properties known from the **G3c** sequent calculus for classical logic, see Troelstra and Schwichtenberg (Cambridge University Press, Cambridge, 2000). Moreover, a sequent calculus for falsity entailment in *SIXTEEN*$_3$ in $\mathcal{L}^*_{tf}$ can be obtained from GL^* not by changing any of the inference rules of the system but simply by appropriately changing the notion of a provable formula.

Y. Shramko and H. Wansing, *Truth and Falsehood*, Trends in Logic 36,
DOI: 10.1007/978-94-007-0907-2_6,

6.1 Standard Sequent Systems for Logics Related to $SIXTEEN_3$

In the present chapter, we will define various standard and non-standard sequent calculi for logics related to the trilattice $SIXTEEN_3$.[1] We will first consider cut-free sequent calculi G_B and G_T for Odintsov's extensions $\vdash_B$ and $\vdash_T$ of the basic logic $\vdash_{\text{base}}$. In this case, the logical object language is the implication-free language $\mathscr{L}_{tf}$. Later these proof systems will be extended to sequent calculi for the logics L_B and L_T in the language $\mathscr{L}_{tf}^{\rightarrow_t}$. Recall that the intersection of $\vdash_T$ and $\vdash_B$ coincides with truth-entailment in $SIXTEEN_3 :\models_t = \vdash_T \cap \vdash_B$. In this sense, G_B and G_L are sequent systems for trilattice logics. Recall also that $\{A \mid \forall v(v(A) = \mathbf{TB}), A \in \mathscr{L}_{tf}^{\rightarrow_t}\} = L^t = L_B \cap L_T$, so that L_B and L_T are closely related to HL^t and hence may be called trilattice logics as well. The sequent calculi G_B and G_L can be obtained in a completely analogous way. Therefore, we will concentrate on G_B. Cut-free sequent calculi for the logic $\vdash_{\text{base}}$ underlying both $\vdash_B$ and $\vdash_L$ have not been obtained so far. Constructing a cut-free sequent calculus for $\vdash_{\text{base}}$ might turn out to be difficult. We will use a translation into the conjunction/disjunction-fragment of classical logic to show that (cut) is an admissible rule of the standard sequent calculus G_B for $\vdash_B$. This translation is suitable for $\vdash_B$ but not for $\vdash_{\text{base}}$. Also, no cut-free sequent calculus for L_{base} is known.

As before, we will use Greek capital letters $\Gamma, \Delta, \ldots$ to represent finite (possibly empty) sets of formulas. The symbol $\stackrel{\circ}{=}$ is used to denote the syntactic identity of symbols and sets of symbols. For any $\sharp \in \{\sim_t, \sim_f, \sim_f\sim_t, \sim_t\sim_f\}$, the expression $\sharp\Gamma$ denotes the set $\{\sharp C \mid C \in \Gamma\}$. An expression of the form $\Gamma \Rightarrow \Delta$ is called a *sequent*. We write $L \vdash S$ to express that the sequent S is provable in the sequent calculus L. A sequent calculus G_B for $\vdash_B$ is introduced below. The $\{\wedge_f, \vee_f\}$-free part of G_B is similar to the sequent calculus L16 discussed in [139].

Definition 6.1 Let $\sim_e$ be $\sim_t$ or $\sim_f\sim_t$. The initial sequents of G_B are of the form:

$$p \Rightarrow p \quad \sim_t p \Rightarrow \sim_t p \quad \sim_f p \Rightarrow \sim_f p \quad \sim_f\sim_t p \Rightarrow \sim_f\sim_t p$$

for any propositional variable p.

The structural inference rules of G_B are of the form:

$$\frac{\Gamma \Rightarrow \Delta, A \quad A, \Sigma \Rightarrow \Pi}{\Gamma, \Sigma \Rightarrow \Delta, \Pi}\text{(cut)} \quad \frac{\Gamma \Rightarrow \Delta}{A, \Gamma \Rightarrow \Delta}\text{(w} - \text{l)} \quad \frac{\Gamma \Rightarrow \Delta}{\Gamma \Rightarrow \Delta, A}\text{(w} - \text{r)}.$$

[1] The presentation is based on [141, 272].

The logical inference rules of G_B are of the form:

$$\frac{A, B, \Gamma \Rightarrow \Delta}{A \wedge_t B, \Gamma \Rightarrow \Delta}(\wedge_t \mathrm{l}) \qquad \frac{\Gamma \Rightarrow \Delta, A \quad \Gamma \Rightarrow \Delta, B}{\Gamma \Rightarrow \Delta, A \wedge_t B}(\wedge_t \mathrm{r})$$

$$\frac{A, \Gamma \Rightarrow \Delta \quad B, \Gamma \Rightarrow \Delta}{A \wedge_t B, \Gamma \Rightarrow \Delta}(\vee_t \mathrm{l}) \qquad \frac{\Gamma \Rightarrow \Delta, A, B}{\Gamma \Rightarrow \Delta, A \vee_t B}(\vee_t \mathrm{r})$$

$$\frac{A, \Gamma \Rightarrow \Delta \quad B, \Gamma \Rightarrow \Delta}{A \wedge_f B, \Gamma \Rightarrow \Delta}(\wedge_f \mathrm{l}) \qquad \frac{\Gamma \Rightarrow \Delta, A, B}{\Gamma \Rightarrow \Delta, A \wedge_f B}(\wedge_f \mathrm{r})$$

$$\frac{A, B, \Gamma \Rightarrow \Delta}{A \vee_f B, \Gamma \Rightarrow \Delta}(\vee_f \mathrm{l}) \qquad \frac{\Gamma \Rightarrow \Delta, A \quad \Gamma \Rightarrow \Delta, B}{\Gamma \Rightarrow \Delta, A \vee_f B}(\vee_f \mathrm{r})$$

$$\frac{A, \Gamma \Rightarrow \Delta}{\sim_t \sim_t A, \Gamma \Rightarrow \Delta}(\sim_t \sim_t \mathrm{l}) \qquad \frac{\Gamma \Rightarrow \Delta, A}{\Gamma \Rightarrow \Delta, \sim_t \sim_t A}(\sim_t \sim_t \mathrm{r})$$

$$\frac{A, \Gamma \Rightarrow \Delta}{\sim_f \sim_f A, \Gamma \Rightarrow \Delta}(\sim_f \sim_f \mathrm{l}) \qquad \frac{\Gamma \Rightarrow \Delta, A}{\Gamma \Rightarrow \Delta, \sim_f \sim_f A}(\sim_f \sim_f \mathrm{r})$$

$$\frac{\sim_t \sim_f A, \Gamma \Rightarrow \Delta}{\sim_f \sim_t A, \Gamma \Rightarrow \Delta}(\sim_f \sim_t \mathrm{l}) \qquad \frac{\Gamma \Rightarrow \Delta, \sim_t \sim_f A}{\Gamma \Rightarrow \Delta, \sim_f \sim_t A}(\sim_f \sim_t \mathrm{r})$$

$$\frac{\sim_f \sim_t A, \Gamma \Rightarrow \Delta}{\sim_t \sim_f A, \Gamma \Rightarrow \Delta}(\sim_t \sim_f \mathrm{l}) \qquad \frac{\Gamma \Rightarrow \Delta, \sim_f \sim_t A}{\Gamma \Rightarrow \Delta, \sim_t \sim_f A}(\sim_t \sim_f \mathrm{r})$$

$$\frac{\sim_f A, \Gamma \Rightarrow \Delta}{\sim_t \sim_f \sim_t A, \Gamma \Rightarrow \Delta}(\sim_t \sim_f \sim_t \mathrm{l}) \qquad \frac{\Gamma \Rightarrow \Delta, \sim_f A}{\Gamma \Rightarrow \Delta, \sim_t \sim_f \sim_t A}(\sim_t \sim_f \sim_t \mathrm{r})$$

$$\frac{\sim_t A, \Gamma \Rightarrow \Delta}{\sim_f \sim_t \sim_f A, \Gamma \Rightarrow \Delta}(\sim_f \sim_t \sim_f \mathrm{l}) \qquad \frac{\Gamma \Rightarrow \Delta, \sim_t A}{\Gamma \Rightarrow \Delta, \sim_f \sim_t \sim_f A}(\sim_f \sim_t \sim_f \mathrm{r})$$

$$\frac{\sim_e A, \Gamma \Rightarrow \Delta \quad \sim_e B, \Gamma \Rightarrow \Delta}{\sim_e (A \wedge_t B), \Gamma \Rightarrow \Delta}(\sim_e \wedge_t \mathrm{l}) \qquad \frac{\Gamma \Rightarrow \Delta, \sim_e A, \sim_e B}{\Gamma \Rightarrow \Delta, \sim_e (A \wedge_t B)}(\sim_e \wedge_t \mathrm{r})$$

$$\frac{\sim_e A, \sim_e B, \Gamma \Rightarrow \Delta}{\sim_e (A \vee_t B), \Gamma \Rightarrow \Delta}(\sim_e \vee_t \mathrm{l}) \qquad \frac{\Gamma \Rightarrow \Delta, \sim_e A \quad \Gamma \Rightarrow \Delta, \sim_e B}{\Gamma \Rightarrow \Delta, \sim_e (A \vee_t B)}(\sim_e \vee_t \mathrm{r})$$

$$\frac{\sim_e A, \sim_e B, \Gamma \Rightarrow \Delta}{\sim_e (A \wedge_f B), \Gamma \Rightarrow \Delta}(\sim_e \wedge_f \mathrm{l}) \qquad \frac{\Gamma \Rightarrow \Delta, \sim_e A \quad \Gamma \Rightarrow \Delta, \sim_e B}{\Gamma \Rightarrow \Delta, \sim_e (A \wedge_f B)}(\sim_e \wedge_f \mathrm{r})$$

$$\frac{\sim_e A, \Gamma \Rightarrow \Delta \quad \sim_e B, \Gamma \Rightarrow \Delta}{\sim_e (A \vee_f B), \Gamma \Rightarrow \Delta}(\sim_e \vee_t \mathrm{l}) \qquad \frac{\Gamma \Rightarrow \Delta, \sim_e A, \sim_e B}{\Gamma \Rightarrow \Delta, \sim_e (A \vee_f B)}(\sim_e \vee_f \mathrm{r})$$

$$\frac{\sim_f A, \Gamma \Rightarrow \Delta \quad \sim_f B, \Gamma \Rightarrow \Delta}{\sim_f (A \wedge_f B), \Gamma \Rightarrow \Delta}(\sim_f \wedge_f \mathrm{l}) \qquad \frac{\Gamma \Rightarrow \Delta, \sim_f A, \sim_f B}{\Gamma \Rightarrow \Delta, \sim_f (A \wedge_f B)}(\sim_f \wedge_f \mathrm{r})$$

$$\frac{\sim_f A,\ \sim_f B, \Gamma \Rightarrow \Delta}{\sim_f (A \vee_f B), \Gamma \Rightarrow \Delta}(\sim_f \vee_f \mathrm{l}) \quad \frac{\Gamma \Rightarrow \Delta,\ \sim_f A \quad \Gamma \Rightarrow \Delta,\ \sim_f B}{\Gamma \Rightarrow \Delta,\ \sim_f (A \vee_f B)}(\sim_f \vee_f \mathrm{r})$$

$$\frac{\sim_f A,\ \sim_f B, \Gamma \Rightarrow \Delta}{\sim_f (A \wedge_t B), \Gamma \Rightarrow \Delta}(\sim_f \wedge_t \mathrm{l}) \quad \frac{\Gamma \Rightarrow \Delta,\ \sim_f A \quad \Gamma \Rightarrow \Delta,\ \sim_f B}{\Gamma \Rightarrow \Delta,\ \sim_f (A \wedge_t B)}(\sim_f \wedge_t \mathrm{r})$$

$$\frac{\sim_f A, \Gamma \Rightarrow \Delta \quad \sim_f B, \Gamma \Rightarrow \Delta}{\sim_f (A \vee_t B), \Gamma \Rightarrow \Delta}(\sim_f \vee_t \mathrm{l}) \quad \frac{\Gamma \Rightarrow \Delta,\ \sim_f A,\ \sim_f B}{\Gamma \Rightarrow \Delta, \sim_f (A \vee_t B)}(\sim_f \vee_t \mathrm{r}).$$

Definition 6.2 A sequent calculus *LK* for the conjunction/disjunction-fragment of classical logic is defined as the $\{\wedge_t, \vee_t\}$-fragment of G_B.

Note that the $\{\wedge_t, \vee_t, \sim_t\}$-fragment of G_B is a sequent calculus for Belnap and Dunn's four-valued logic [22, 23, 75] and that the inference rules $(\sim_t \wedge_f \mathrm{l})$, $(\sim_t \wedge_f \mathrm{r})$, $(\sim_t \vee_f \mathrm{l})$ and $(\sim_t \vee_f \mathrm{r})$ appear in Arieli and Avron's bilattice logic [7] if $\wedge_f$ and $\vee_f$ respectively are read as the (multiplicative) conjunction and disjunction connectives $*$ and $+$ used in [7]. Therefore, G_B may be regarded as a natural extension and generalization of Belnap and Dunn's logic and Arieli and Avron's logic.

A sequent calculus G_T for $\vdash_T$ is obtained from G_B by replacing the inference rules $\{(\wedge_f \mathrm{l}), (\wedge_f \mathrm{r}), (\vee_f \mathrm{l}), (\vee_f \mathrm{r})\}$ by the following inference rules:

$$\frac{A, B, \Gamma \Rightarrow \Delta}{A \wedge_f B, \Gamma \Rightarrow \Delta} \quad \frac{\Gamma \Rightarrow \Delta, A \quad \Gamma \Rightarrow \Delta, B}{\Gamma \Rightarrow \Delta, A \wedge_f B}$$

$$\frac{A, \Gamma \Rightarrow \Delta \quad B, \Gamma \Rightarrow \Delta}{A \vee_f B, \Gamma \Rightarrow \Delta} \quad \frac{\Gamma \Rightarrow \Delta, A, B}{\Gamma \Rightarrow \Delta, A \vee_f B}.$$

Note that G_T is an extension of the $\rightarrow$-free fragment of Arieli and Avron's logic.

In what follows, if L is a sequent calculus, then we say that L is cut-free iff it does not contain (cut) as a primitive rule. If L is cut-free, we denote this by $L-$ (cut).

Proposition 6.1 *The following rules are derivable in cut-free G_B:*

$$\frac{\sim_t \sim_f A, \Gamma \Rightarrow \Delta \quad \sim_t \sim_f B, \Gamma \Rightarrow \Delta}{\sim_t \sim_f (A \wedge_t B), \Gamma \Rightarrow \Delta}(\sim_t \sim_f \wedge_t \mathrm{l}) \quad \frac{\Gamma \Rightarrow \Delta, \sim_t \sim_f A, \sim_t \sim_f B}{\Gamma \Rightarrow \Delta, \sim_t \sim_f (A \wedge_t B)}(\sim_t \sim_f \wedge_t \mathrm{r})$$

$$\frac{\sim_t \sim_f A, \sim_t \sim_f B, \Gamma \Rightarrow \Delta}{\sim_t \sim_f (A \vee_t B), \Gamma \Rightarrow \Delta}(\sim_t \sim_f \vee_t \mathrm{l}) \quad \frac{\Gamma \Rightarrow \Delta, \sim_t \sim_f A \quad \Gamma \Rightarrow \Delta, \sim_t \sim_f B}{\Gamma \Rightarrow \Delta, \sim_t \sim_f (A \vee_t B)}(\sim_t \sim_f \vee_t \mathrm{r})$$

$$\frac{\sim_t \sim_f A, \sim_t \sim_f B, \Gamma \Rightarrow \Delta}{\sim_t \sim_f (A \wedge_f B), \Gamma \Rightarrow \Delta}(\sim_t \sim_f \wedge_f \mathrm{l}) \quad \frac{\Gamma \Rightarrow \Delta, \sim_t \sim_f A \quad \Gamma \Rightarrow \Delta, \sim_t \sim_f B}{\Gamma \Rightarrow \Delta, \sim_t \sim_f (A \wedge_f B)}(\sim_t \sim_f \wedge_f \mathrm{r})$$

$$\frac{\sim_t \sim_f A, \Gamma \Rightarrow \Delta \quad \sim_t \sim_f B, \Gamma \Rightarrow \Delta}{\sim_t \sim_f (A \vee_f B), \Gamma \Rightarrow \Delta}(\sim_t \sim_f \vee_f \mathrm{l}) \quad \frac{\Gamma \Rightarrow \Delta, \sim_t \sim_f A, \sim_t \sim_f B}{\Gamma \Rightarrow \Delta, \sim_t \sim_f (A \vee_f B)}(\sim_t \sim_f \vee_f \mathrm{r}).$$

Proof By using $(\sim_f\sim_t\mathrm{l})$, $(\sim_f\sim_t\mathrm{r})$, $(\sim_t\sim_f\mathrm{l})$, and $(\sim_t\sim_f\mathrm{r})$. □

Proposition 6.2 *Let $\sim_b$ be either $\sim_f\sim_t$ or $\sim_t\sim_f$. The following rules are derivable in cut-free G_B:*

$$\frac{A, \Gamma \Rightarrow \Delta}{\sim_b\sim_b A, \Gamma \Rightarrow \Delta}(\sim_b\sim_b\mathrm{l}) \quad \frac{\Gamma \Rightarrow \Delta, A}{\Gamma \Rightarrow \Delta, \sim_b\sim_b A}(\sim_b\sim_b\mathrm{r}).$$

Proof By using $(\sim_f\sim_f\mathrm{r})$, $(\sim_f\sim_f\mathrm{l})$, $(\sim_t\sim_t\mathrm{r})$, $(\sim_t\sim_t\mathrm{l})$, $(\sim_t\sim_f\sim_t\mathrm{r})$, $(\sim_t\sim_f\sim_t\mathrm{l})$, $(\sim_f\sim_t\sim_f\mathrm{r})$, and $(\sim_f\sim_t\sim_f\mathrm{l})$. □

Proposition 6.3 *Let $\sharp$ be either $\sim_t, \sim_f, \sim_f\sim_t$ or $\sim_t\sim_f$. The following rules are admissible in cut-free G_B:*

$$\frac{\sharp\sharp A, \Gamma \Rightarrow \Delta}{A, \Gamma \Rightarrow \Delta}(\sharp\sharp\mathrm{l}^{-1}) \quad \frac{\Gamma \Rightarrow \Delta, \sharp\sharp A}{\Gamma \Rightarrow \Delta, A}(\sharp\sharp\mathrm{r}^{-1})$$

$$\frac{\sim_t\sim_f\sim_t A, \Gamma \Rightarrow \Delta}{\sim_f A, \Gamma \Rightarrow \Delta}(\sim_t\sim_f\sim_t\mathrm{l}^{-1}) \quad \frac{\Gamma \Rightarrow \Delta, \sim_t\sim_f\sim_t A}{\Gamma \Rightarrow \Delta, \sim_f A}(\sim_t\sim_f\sim_t\mathrm{r}^{-1})$$

$$\frac{\sim_f\sim_t\sim_f A, \Gamma \Rightarrow \Delta}{\sim_t A, \Gamma \Rightarrow \Delta}(\sim_f\sim_t\sim_f\mathrm{l}^{-1}) \quad \frac{\Gamma \Rightarrow \Delta, \sim_f\sim_t\sim_f A}{\Gamma \Rightarrow \Delta, \sim_t A}(\sim_f\sim_t\sim_f\mathrm{r}^{-1}).$$

Proof Straightforward. □

Proposition 6.4 *The following rules are admissible in cut-free G_B:*

$$\frac{\Gamma \Rightarrow \Delta}{\sim_t\Delta \Rightarrow \sim_t\Gamma}(\sim_t\mathrm{twist}) \quad \frac{\Gamma \Rightarrow \Delta}{\sim_f\Gamma \Rightarrow \sim_f\Delta}(\sim_f\mathrm{invol}).$$

Proof We show only the case for $(\sim_t\mathrm{twist})$. We show that if $G_B - (\mathrm{cut}) \vdash \Gamma \Rightarrow \Delta$, then $G_B - (\mathrm{cut}) \vdash \sim_t\Delta \Rightarrow \sim_t\Gamma$. We prove this by induction on a cut-free proof π of $\Gamma \Rightarrow \Delta$ and distinguish the cases according to the final inference of π. We show some cases.

Case $(\sim_f\sim_t\wedge_t\mathrm{l})$: The last inference rule applied in π is of the form:

$$\frac{\sim_f\sim_t A, \Gamma \Rightarrow \Delta \quad \sim_f\sim_t B, \Gamma \Rightarrow \Delta}{\sim_f\sim_t(A\wedge_t B), \Gamma \Rightarrow \Delta}(\sim_f\sim_t\wedge_t\mathrm{l}).$$

By the induction hypothesis, we have $G_B - (\mathrm{cut}) \vdash \sim_t\Delta \Rightarrow \sim_t\sim_f\sim_t A, \sim_t\Gamma$ and $G_B - (\mathrm{cut}) \vdash \sim_t\Delta \Rightarrow \sim_t\sim_f\sim_t B, \sim_t\Gamma$. We then obtain:

$$\dfrac{\dfrac{\dfrac{\vdots}{\sim_t\Delta \Rightarrow \sim_t\sim_f\sim_t A, \sim_t\Gamma}}{\sim_t\Delta \Rightarrow \sim_f A, \sim_t\Gamma}(\sim_t\sim_f\sim_t\mathrm{r}^{-1}) \quad \dfrac{\dfrac{\vdots}{\sim_t\Delta \Rightarrow \sim_t\sim_f\sim_t B, \sim_t\Gamma}}{\sim_t\Delta \Rightarrow \sim_f B, \sim_t\Gamma}(\sim_t\sim_f\sim_t\mathrm{r}^{-1})}{\dfrac{\sim_t\Delta \Rightarrow \sim_f(A\wedge_t B), \sim_t\Gamma}{\sim_t\Delta \Rightarrow \sim_t\sim_f\sim_t(A\wedge_t B), \sim_t\Gamma}(\sim_t\sim_f\sim_t\mathrm{r})}(\sim_f\wedge_t\mathrm{r})$$

where $(\sim_t\sim_f\sim_t\mathrm{r}^{-1})$ is admissible in cut-free G_B by Proposition 6.3.

Case $(\sim_f\sim_t\sim_f\mathrm{r})$: The last inference rule applied in π is of the form:

$$\frac{\Gamma \Rightarrow \Delta,\ \sim_t A}{\Gamma \Rightarrow \Delta,\ \sim_f\sim_t\sim_f A}(\sim_f\sim_t\sim_f\mathrm{r}).$$

By the induction hypothesis, we have $G_B - (\mathrm{cut}) \vdash\ \sim_t\Delta,\ \sim_t\sim_t A \Rightarrow\ \sim_t\Gamma$. We then have:

$$\dfrac{\dfrac{\dfrac{\dfrac{\vdots}{\sim_t\Delta, \sim_t\sim_t A \Rightarrow \sim_t\Gamma}}{\sim_t\Delta, A \Rightarrow \sim_t\Gamma}(\sim_t\sim_t\mathrm{l}^{-1})}{\sim_t\Delta, \sim_f\sim_f A \Rightarrow \sim_t\Gamma}(\sim_f\sim_f\mathrm{l})}{\sim_t\Delta, \sim_t\sim_f\sim_t\sim_f A \Rightarrow \sim_t\Gamma}(\sim_t\sim_f\sim_t\mathrm{l})$$

where $(\sim_t\sim_t\mathrm{l}^{-1})$ is admissible in cut-free G_B by Proposition 6.3.

Case $(\sim_f\sim_t\mathrm{l})$: The last inference rule applied in π is of the form:

$$\frac{\sim_t\sim_f A, \Gamma \Rightarrow \Delta}{\sim_f \sim_t A, \Gamma \Rightarrow \Delta}(\sim_f\sim_t\mathrm{l}).$$

By the induction hypothesis, we have $G_B - (\mathrm{cut}) \vdash\ \sim_t\Delta \Rightarrow\ \sim_t \sim_t \sim_f A,\ \sim_t\Gamma$. We then obtain:

$$\dfrac{\dfrac{\dfrac{\vdots}{\sim_t\Delta \Rightarrow \sim_t\sim_t\sim_f A, \sim_t\Gamma}}{\sim_t\Delta \Rightarrow \sim_f A, \sim_t\Gamma}(\sim_t\sim_t\mathrm{l}^{-1})}{\sim_t\Delta \Rightarrow \sim_t\sim_f\sim_t A, \sim_t\Gamma}(\sim_t\sim_f\sim_t\mathrm{r})$$

where $(\sim_t\sim_t\mathrm{l}^{-1})$ is admissible in cut-free G_B by Proposition 6.3. □

Proposition 6.5 *For any $\mathscr{L}_{tf}$-formula A,*

1. $G_B - (\mathrm{cut}) \vdash\ \sim_t A \Rightarrow\ \sim_t A$ *and* 2. $G_B - (\mathrm{cut}) \vdash\ \sim_f A \Rightarrow\ \sim_f A$.

Proof By induction on A. We show only the following case for (1).

Case $(A \doteq \sim_f B)$: By the induction hypothesis, we have $G_B - (\mathrm{cut}) \vdash\ \sim_t B \Rightarrow\ \sim_t B$. We then have:

$$\cfrac{\cfrac{\cfrac{\begin{matrix}\vdots \\ \sim_t B \Rightarrow \sim_t B\end{matrix}}{\sim_f\sim_t B \Rightarrow \sim_f\sim_t B}\ (\sim_f\text{invol})}{\sim_t\sim_f B \Rightarrow \sim_f\sim_t B}\ (\sim_t\sim_f\text{l})}{\sim_t\sim_f B \Rightarrow \sim_t\sim_f B}\ (\sim_f\sim_t\text{r})$$

where ($\sim_f$invol) is admissible in cut-free G_B by Proposition 6.4. □

Proposition 6.6 *For any* $\mathscr{L}_{tf}$*-formula* A, $G_B - (\text{cut}) \vdash A \Rightarrow A$.

Proof By induction on A. We use Proposition 6.5. □

Let Γ be a non-empty set $\{C_1, C_2, \ldots, C_n\} (n \geq 1)$ of formulas. The expressions $\bigwedge_t \Gamma$ and $\bigvee_t \Gamma$ stand for $C_1 \wedge_t C_2 \wedge_t \ldots \wedge_t C_n$ and $C_1 \vee_t C_2 \vee_t \ldots \vee_t C_n$, respectively. The following theorem shows that G_B and $\vdash_B$ coincide for first-degree consequences.

Theorem 6.1 *Let* A *and* B *be* $\mathscr{L}_{tf}$*-formulas. Then* $\vdash_B A \Rightarrow B$ *iff* $G_B \vdash A \Rightarrow B$.

Proof Left-to-right: Straightforward. For $\sim_t\sim_f A \Leftrightarrow \sim_f\sim_t A$ we use Proposition 6.6 and the rules ($\sim_f\sim_t$r) and ($\sim_t\sim_f$r).

Right-to-left: Let Γ and Δ be non-empty sets of formulas. We show that $G_B \vdash \bigwedge_t \Delta \Rightarrow \bigvee_t \Gamma$ implies $\vdash_B \bigwedge_t \Delta \Rightarrow \bigvee_t \Gamma$. By the characterization of $\vdash_B$ in terms of $\models_B$ and Proposition 3.2 from Chap. 3 (or Theorem 5.1 from Chap. 5), it is enough to show that

$$(*)\ \forall v(\mathbf{B} \in v(\textstyle\bigwedge_t \Delta) \text{ implies } \mathbf{B} \in v(\bigvee_t \Gamma)).$$

We prove $(*)$ by induction on proofs in G_B. Most of the cases are obvious. Suppose, for example, that the last inference rule applied in the proof is ($\vee_f$l). By the induction hypothesis, $\forall v(\mathbf{B} \in v(A \wedge_t B \wedge_t \bigwedge_t \Gamma)$ implies $\mathbf{B} \in v(\bigvee_t \Delta))$. Then $\forall v(\mathbf{B} \in v((A \vee_f B) \wedge_t \bigwedge_t \Gamma)$ implies $\mathbf{B} \in v(\bigvee_t \Delta))$, because for every $x, y \in \mathbf{16}$, $\mathbf{B} \in x \sqcap_f y$ iff ($\mathbf{B} \in x$ and $\mathbf{B} \in y$).

If the last rule applied in the proof is ($\sim_f \vee_t$ r), by the induction hypothesis, we have $\forall v(\mathbf{B} \in v(\bigwedge_t \Gamma)$ implies $\mathbf{B} \in v(\bigvee_t \Delta) \vee_t \sim_f A \vee_t \sim_f B))$. By Propositions 3.2 and 3.4 from Chap. 3 (or Theorem 5.1 from Chap. 5) and the definition of valuations,

$$\begin{aligned}\mathbf{B} \in v(\sim_f(A \vee_t B)) &\text{ iff}\\ \mathbf{T} \in v(A \vee_t B) &\text{ iff}\\ \mathbf{T} \in v(A) \text{ or } \mathbf{T} \in v(B) &\text{ iff}\\ \mathbf{B} \in v(\sim_f A) \text{ or } \mathbf{B} \in v(\sim_f B)&.\end{aligned}$$

Therefore, $\forall v(\mathbf{B} \in v(\bigwedge_t \Gamma)$ implies $\mathbf{B} \in v(\bigvee_t \Delta) \vee_t \sim_f(A \vee_t B))$. □

In order to show the cut-elimination theorem for G_B, we introduce an embedding of G_B into *LK*.

Definition 6.3 Let $\sim_e$ be either $\sim_t$ or $\sim_f\sim_t$. Starting from the set $Atom = \Phi$ of propositional variables, we define the sets $\Phi' := \{p' \mid p \in \Phi\}, \Phi'' := \{p'' \mid p \in \Phi\}$ and $\Phi''' := \{p''' \mid p \in \Phi\}$ of propositional variables. The language $\mathscr{L}_{LK}$ of LK is obtained from $\mathscr{L}_{tf}$ by adding $\{\Phi', \Phi'', \Phi'''\}$ and deleting $\{\wedge_f, \vee_f, \sim_t, \sim_f\}$.

The mapping f from $\mathscr{L}_{tf}$ to $\mathscr{L}_{LK}$ is defined as follows:

1. $f(p) := p, f(\sim_t p) := p' \in \Phi', f(\sim_f p) := p'' \in \Phi''$ and $f(\sim_t\sim_f p) = f(\sim_f\sim_t p) := p''' \in \Phi'''$ for any $p \in \Phi$,
2. $f(A \circ B) := f(A) \circ f(B)$ where $\circ \in \{\wedge_t, \vee_t\}$,
3. $f(A\wedge_f B) := f(A)\vee_t f(B)$,
4. $f(A\vee_f B) := f(A)\wedge_t f(B)$,
5. $f(\sim_t\sim_t A) := f(A)$,
6. $f(\sim_f\sim_f A) := f(A)$,
7. $f(\sim_t\sim_f A) := f(\sim_f\sim_t A)$,
8. $f(\sim_t\sim_f\sim_t A) := f(\sim_f A)$,
9. $f(\sim_f\sim_t\sim_f A) := f(\sim_t A)$,
10. $f(\sim_e(A\wedge_t B)) := f(\sim_e A)\vee_t f(\sim_e B)$,
11. $f(\sim_e(A\vee_t B)) := f(\sim_e A)\wedge_t f(\sim_e B)$,
12. $f(\sim_e(A\wedge_f B)) := f(\sim_e A)\vee_t f(\sim_e B)$,
13. $f(\sim_e(A\vee_f B)) := f(\sim_e A)\wedge_t f(\sim_e B)$,
14. $f(\sim_f(A\wedge_f B)) := f(\sim_f A)\wedge_t f(\sim_f B)$,
15. $f(\sim_f(A\vee_f B)) := f(\sim_f A)\vee_t f(\sim_f B)$,
16. $f(\sim_f(A\wedge_t B)) := f(\sim_f A)\wedge_t f(\sim_f B)$,
17. $f(\sim_f(A\vee_t B)) := f(\sim_f A)\vee_t f(\sim_f B)$.

Let Γ be a set of formulas in $\mathscr{L}_{tf}$. Then, $f(\Gamma)$ denotes the result of replacing every occurrence of a formula A in Γ by an occurrence of $f(A)$.

A translation similar to f has been used by Gurevich [128] and Rautenberg [204] to embed Nelson's constructive logic **N3** [2, 180] into intuitionistic logic. The introduction of new atoms may also be used to embed Nelson's constructive paraconsistent logic **N4** into positive intuitionistic logic, cf. also [267].

Theorem 6.2 *Let Δ, Γ be sets of formulas in $\mathscr{L}_{tf}$, and f be the mapping defined in Definition 6.3.*

1. *If $G_B \vdash \Gamma \Rightarrow \Delta$, then $LK \vdash f(\Gamma) \Rightarrow f(\Delta)$.*
2. *If $LK - (cut) \vdash f(\Gamma) \Rightarrow f(\Delta)$, then $G_B - (cut) \vdash \Gamma \Rightarrow \Delta$.*

Proof Since the second claim can be proved similarly, we only show the first claim by induction on a proof π of $\Gamma \Rightarrow \Delta$ in G_B. We distinguish the cases according to the last inference of π. We show some cases.

Case $(\sim_f\sim_t p \Rightarrow \sim_t\sim_f p)$: π is of the form $\sim_f\sim_t p \Rightarrow \sim_t\sim_f p$ where p is a propositional variable. By the definition of f, we have $f(\sim_f\sim_t p) = f(\sim_t\sim_f p) = p''' \in \Phi'''$, and hence $LK \vdash p''' \Rightarrow p'''$.

Case ($\sim_t\sim_f\sim_t$l): The last inference rule applied in π is of the form:

$$\frac{\sim_f A, \Sigma \Rightarrow \Delta}{\sim_t\sim_f\sim_t A, \Sigma \Rightarrow \Delta}\ (\sim_t\sim_f\sim_t\text{l}).$$

By the induction hypothesis, we have $LK \vdash f(\sim_f A), f(\Sigma) \Rightarrow f(\Delta)$, and hence $LK \vdash f(\sim_t \sim_f \sim_t A), f(\Sigma) \Rightarrow f(\Delta)$ by the definition of f.

Case ($\sim_f\sim_t\wedge_t$l): The last inference rule applied in π is of the form:

$$\frac{\sim_f\sim_t A, \Sigma \Rightarrow \Delta \quad \sim_f\sim_t B, \Sigma \Rightarrow \Delta}{\sim_f\sim_t(A\wedge_t B), \Sigma \Rightarrow \Delta}\ (\sim_f\sim_t\wedge_t\text{l}).$$

By the induction hypothesis, we have $LK \vdash f(\sim_f \sim_t A), f(\Sigma) \Rightarrow f(\Delta)$ and $LK \vdash f(\sim_f \sim_t B), f(\Sigma) \Rightarrow f(\Delta)$. Thus, we obtain

$$\frac{\begin{matrix}\vdots \\ f(\sim_f \sim_t A), f(\Sigma) \Rightarrow f(\Delta)\end{matrix} \quad \begin{matrix}\vdots \\ f(\sim_f \sim_t B), f(\Sigma) \Rightarrow f(\Delta)\end{matrix}}{f(\sim_f \sim_t A) \vee_t f(\sim_f \sim_t B), f(\Sigma) \Rightarrow f(\Delta)}\ (\vee_t\text{l}).$$

Therefore, $LK \vdash f(\sim_f\sim_t(A\wedge_t B)), f(\Sigma) \Rightarrow f(\Delta)$ by the definition of f. □

We may then straightforwardly prove the cut-elimination theorem for G_B.[2]

Theorem 6.3 *The rule* (cut) *is admissible in cut-free* G_B.

Proof Suppose $G_B \vdash \Gamma \Rightarrow \Delta$. Then, we have $LK \vdash f(\Gamma) \Rightarrow f(\Delta)$ by Theorem 6.2 (1), and hence $LK - (\text{cut}) \vdash f(\Gamma) \Rightarrow f(\Delta)$ by the well-known cut-elimination theorem for LK. By Theorem 6.2 (2), it follows that $G_B - (\text{cut}) \vdash \Gamma \Rightarrow \Delta$. □

Theorem 6.4 G_B *is decidable.*

Proof By Theorem 6.2, the provability in G_B can be transformed into provability in LK. The translation from the formulas of G_B into the corresponding formulas of LK can be performed in finitely many steps. It follows that since LK is decidable, G_B is also decidable. □

Definition 6.4 Let $\sharp$ be a unary (negation-like) connective. A sequent calculus L is called *explosive* with respect to $\sharp$ iff for any formulas A and B, the sequent $A, \sharp A \Rightarrow B$ is provable in L. The system L is called *paraconsistent* with respect to $\sharp$ iff it is not explosive with respect to $\sharp$.

Proposition 6.7 *Let* $\sharp$ *be* $\sim_t, \sim_f, \sim_t\sim_f$ *or* $\sim_f\sim_t$. *Then,* G_B *is paraconsistent with respect to* $\sharp$.

---

[2] Note that cut-elimination for (variants of) G_B can be shown by using other methods as well. In [142], semantical proofs based on Maehara's method and Schütte's method were obtained for some modifications and extensions of G_B. A purely syntactical, direct proof of the cut-elimination theorem for G_B or its modifications is likely to exist, but is not yet available.

Proof Consider a sequent $p, \sharp p \Rightarrow q$ where p and q are distinct propositional variables. Then, the unprovability of this sequent is guaranteed by Theorem 6.3. □

6.2 Alternative Sequent Calculi

The sequent system G_B does not satisfy the subformula property. In the following, a four-place (horizontal) matrix sequent calculus F_B, which enjoys the subformula property, is introduced. This calculus may be regarded as a natural extension or generalization of the subformula calculi discussed in [138, 141]. Moreover, F_B is one way of implementing in a proof system Odintsov's idea of using a 2×2-matrix for representing elements of **16**. A similar four-place display sequent calculus for Nelson's constructive paraconsistent logic **N4** is studied in [143].

A *four-place sequent* of F_B is of the form:

$$\Gamma_1 : \Gamma_2 : \Gamma_3 : \Gamma_4', \Gamma_4'' \Rightarrow \Delta_1 : \Delta_2 : \Delta_3 : \Delta_4', \Delta_4''$$

which corresponds to the standard sequent of the form:

$$\Gamma_1, {\sim_t}\Gamma_2, {\sim_f}\Gamma_3, {\sim_f}{\sim_t}\Gamma_4', {\sim_t}{\sim_f}\Gamma_4'' \Rightarrow \Delta_1, {\sim_t}\Delta_2, {\sim_f}\Delta_3, {\sim_f}{\sim_t}\Delta_4', {\sim_t}{\sim_f}\Delta_4''.$$

Some of the logical inference rules in G_B for negated compound formulas block the subformula property. By introducing four-place sequents, these inference rules can be transformed into so-called separated inference rules in F_B, which exhibit only one connective.

The expression $\vec{\Gamma}$ is used to denote $\Gamma_1 : \Gamma_2 : \Gamma_3 : \Gamma_4', \Gamma_4''$. The part Γ_4', Γ_4'' in the expression $\Gamma_1 : \Gamma_2 : \Gamma_3 : \Gamma_4', \Gamma_4''$ will sometimes be denoted as Γ_4. The expression $i : A, \vec{\Gamma}$ or $\vec{\Gamma}, i : A$ for each $i \in \{1,2,3,4\}$ will be used to denote the place-wise union of $\vec{\Gamma}$ and $\{A\}$, for example, $3 : A, \vec{\Gamma}$ denotes $\Gamma_1 : \Gamma_2 : A, \Gamma_3 : \Gamma_4$.

Definition 6.5 Let $a \in \{1,2,3,4\}$ and $e \in \{2,4\}$. The initial sequents of F_B are of the form $a : p \Rightarrow a : p$ for any propositional variable p. The inference rules of F_B are of the form:

$$\frac{\Gamma : \Gamma_2 : \Gamma_3 : \Gamma_4 \Rightarrow \Delta, A : \Delta_2 : \Delta_3 : \Delta_4 \quad A, \Sigma : \Gamma_2 : \Gamma_3 : \Gamma_4 \Rightarrow \Pi : \Delta_2 : \Delta_3 : \Delta_4}{\Gamma, \Sigma : \Gamma_2 : \Gamma_3 : \Gamma_4 \Rightarrow \Delta, \Pi : \Delta_2 : \Delta_3 : \Delta_4}\ (\text{cut}1)$$

$$\frac{\Gamma_1 : \Gamma : \Gamma_3 : \Gamma_4 \Rightarrow \Delta_1 : \Delta, A : \Delta_3 : \Delta_4 \quad \Gamma_1 : A, \Sigma : \Gamma_3 : \Gamma_4 \Rightarrow \Delta_1 : \Pi : \Delta_3 : \Delta_4}{\Gamma_1 : \Gamma, \Sigma : \Gamma_3 : \Gamma_4 \Rightarrow \Delta_1 : \Delta, \Pi : \Delta_3 : \Delta_4}\ (\text{cut}2)$$

$$\frac{\Gamma_1 : \Gamma_2 : \Gamma : \Gamma_4 \Rightarrow \Delta_1 : \Delta_2 : \Delta, A : \Delta_4 \quad \Gamma_1 : \Gamma_2 : A, \Sigma : \Gamma_4 \Rightarrow \Delta_1 : \Delta_2 : \Pi : \Delta_4}{\Gamma_1 : \Gamma_2 : \Gamma, \Sigma : \Gamma_4 \Rightarrow \Delta_1 : \Delta_2 : \Delta, \Pi : \Delta_4}\ (\text{cut}3)$$

$$\frac{\Gamma_1 : \Gamma_2 : \Gamma_3 : \Gamma \Rightarrow \Delta_1 : \Delta_2 : \Delta_3 : \Delta, A \quad \Gamma_1 : \Gamma_2 : \Gamma_3 : A, \Sigma \Rightarrow \Delta_1 : \Delta_2 : \Delta_3 : \Pi}{\Gamma_1 : \Gamma_2 : \Gamma_3 : \Gamma, \Sigma \Rightarrow \Delta_1 : \Delta_2 : \Delta_3 : \Delta, \Pi}\ (\text{cut}4)$$

$$\frac{\vec{\Gamma} \Rightarrow \vec{\Delta}}{a : A, \vec{\Gamma} \Rightarrow \vec{\Delta}} \quad \frac{\vec{\Gamma} \Rightarrow \vec{\Delta}}{\vec{\Gamma} \Rightarrow \vec{\Delta}, a : A}$$

$$\frac{1 : A, 1 : B, \vec{\Gamma} \Rightarrow \vec{\Delta}}{1 : A \wedge_t B, \vec{\Gamma} \Rightarrow \vec{\Delta}} \quad \frac{\vec{\Gamma} \Rightarrow \vec{\Delta}, 1 : A \quad \vec{\Gamma} \Rightarrow \vec{\Delta}, 1 : B}{\vec{\Gamma} \Rightarrow \vec{\Delta}, 1 : A \wedge_t B}$$

$$\frac{1 : A, \vec{\Gamma} \Rightarrow \vec{\Delta} \quad 1 : B, \vec{\Gamma} \Rightarrow \vec{\Delta}}{1 : A \vee_t B, \vec{\Gamma} \Rightarrow \vec{\Delta}} \quad \frac{\vec{\Gamma} \Rightarrow \vec{\Delta}, 1 : A, 1 : B}{\vec{\Gamma} \Rightarrow \vec{\Delta}, 1 : A \vee_t B}$$

$$\frac{1 : A, \vec{\Gamma} \Rightarrow \vec{\Delta} \quad 1 : B, \vec{\Gamma} \Rightarrow \vec{\Delta}}{1 : A \wedge_f B, \vec{\Gamma} \Rightarrow \vec{\Delta}} \quad \frac{\vec{\Gamma} \Rightarrow \vec{\Delta}, 1 : A, 1 : B}{\vec{\Gamma} \Rightarrow \vec{\Delta}, 1 : A \wedge_f B}$$

$$\frac{1 : A, 1 : B, \vec{\Gamma} \Rightarrow \vec{\Delta}}{1 : A \vee_f B, \vec{\Gamma} \Rightarrow \vec{\Delta}} \quad \frac{\vec{\Gamma} \Rightarrow \vec{\Delta}, 1 : A \quad \vec{\Gamma} \Rightarrow \vec{\Delta}, 1 : B}{\vec{\Gamma} \Rightarrow \vec{\Delta}, 1 : A \vee_f B}$$

$$\frac{1 : A, \vec{\Gamma} \Rightarrow \vec{\Delta}}{2 : \sim_t A, \vec{\Gamma} \Rightarrow \vec{\Delta}} \quad \frac{\vec{\Gamma} \Rightarrow \vec{\Delta}, 1 : A}{\vec{\Gamma} \Rightarrow \vec{\Delta}, 2 : \sim_t A}$$

$$\frac{1 : A, \vec{\Gamma} \Rightarrow \vec{\Delta}}{3 : \sim_f A, \vec{\Gamma} \Rightarrow \vec{\Delta}} \quad \frac{\vec{\Gamma} \Rightarrow \vec{\Delta}, 1 : A}{\vec{\Gamma} \Rightarrow \vec{\Delta}, 3 : \sim_f A}$$

$$\frac{3 : A, \vec{\Gamma} \Rightarrow \vec{\Delta}}{4 : \sim_t A, \vec{\Gamma} \Rightarrow \vec{\Delta}} \quad \frac{\Gamma \Rightarrow \vec{\Delta}, 3 : A}{\vec{\Gamma} \Rightarrow \vec{\Delta}, 4 : \sim_t A}$$

$$\frac{2 : A, \vec{\Gamma} \Rightarrow \vec{\Delta}}{4 : \sim_f A, \vec{\Gamma} \Rightarrow \vec{\Delta}} \quad \frac{\Gamma \Rightarrow \vec{\Delta}, 2 : A}{\vec{\Gamma} \Rightarrow \vec{\Delta}, 4 : \sim_f A}$$

$$\frac{e : A, \vec{\Gamma} \Rightarrow \vec{\Delta} \quad e : B, \vec{\Gamma} \Rightarrow \vec{\Delta}}{e : A \wedge_t B, \vec{\Gamma} \Rightarrow \vec{\Delta}} \quad \frac{\vec{\Gamma} \Rightarrow e : A, e : B, \vec{\Delta}}{\vec{\Gamma} \Rightarrow e : A \wedge_t B, \vec{\Delta}}$$

$$\frac{e : A, e : B, \vec{\Gamma} \Rightarrow \vec{\Delta}}{e : A \vee_t B, \vec{\Gamma} \Rightarrow \vec{\Delta}} \quad \frac{\vec{\Gamma} \Rightarrow \vec{\Delta}, e : A \quad \vec{\Gamma} \Rightarrow \vec{\Delta}, e : B}{\vec{\Gamma} \Rightarrow \vec{\Delta}, e : A \vee_t B}$$

$$\frac{e : A, e : B, \vec{\Gamma} \Rightarrow \vec{\Delta}}{e : A \wedge_f B, \vec{\Gamma} \Rightarrow \vec{\Delta}} \quad \frac{\vec{\Gamma} \Rightarrow \vec{\Delta}, e : A \quad \vec{\Gamma} \Rightarrow \vec{\Delta}, e : B}{\vec{\Gamma} \Rightarrow \vec{\Delta}, e : A \wedge_f B}$$

$$\frac{e : A, \vec{\Gamma} \Rightarrow \vec{\Delta} \quad e : B, \vec{\Gamma} \Rightarrow \vec{\Delta}}{e : A \vee_f B, \vec{\Gamma} \Rightarrow \vec{\Delta}} \quad \frac{\vec{\Gamma} \Rightarrow \vec{\Delta}, e : A, B}{\vec{\Gamma} \Rightarrow \vec{\Delta}, e : A \vee_f B}$$

$$\frac{3:A,\vec{\Gamma}\Rightarrow\vec{\Delta}\quad 3:B,\vec{\Gamma}\Rightarrow\vec{\Delta}}{3:A\wedge_f B,\vec{\Gamma}\Rightarrow\vec{\Delta}}\quad\frac{\vec{\Gamma}\Rightarrow\vec{\Delta},3:A,3:B}{\vec{\Gamma}\Rightarrow\vec{\Delta},3:A\wedge_f B}$$

$$\frac{3:A,3:B,\vec{\Gamma}\Rightarrow\vec{\Delta}}{3:A\vee_f B,\vec{\Gamma}\Rightarrow\vec{\Delta}}\quad\frac{\vec{\Gamma}\Rightarrow\vec{\Delta},3:A\quad\vec{\Gamma}\Rightarrow\vec{\Delta},3:B}{\vec{\Gamma}\Rightarrow\vec{\Delta},3:A\vee_f B}$$

$$\frac{3:A,3:B,\vec{\Gamma}\Rightarrow\vec{\Delta}}{3:A\wedge_t B,\vec{\Gamma}\Rightarrow\vec{\Delta}}\quad\frac{\vec{\Gamma}\Rightarrow\vec{\Delta},3:A\quad\vec{\Gamma}\Rightarrow\vec{\Delta},3:B}{\vec{\Gamma}\Rightarrow\vec{\Delta},3:A\wedge_t B}$$

$$\frac{3:A,\vec{\Gamma}\Rightarrow\vec{\Delta}\quad 3:B,\vec{\Gamma}\Rightarrow\vec{\Delta}}{3:A\vee_t B,\vec{\Gamma}\Rightarrow\vec{\Delta}}\quad\frac{\vec{\Gamma}\Rightarrow\vec{\Delta},3:A,3:B}{\vec{\Gamma}\Rightarrow\vec{\Delta},3:A\vee_t B}.$$

The system F_B^- is defined as the cut-free system $\mathrm{F}_B - \{(\mathrm{cut1}), (\mathrm{cut2}), (\mathrm{cut3}), (\mathrm{cut4})\}$.

Next, we consider the correspondence between F_B and G_B. For example, the following inference rules in F_B:

$$\frac{3:A,3:B,\vec{\Gamma}\Rightarrow\vec{\Delta}}{3:A\wedge_t B,\vec{\Gamma}\Rightarrow\vec{\Delta}}\quad\frac{1:A,\vec{\Gamma}\Rightarrow\vec{\Delta}}{2:{\sim_t}A,\vec{\Gamma}\Rightarrow\vec{\Delta}}$$

correspond to the following inference rules in G_B, respectively:

$$\frac{{\sim_f}A,{\sim_f}B,\Gamma\Rightarrow\Delta}{{\sim_f}(A\wedge_t B),\Gamma\Rightarrow\Delta}({\sim_f}\wedge_t\mathrm{l})\qquad\frac{A,\Gamma\Rightarrow\Delta}{{\sim_t}{\sim_t}A,\Gamma\Rightarrow\Delta}({\sim_t}{\sim_t}\mathrm{l}).$$

Theorem 6.5 *Let* $\Gamma_1,\Gamma_2,\Gamma_3,\Gamma_4',\Gamma_4'',\Delta_1,\Delta_2,\Delta_3,\Delta_4'$, *and* Δ_4'' *be sets of* $\mathscr{L}_{tf}$-*formulas.*

1. *If* $\mathrm{F}_B\vdash\Gamma_1:\Gamma_2:\Gamma_3:\Gamma_4',\Gamma_4''\rightarrow\Delta_1:\Delta_2:\Delta_3:\Delta_4',\Delta_4''$, *then* $G_B\vdash\Gamma_1,{\sim_t}\Gamma_2,{\sim_f}\Gamma_3,{\sim_f}{\sim_t}\Gamma_4',{\sim_t}{\sim_f}\Gamma_4''\Rightarrow\Delta_1,{\sim_t}\Delta_2,{\sim_f}\Delta_3,{\sim_f}{\sim_t}\Delta_4',{\sim_t}{\sim_f}\Delta_4''$.
2. *If* $G_B-(\mathrm{cut})\vdash\Gamma_1,{\sim_t}\Gamma_2,{\sim_f}\Gamma_3,{\sim_f}{\sim_t}\Gamma_4',{\sim_t}{\sim_f}\Gamma_4''\Rightarrow\Delta_1,{\sim_t}\Delta_2,{\sim_f}\Delta_3,{\sim_f}{\sim_t}\Delta_4'$, ${\sim_t}{\sim_f}\Delta_4''$, *then* $\mathrm{F}_B^-\vdash\Gamma_1:\Gamma_2:\Gamma_3:\Gamma_4',\Gamma_4''\Rightarrow\Delta_1:\Delta_2:\Delta_3:\Delta_4',\Delta_4''$.

Proof Obvious by the definition of F_B. □

Theorem 6.6 *The rules* (cut1), (cut2), (cut3) *and* (cut4) *are admissible in cut-free* F_B.

Proof Suppose $\mathrm{F}_B\vdash\Gamma_1:\Gamma_2:\Gamma_3:\Gamma_4',\Gamma_4''\rightarrow\Delta_1:\Delta_2:\Delta_3:\Delta_4',\Delta_4''$. Then, $G_B\vdash\Gamma_1,{\sim_t}\Gamma_2,{\sim_f}\Gamma_3,{\sim_f}{\sim_t}\Gamma_4',{\sim_t}{\sim_f}\Gamma_4''\Rightarrow\Delta_1,{\sim_t}\Delta_2,{\sim_f}\Delta_3,{\sim_f}{\sim_t}\Delta_4',{\sim_t}{\sim_f}\Delta_4''$ by Theorem 6.5 (1), and hence $G_B-(\mathrm{cut})\vdash\Gamma_1,{\sim_t}\Gamma_2,{\sim_f}\Gamma_3,{\sim_f}{\sim_t}\Gamma_4',{\sim_t}{\sim_f}\Gamma_4''\Rightarrow\Delta_1,{\sim_t}\Delta_2,{\sim_f}\Delta_3,{\sim_f}{\sim_t}\Delta_4',{\sim_t}{\sim_f}\Delta_4''$ by Theorem 6.3. Therefore $\mathrm{F}_B^-\vdash\Gamma_1:\Gamma_2:\Gamma_3:\Gamma_4',\Gamma_4''\Rightarrow\Delta_1:\Delta_2:\Delta_3:\Delta_4',\Delta_4''$ by Theorem 6.5 (2). □

Corollary 6.1 *If a four-place sequent S is provable in* F_B, *then there is a proof* π *of S such that all formulas that appear in* π *are subformulas of some formula in S.*

Corollary 6.2 *Let X and Y both be non-empty subsets of* $\{\sim_t, \sim_f, \wedge_t, \wedge_f, \vee_t, \vee_f\}$, *and let X be a proper subset of Y. Then* $\vdash_B$ *restricted to Y is a conservative extension of* $\vdash_B$ *restricted to X.*

Next, we will introduce yet another sequent calculus for $\vdash_B$. This quadruple (vertical) matrix sequent calculus Q_B may be regarded as a natural extension and generalization of the dual calculi discussed in [137, 140]. The system Q_B, however, does not enjoy the subformula property. It is considered here because it reflects the co-ordinate valuations from Chap. 5 by making use of four kinds of sequents: $(\Gamma \Rightarrow_n \Delta), (\Gamma \Rightarrow_t \Delta), (\Gamma \Rightarrow_f \Delta)$ and $(\Gamma \Rightarrow_b \Delta)$ which, respectively, correspond to the standard sequents: $(\Gamma \Rightarrow \Delta)(\sim_t\Gamma \Rightarrow \sim_t\Delta), (\sim_f\Gamma \Rightarrow \sim_f\Delta)$ and $(\sim_f\sim_t\Gamma', \sim_t\sim_f\Gamma'' \Rightarrow \sim_f\sim_t\Delta', \sim_t\sim_f\Delta'')$, where $(\Gamma \doteq \Gamma', \Gamma'')$ and $(\Delta \doteq \Delta', \Delta'')$.

Definition 6.6 Let $\Rightarrow_*$ be either $\Rightarrow_n, \Rightarrow_t, \Rightarrow_f$ or $\Rightarrow_b$, and let $\Rightarrow_e$ be either $\Rightarrow_t$ or $\Rightarrow_b$. The initial sequents of Q_B are of the form $p \Rightarrow_* p$ for any propositional variable p. The inference rules of Q_B are of the form:

$$\frac{\Gamma \Rightarrow_* \Delta, A \quad A, \Sigma \Rightarrow_* \Pi}{\Gamma, \Sigma \Rightarrow_* \Delta, \Pi}\,(\mathrm{cut} : *)$$

$$\frac{\sim_t\Gamma_1, \Gamma_2 \Rightarrow_n \sim_t\Delta_1, \Delta_2}{\Gamma_1, \sim_t\Gamma_2 \Rightarrow_t \Delta_1, \sim_t\Delta_2} \quad \frac{\sim_t\Gamma_1, \Gamma_2 \Rightarrow_t \sim_t\Delta_1, \Delta_2}{\Gamma_1, \sim_t\Gamma_2 \Rightarrow_n \Delta_1, \sim_t\Delta_2}$$

$$\frac{\sim_f\Gamma_1, \Gamma_2 \Rightarrow_n \sim_f\Delta_1, \Delta_2}{\Gamma_1, \sim_f\Gamma_2 \Rightarrow_f \Delta_1, \sim_f\Delta_2} \quad \frac{\sim_f\Gamma_1, \Gamma_2 \Rightarrow_f \sim_f\Delta_1, \Delta_2}{\Gamma_1, \sim_f\Gamma_2 \Rightarrow_n \Delta_1, \sim_f\Delta_2}$$

$$\frac{\sim_f\sim_t\Gamma_1, \sim_t\sim_f\Gamma_2, \Gamma_3, \Gamma_4 \Rightarrow_n \sim_f\sim_t\Delta_1, \sim_t\sim_f\Delta_2, \Delta_3, \Delta_4}{\Gamma_1, \Gamma_2, \sim_f\sim_t\Gamma_3, \sim_t\sim_f\Gamma_4 \Rightarrow_b \Delta_1, \Delta_2, \sim_f\sim_t\Delta_3, \sim_t\sim_f\Delta_4}$$

$$\frac{\sim_f\sim_t\Gamma_1, \sim_t\sim_f\Gamma_2, \Gamma_3, \Gamma_4 \Rightarrow_b \sim_f\sim_t\Delta_1, \sim_t\sim_f\Delta_2, \Delta_3, \Delta_4}{\Gamma_1, \Gamma_2, \sim_f\sim_t\Gamma_3, \sim_t\sim_f\Gamma_4 \Rightarrow_n \Delta_1, \Delta_2, \sim_f\sim_t\Delta_3, \sim_t\sim_f\Delta_4}$$

$$\frac{\sim_t\Gamma_1, \Gamma_2 \Rightarrow_f \sim_t\Delta_1, \Delta_2}{\Gamma_1, \sim_t\Gamma_2 \Rightarrow_b \Delta_1, \sim_t\Delta_2} \quad \frac{\sim_t\Gamma_1, \Gamma_2 \Rightarrow_b \sim_t\Delta_1, \Delta_2}{\Gamma_1, \sim_t\Gamma_2 \Rightarrow_f \Delta_1, \sim_t\Delta_2}$$

$$\frac{\sim_f\Gamma_1, \Gamma_2 \Rightarrow_t \sim_f\Delta_1, \Delta_2}{\Gamma_1, \sim_f\Gamma_2 \Rightarrow_b \Delta_1, \sim_f\Delta_2} \quad \frac{\sim_f\Gamma_1, \Gamma_2 \Rightarrow_b \sim_f\Delta_1, \Delta_2}{\Gamma_1, \sim_f\Gamma_2 \Rightarrow_t \Delta_1, \sim_f\Delta_2}$$

$$\frac{\sim_f\sim_t\Gamma_1, \Gamma_2 \Rightarrow_t \sim_f\sim_t\Delta_1, \Delta_2}{\Gamma_1, \sim_t\sim_f\Gamma_2 \Rightarrow_f \Delta_1, \sim_t\sim_f\Delta_2} \quad \frac{\sim_f\sim_t\Gamma_1, \Gamma_2 \Rightarrow_f \sim_f\sim_t\Delta_1, \Delta_2}{\Gamma_1, \sim_t\sim_f\Gamma_2 \Rightarrow_t \Delta_1, \sim_t\sim_f\Delta_2}$$

$$\frac{\Gamma \Rightarrow_* \Delta}{A, \Gamma \Rightarrow_* \Delta} \quad \frac{\Gamma \Rightarrow_* \Delta}{\Gamma \Rightarrow_* \Delta, A}$$

$$\frac{A, B, \Gamma \Rightarrow_n \Delta}{A \wedge_t B, \Gamma \Rightarrow_n \Delta} \quad \frac{\Gamma \Rightarrow_n \Delta, A \quad \Gamma \Rightarrow_n \Delta, B}{\Gamma \Rightarrow_n \Delta, A \wedge_t B}$$

$$\frac{A, \Gamma \Rightarrow_n \Delta \quad B, \Gamma \Rightarrow_n \Delta}{A \vee_t B, \Gamma \Rightarrow_n \Delta} \quad \frac{\Gamma \Rightarrow_n \Delta, A, B}{\Gamma \Rightarrow_n \Delta, A \vee_t B}$$

$$\frac{A, \Gamma \Rightarrow_n \Delta \quad B, \Gamma \Rightarrow_n \Delta}{A \wedge_f B, \Gamma \Rightarrow_n \Delta} \quad \frac{\Gamma \Rightarrow_n \Delta, A, B}{\Gamma \Rightarrow_n \Delta, A \wedge_f B}$$

$$\frac{A, B, \Gamma \Rightarrow_n \Delta}{A \vee_f B, \Gamma \Rightarrow_n \Delta} \quad \frac{\Gamma \Rightarrow_n \Delta, A \quad \Gamma \Rightarrow_n \Delta, B}{\Gamma \Rightarrow_n \Delta, A \vee_f B}$$

$$\frac{A, \Gamma \Rightarrow_e \Delta \quad B, \Gamma \Rightarrow_e \Delta}{A \wedge_t B, \Gamma \Rightarrow_e \Delta} \quad \frac{\Gamma \Rightarrow_e \Delta, A, B}{\Gamma \Rightarrow_e \Delta, A \wedge_t B}$$

$$\frac{A, B, \Gamma \Rightarrow_e \Delta}{A \vee_t B, \Gamma \Rightarrow_e \Delta} \quad \frac{\Gamma \Rightarrow_e \Delta, A \quad \Gamma \Rightarrow_e \Delta, B}{\Gamma \Rightarrow_e \Delta, A \vee_t B}$$

$$\frac{A, B, \Gamma \Rightarrow_e \Delta}{A \wedge_f B, \Gamma \Rightarrow_e \Delta} \quad \frac{\Gamma \Rightarrow_e \Delta, A \quad \Gamma \Rightarrow_e \Delta, B}{\Gamma \Rightarrow_e \Delta, A \wedge_f B}$$

$$\frac{A, \Gamma \Rightarrow_e \Delta \quad B, \Gamma \Rightarrow_e \Delta}{A \vee_f B, \Gamma \Rightarrow_e \Delta} \quad \frac{\Gamma \Rightarrow_e \Delta, A, B}{\Gamma \Rightarrow_e \Delta, A \vee_f B}$$

$$\frac{A, \Gamma \Rightarrow_f \Delta \quad B, \Gamma \Rightarrow_f \Delta}{A \wedge_f B, \Gamma \Rightarrow_f \Delta} \quad \frac{\Gamma \Rightarrow_f \Delta, A, B}{\Gamma \Rightarrow_f \Delta, A \wedge_f B}$$

$$\frac{A, B, \Gamma \Rightarrow_f \Delta}{A \vee_f B, \Gamma \Rightarrow_f \Delta} \quad \frac{\Gamma \Rightarrow_f \Delta, A \quad \Gamma \Rightarrow_f \Delta, B}{\Gamma \Rightarrow_f \Delta, A \vee_f B}$$

$$\frac{A, B, \Gamma \Rightarrow_f \Delta}{A \wedge_t B, \Gamma \Rightarrow_f \Delta} \quad \frac{\Gamma \Rightarrow_f \Delta, A \quad \Gamma \Rightarrow_f \Delta, B}{\Gamma \Rightarrow_f \Delta, A \wedge_t B}$$

$$\frac{A, \Gamma \Rightarrow_f \Delta \quad B, \Gamma \Rightarrow_f \Delta}{A \vee_t B, \Gamma \Rightarrow_f \Delta} \quad \frac{\Gamma \Rightarrow_f \Delta, A, B}{\Gamma \Rightarrow_f \Delta, A \vee_t B}.$$

Next, we consider the correspondence between Q_B and G_B. For example, the following inference rules in Q_B:

$$\frac{A, \Gamma \Rightarrow_f \Delta \quad B, \Gamma \Rightarrow_f \Delta}{A \wedge_f B, \Gamma \Rightarrow_f \Delta}$$

$$\frac{\sim_f \sim_t \Gamma_1, \sim_t \sim_f \Gamma_2, \Gamma_3, \Gamma_4 \Rightarrow_n \sim_f \sim_t \Delta_1, \sim_t \sim_f \Delta_2, \Delta_3, \Delta_4}{\Gamma_1, \Gamma_2, \sim_f \sim_t \Gamma_3, \sim_t \sim_f \Gamma_4 \Rightarrow_b \Delta_1, \Delta_2, \sim_f \sim_t \Delta_3, \sim_t \sim_f \Delta_4}$$

correspond to the inference rule ($\sim_f \wedge_f l$) in G_B and the inference rules ($\sim_b \sim_b l$) and ($\sim_b \sim_b r$) (appearing in Proposition 6.2) in G_B, respectively.

Theorem 6.7 *Let* Γ *and* Δ *be sets of* $\mathscr{L}_{tf}$*-formulas,* Γ *be* (Γ', Γ'')*, and* Δ *be* (Δ', Δ'').

1. *If* $\mathrm{Q}_B \vdash \Gamma \Rightarrow_* \Delta$ *where* $* \in \{n, t, f, b\}$*, then*
   (a) $G_B \vdash \Gamma \Rightarrow \Delta$ *if* $* = n$,
   (b) $G_B \vdash {\sim_t}\Gamma \Rightarrow {\sim_t}\Delta$ *if* $* = t$,
   (c) $G_B \vdash {\sim_f}\Gamma \Rightarrow {\sim_f}\Delta$ *if* $* = f$ *or*
   (d) $G_B \vdash {\sim_f}{\sim_t}\Gamma', {\sim_t}{\sim_f}\Gamma'' \Rightarrow {\sim_f}{\sim_t}\Delta', {\sim_t}{\sim_f}\Delta''$ if $* = b$.
2. *If* $G_B - (\mathrm{cut}) \vdash \Gamma \Rightarrow \Delta$*, then* $\mathrm{Q}_B - (\mathrm{cut} : *) \vdash \Gamma \Rightarrow_n \Delta$.

Proof Obvious by the definition of Q_B. □

Theorem 6.8 *The rule* (cut:$*$) *is admissible in cut-free* Q_B.

Proof By Theorems 6.7 and 6.3. □

6.3 Extensions

Odintsov obtained the axiomatizations L_{base}, L_B and L_T by extending $\vdash_{\mathrm{base}}, \vdash_B$ and $\vdash_T$, respectively. The language $\mathscr{L}_{tf}^{\rightarrow_t}$ of L_{base}, L_B, and L_T is obtained from that of $\vdash_{\mathrm{base}}$ by adding $\rightarrow_t$ (material implication) and $\neg$ (classical negation), the latter being defined by $\neg A := A \rightarrow_t {\sim_t}(p \rightarrow_t p)$, for some fixed propositional variable p. We will define a standard Gentzen-style sequent calculus GL_B for L_B and prove the cut-elimination theorem for GL_B. Although a four-place sequent calculus FL_B and a quadruple sequent calculus QL_B can also be obtained for L_B in a similar way to how it was obtained in the previous section, detailed discussions of these calculi are omitted here because of the tedious and complex constructions of these calculi. The cut-elimination theorem for FL_B and QL_B can also be proved in a similar way to how it was proved in Sect. 6.2.

For the remainder of this section, we will use the symbol $\rightarrow$ instead of $\rightarrow_t$. The expression $A \leftrightarrow B$ is an abbreviation of $(A \rightarrow B)\wedge_t(B \rightarrow A)$. We repeat the definition of the axiom system L_B.

Definition 6.7 Suppose that the symbols $\vdash$ and $\dashv\vdash$ in the definition of $\vdash_B$ (Definition 5.5) are replaced by $\rightarrow$ and $\leftrightarrow$, respectively. Let ${\sim}_a$ be either ${\sim_f}$ or ${\sim_e}$ (cf. Definition 6.1). L_B is obtained from $\vdash_B$ by deleting the rules

$$\frac{A \vdash B \quad B \vdash C}{A \vdash C} \quad \frac{A \vdash B \quad A \vdash C}{A \vdash B \wedge_t C} \quad \frac{A \vdash C \quad B \vdash C}{A \vee_t B \vdash C}$$

and adding the following axiom schemes and inference rule:

1. $A \to (B \to A)$,
2. $(A \to (B \to C)) \to ((A \to B) \to (A \to C))$,
3. $(A \to B) \to ((A \to C) \to (A \to (B \wedge_t C)))$,
4. $(A \to C) \to ((B \to C) \to ((A \vee_t B) \to C))$,
5. $A \vee_t (A \to B)$,
6. $\neg{\sim_a} A \leftrightarrow {\sim_a}\neg A$,
7. $(A \to B) \leftrightarrow (\neg A \vee_t B)$,
8. ${\sim_e}(A \to B) \leftrightarrow ({\sim_e}\neg A \wedge_t {\sim_e} B)$,
9. ${\sim_f}(A \to B) \leftrightarrow ({\sim_f} A \to {\sim_f} B)$,

$$\frac{A \quad A \to B}{B}$$

Definition 6.8 The sequent calculus GL_B for L_B is obtained from G_B by adding the following inference rules:

$$\frac{\Gamma \Rightarrow \Sigma, A \quad B, \Delta \Rightarrow \Pi}{A \to B, \Gamma, \Delta \Rightarrow \Sigma, \Pi}(\to \mathrm{l}) \qquad \frac{A, \Gamma \Rightarrow \Delta, B}{\Gamma \Rightarrow \Delta, A \to B}(\to \mathrm{r})$$

$$\frac{\Gamma \Rightarrow \Delta, A}{\neg A, \Gamma \Rightarrow \Delta}(\neg \mathrm{l}) \qquad \frac{A, \Gamma \Rightarrow \Delta}{\Gamma \Rightarrow \Delta, \neg A}(\neg \mathrm{r})$$

$$\frac{{\sim_a}\neg A, \Gamma \Rightarrow \Delta}{\neg{\sim_a} A, \Gamma \Rightarrow \Delta}(\neg{\sim_a}\mathrm{l}) \qquad \frac{\Gamma \Rightarrow \Delta, {\sim_a}\neg A}{\Gamma \Rightarrow \Delta, \neg{\sim_a} A}(\neg{\sim_a}\mathrm{r})$$

$$\frac{\neg{\sim_a} A, \Gamma \Rightarrow \Delta}{{\sim_a}\neg A, \Gamma \Rightarrow \Delta}({\sim_a}\neg \mathrm{l}) \qquad \frac{\Gamma \Rightarrow \Delta, \neg{\sim_a} A}{\Gamma \Rightarrow \Delta, {\sim_a}\neg A}({\sim_a}\neg \mathrm{r})$$

$$\frac{\neg A, \Gamma \Rightarrow \Delta}{{\sim_a}\neg{\sim_a} A, \Gamma \Rightarrow \Delta}({\sim_a}\neg{\sim_a}\mathrm{l}) \qquad \frac{\Gamma \Rightarrow \Delta, \neg A}{\Gamma \Rightarrow \Delta, {\sim_a}\neg{\sim_a} A}({\sim_a}\neg{\sim_a}\mathrm{r})$$

$$\frac{{\sim_e}\neg A, {\sim_e} B, \Gamma \Rightarrow \Delta}{{\sim_e}(A \to B), \Gamma \Rightarrow \Delta}({\sim_e} \to \mathrm{l}) \qquad \frac{\Gamma \Rightarrow \Delta, {\sim_e}\neg A \quad \Gamma \Rightarrow \Delta, {\sim_e} B}{\Gamma \Rightarrow \Delta, {\sim_e}(A \to B)}({\sim_e} \to \mathrm{r})$$

$$\frac{\Gamma \Rightarrow \Delta, {\sim_a} A}{{\sim_a}\neg A, \Gamma \Rightarrow \Delta}({\sim_a}\neg \mathrm{l}) \qquad \frac{{\sim_a} A, \Gamma \Rightarrow \Delta}{\Gamma \Rightarrow \Delta, {\sim_a}\neg A}({\sim_a}\neg \mathrm{r})$$

$$\frac{\Gamma \Rightarrow \Sigma, {\sim_f} A \quad {\sim_f} B, \Delta \Rightarrow \Pi}{{\sim_f}(A \to B), \Gamma, \Delta \Rightarrow \Sigma, \Pi}({\sim_f} \to \mathrm{l}) \qquad \frac{{\sim_f} A, \Gamma \Rightarrow \Delta, {\sim_f} B}{\Gamma \Rightarrow \Delta, {\sim_f}(A \to B)}({\sim_f} \to \mathrm{r}).$$

Note that a four-place sequent calculus FL_B and a quadruple sequent calculus QL_B can be obtained from GL_B in a natural way. For example, the inference rule:

$$\frac{\Gamma \Rightarrow \Delta, {\sim_t}\neg A \quad \Gamma \Rightarrow \Delta, {\sim_t}B}{\Gamma \Rightarrow \Delta, {\sim_t}(A \rightarrow B)}({\sim_t} \rightarrow \mathrm{r})$$

in GL_B gives rise to the corresponding inference rules:

$$\frac{\vec{\Gamma} \Rightarrow \vec{\Delta}, 2 : \neg A \quad \vec{\Gamma} \Rightarrow \vec{\Delta}, 2 : B}{\vec{\Gamma} \Rightarrow \vec{\Delta}, 2 : A \rightarrow B} \quad \frac{\Gamma \Rightarrow_t \Delta, \neg A \quad \Gamma \Rightarrow_t \Delta, B}{\Gamma \Rightarrow_t \Delta, A \rightarrow B}$$

in FL_B and QL_B, respectively.

Proposition 6.8 *The rules presented in Proposition 6.3 and the following rules are admissible in cut-free* GL_B:

$$\frac{{\sim_a}\neg{\sim_a}A, \Gamma \Rightarrow \Delta}{\neg A, \Gamma \Rightarrow \Delta}({\sim_a}\neg{\sim_a}\mathrm{l}^{-1}) \quad \frac{\Gamma \Rightarrow \Delta, {\sim_a}\neg{\sim_a}A}{\Gamma \Rightarrow \Delta, \neg A}({\sim_a}\neg{\sim_a}\mathrm{r}^{-1}).$$

We cannot show the admissibility of (${\sim_t}$twist) and (${\sim_f}$invol) (Proposition 6.4) for GL_B because of the presence of $\rightarrow$, and hence we have to find another way (without Proposition 6.4) of proving that $\mathrm{GL}_B - (\mathrm{cut}) \vdash A \Rightarrow A$ for any $\mathscr{L}_{tf}^{\rightarrow_t}$-formula A. We present such a way below.

Proposition 6.9 *Let* ${\sim_d}$ *be either* ${\sim_t}{\sim_t}$, ${\sim_f}{\sim_f}$, *or* ${\sim_b}{\sim_b}$. *The following sequents are provable in cut-free* GL_B, *for any* $\mathscr{L}_{tf}^{\rightarrow_t}$*-formula A:*

1. ${\sim_t}A \Rightarrow {\sim_t}A$, 2. ${\sim_f}A \Rightarrow {\sim_f}A$, 3. ${\sim_b}A \Rightarrow {\sim_b}A$, 4. ${\sim_d}A \Rightarrow {\sim_d}A$.

Proof By simultaneous induction for (1)–(4) on the structure of A.

(1) We show some cases.

(Case $A \mathrel{\hat{=}} {\sim_f}B$): By the induction hypothesis for (3), we obtain $\mathrm{GL}_B - (\mathrm{cut}) \vdash {\sim_t}{\sim_f}B \Rightarrow {\sim_t}{\sim_f}B$.

(Case $A \mathrel{\hat{=}} {\sim_t}B$): By the induction hypothesis for (4), we obtain $\mathrm{GL}_B - (\mathrm{cut}) \vdash {\sim_t}{\sim_t}B \Rightarrow {\sim_t}{\sim_t}B$.

(Case $A \mathrel{\hat{=}} B \rightarrow C$): We show $\mathrm{GL}_B - (\mathrm{cut}) \vdash {\sim_t}(B \rightarrow C) \Rightarrow {\sim_t}(B \rightarrow C)$. By the induction hypothesis for (4), we have $\mathrm{GL}_B - (\mathrm{cut}) \vdash {\sim_t}{\sim_t}B \Rightarrow {\sim_t}{\sim_t}B$, and by the induction hypothesis for (1), we have $\mathrm{GL}_B - (\mathrm{cut}) \vdash {\sim_t}C \Rightarrow {\sim_t}C$. Then we have:

$$
\dfrac{
\dfrac{\dfrac{\dfrac{\dfrac{\vdots}{\sim_t\sim_t B \Rightarrow \sim_t\sim_t B}}{B \Rightarrow \sim_t\sim_t B}\,(\sim_t\sim_t\mathrm{l}^{-1})}{B \Rightarrow B}\,(\sim_t\sim_t\mathrm{r}^{-1})}{\dfrac{B, \sim_t C \Rightarrow B}{\sim_t(B \rightarrow C) \Rightarrow B}\,(\sim_t \rightarrow \mathrm{l})}\,(\text{w-l})
\qquad
\dfrac{\dfrac{\dfrac{\vdots}{\sim_t C \Rightarrow \sim_t C}}{B, \sim_t C \Rightarrow \sim_t C}\,(\text{w-l})}{\sim_t(B \rightarrow C) \Rightarrow \sim_t C}\,(\sim_t \rightarrow \mathrm{l})
}{\sim_t(B \rightarrow C) \Rightarrow \sim_t(B \rightarrow C)}\,(\sim_t \rightarrow \mathrm{r})
$$

where $(\sim_t\sim_t l^{-1})$ and $(\sim_t\sim_t r^{-1})$ are admissible in cut-free GL_B by Proposition 6.8.

(2) Similar to (1).

(3) We show only the case for $\sim_b = \sim_f\sim_t$ since the case for $\sim_b = \sim_t\sim_f$ can be obtained similarly. We show some cases.

(Case $A \mathrel{\mathring{=}} \sim_t B$): We show $\mathrm{GL}_B - (\mathrm{cut}) \vdash \sim_f\sim_t\sim_t B \Rightarrow \sim_f\sim_t\sim_t B$. By the induction hypothesis for (2), we have $\mathrm{GL}_B - (\mathrm{cut}) \vdash \sim_f B \Rightarrow \sim_f B$. Then we obtain:

$$
\dfrac{\dfrac{\dfrac{\vdots}{\sim_f B \Rightarrow \sim_f B}}{\sim_f\sim_t\sim_t B \Rightarrow \sim_f B}\,(\sim_f\sim_t\sim_t\mathrm{l})}{\sim_f\sim_t\sim_t B \Rightarrow \sim_f\sim_t\sim_t B}\,(\sim_f\sim_t\sim_t\mathrm{r}).
$$

(Case $A \mathrel{\mathring{=}} B \wedge_t C$): We show $\mathrm{GL}_B - (\mathrm{cut}) \vdash \sim_f\sim_t(B \wedge_t C) \Rightarrow \sim_f\sim_t(B \wedge_t C)$. By the induction hypothesis for (3), we have $\mathrm{GL}_B - (\mathrm{cut}) \vdash \sim_f\sim_t B \Rightarrow \sim_f\sim_t B$ and $\mathrm{GL}_B - (\mathrm{cut}) \vdash \sim_f\sim_t C \Rightarrow \sim_f\sim_t C$. Then we obtain:

$$
\dfrac{
\dfrac{\dfrac{\vdots}{\sim_f\sim_t B \Rightarrow \sim_f\sim_t B}}{\sim_f\sim_t B \Rightarrow \sim_f\sim_t(B \wedge_t C)}\,(\sim_f\sim_t\mathrm{r1})
\qquad
\dfrac{\dfrac{\vdots}{\sim_f\sim_t C \Rightarrow \sim_f\sim_t C}}{\sim_f\sim_t C \Rightarrow \sim_f\sim_t(B \wedge_t C)}\,(\sim_f\sim_t\mathrm{r2})
}{\sim_f\sim_t(B \wedge_t C) \Rightarrow \sim_f\sim_t(B \wedge_t C)}\,(\sim_f\sim_t\wedge_t\mathrm{l}).
$$

(4) We show some cases.

(Case $A \mathrel{\mathring{=}} p$ for a propositional variable p): We can obtain $\mathrm{GL}_B - (\mathrm{cut}) \vdash \sim_d p \Rightarrow \sim_d p$ from the initial sequent $p \Rightarrow p$ by applying $(\sim_d l)$ and $(\sim_d r)$.

(Case $A \mathrel{\mathring{=}} B \rightarrow C$): We show $\mathrm{GL}_B - (\mathrm{cut}) \vdash \sim_d(B \rightarrow C) \Rightarrow \sim_d(B \rightarrow C)$. By the induction hypothesis for (4), we have $\mathrm{GL}_B - (\mathrm{cut}) \vdash \sim_d B \Rightarrow \sim_d B$ and $\mathrm{GL}_B - (\mathrm{cut}) \vdash \sim_d C \Rightarrow \sim_d C$. Then we obtain:

$$
\dfrac{\dfrac{\dfrac{\dfrac{\dfrac{\dfrac{\vdots}{\sim_d B \Rightarrow \sim_d B}}{B \Rightarrow \sim_d B}\,(\sim_d \mathrm{l}^{-1})}{B \Rightarrow B}\,(\sim_d \mathrm{r}^{-1}) \qquad \dfrac{\dfrac{\dfrac{\vdots}{\sim_d C \Rightarrow \sim_d C}}{C \Rightarrow \sim_d C}\,(\sim_d \mathrm{l}^{-1})}{C \Rightarrow C}\,(\sim_d \mathrm{r}^{-1})}{B, B \rightarrow C \Rightarrow C}\,(\rightarrow \mathrm{l})}{B \rightarrow C \Rightarrow B \rightarrow C}\,(\rightarrow \mathrm{r})}{\sim_d (B \rightarrow C) \Rightarrow B \rightarrow C}\,(\sim_d \mathrm{l})}{\sim_d (B \rightarrow C) \Rightarrow \sim_d (B \rightarrow C)}\,(\sim_d \mathrm{r})
$$

where $(\sim_d \mathrm{l}^{-1})$ and $(\sim_d \mathrm{r}^{-1})$ are admissible in cut-free GL_B by Proposition 6.8. □

Proposition 6.10 *The following sequents are provable in cut-free* GL_B *for any* $\mathscr{L}_{tf}^{\rightarrow_t}$*-formula* A: 1. $\sim_f \sim_t A \Rightarrow \sim_t \sim_f A$, 2. $\sim_t \sim_f A \Rightarrow \sim_f \sim_t A$.

Proof By induction on A. We use Proposition 6.9. □

Proposition 6.11 *For any* $\mathscr{L}_{tf}^{\rightarrow_t}$*-formula* A, $\mathrm{GL}_B - (\mathrm{cut}) \vdash A \Rightarrow A$.

Proof By induction on A. We use Proposition 6.9 (1) and (2) for the cases $A \triangleq \sim_t B$ and $A \triangleq \sim_f B$. □

Let the expressions $\bigwedge_t \Gamma$ and $\bigvee_t \Gamma$ stand for the conjunction $C_1 \wedge_t C_2 \wedge_t \cdots \wedge_t C_n$ and the disjunction $C_1 \vee_t C_2 \vee_t \cdots \vee_t C_n$, respectively, if $\Gamma \triangleq \{C_1, C_2, \ldots, C_n\}$ $(n \geq 1)$ and for $p \rightarrow p$ and $\neg(p \rightarrow p)$ (for some propositional variable p), respectively, if $\Gamma \triangleq \emptyset$.

Theorem 6.9 *Let* A *be an* $\mathscr{L}_{tf}^{\rightarrow_t}$*-formula. Then* $\mathrm{GL}_B \vdash \Rightarrow A$ *iff* $\mathrm{L}_B \vdash A$.

Proof Right-to-left: If A is an axiom of L_B, then $\mathrm{GL}_B - (\mathrm{cut}) \vdash \Rightarrow A$. Moreover, using (cut) it can be shown that in GL_B, $\{\Rightarrow A, \Rightarrow A \rightarrow B\} \vdash \Rightarrow B$.

Left-to-right: In Chap. 5 we have seen that for every $\mathscr{L}_{tf}^{\rightarrow_t}$-formula A:

$$L_T = \{A \mid \forall v(v_t(A) = 1)\} = \{A \mid \forall v(v_f(A) = 0)\}.$$

$$L_B = \{A \mid \forall v(v_b(A) = 1)\} = \{A \mid \forall v(v_n(A) = 0)\}.$$

Therefore, it is enough to show that $\mathrm{GL}_B \vdash \Delta \Rightarrow \Gamma$ implies

$$\forall v \big(\mathbf{B} \in v(\textstyle\bigwedge_t \Delta) \text{ implies } \mathbf{B} \in v(\textstyle\bigvee_t \Gamma)\big).$$

We show some cases of the proof by induction on proofs on GL_B.

$(\rightarrow \mathrm{l})$: Suppose $\forall v(\mathbf{B} \in v(\bigwedge_t \Gamma)$ implies $\mathbf{B} \in v(\bigvee_t \Sigma \vee_t A))$ and $\forall v(\mathbf{B} \in v(B \wedge_t \bigwedge_t \Delta)$ implies $\mathbf{B} \in v(\bigvee_t \Pi))$, and let $\mathbf{B} \in v'((A \rightarrow B) \wedge_t \bigwedge_t \Pi \wedge_t \bigwedge_t \Delta)$ for some valuation v'. By the definition of $\sqsupseteq_t$, obviously $\mathbf{B} \in v'(\bigvee_t \Sigma \vee_t \bigvee_t \Pi)$.

($\neg r$): Suppose $\forall v(\mathbf{B} \in v(A \wedge_t \bigwedge_t \Gamma)$ implies $\mathbf{B} \in v(\bigvee_t \Delta))$ and let $\mathbf{B} \in v'(\bigwedge_t \Gamma)$ for some valuation v'. Since $\mathbf{B} \in v'(A \to \sim_f(p \to p)$ iff $\mathbf{B} \notin v'(A), \mathbf{B} \in v'$ $(\bigvee_t \Delta \vee_t \neg A)(\sim_a \neg\sim_a 1)$ with $a = \sim_t$ Suppose $\forall v(\mathbf{B} \in v(\neg A \wedge_t \bigwedge_t \Gamma)$ implies $\mathbf{B} \in v(\bigvee_t \Delta))$ and let $\mathbf{B} \in v'(\sim_t \neg\sim_t A \wedge_t \bigwedge_t \Gamma)$ for some valuation v'. Since by Proposition 3.4,

$$\begin{aligned} \mathbf{B} \in v'(\sim_t \neg\sim_t A) &\text{ iff} \\ \mathbf{F} \in v'(\neg\sim_t A) &\text{ iff} \\ \mathbf{F} \notin v'(\sim_t A) &\text{ iff} \\ \mathbf{B} \notin v'(A), & \end{aligned}$$

it follows that $\mathbf{B} \in v'(\bigvee_t \Delta)$.

($\sim_f \to$ r): Suppose $\forall v(\mathbf{B} \in v(\sim_f A \wedge_t \bigwedge_t \Gamma)$ implies $\mathbf{B} \in v(\bigvee_t \Delta) \vee_t \sim_f B))$ and let $\mathbf{B} \in v'(\bigwedge_t \Gamma)$ for some valuation v'. It is enough to note that by Proposition 3.4 and the definition of $\sqsupseteq_t$,

$$\begin{aligned} \mathbf{B} \in v'(\sim_f(A \to B)) &\text{ iff} \\ \mathbf{T} \in v'(A \to B) &\text{ iff} \\ (\mathbf{T} \in v'(A) \text{ implies } \mathbf{T} \in v'(B)) &\text{ iff} \\ (\mathbf{B} \in v'(\sim_f A) \text{ implies } \mathbf{B} \in v'(\sim_f B)). & \end{aligned}$$

The embedding theorem of GL_B into *LK* with the inference rules for $\to$ and $\neg$ can naturally be shown, and hence the following cut-elimination result can be obtained.

Theorem 6.10 *The rule* (cut) *is admissible in cut-free* GL_B.

The decidability and paraconsistency of GL_B with respect to $\sim_a$ can also be shown.

6.4 Sequent Calculi for Truth Entailment and Falsity Entailment in *SIXTEEN*$_3$

Gentzen introduced sequents as structures $\Delta \vdash \Gamma$, where Δ and Γ are finite sequences of formulas. Various structural inference rules not displaying any logical operations are assumed such that Δ and Γ may be taken to be finite sets, and the intuitive understanding of $\Delta \vdash \Gamma$ is expressed by the formula $\bigwedge\Delta \to \bigvee\Gamma$. If this translation of sequents is a translation into classical logic and $\Delta = \{A_1, \ldots, A_n\}$, then the translation of $\Delta \vdash \Gamma$ is equivalent with $\bigvee\{\neg A_1, \ldots, \neg A_n\} \vee \bigvee\Gamma$. In other words, $\Delta \vdash \Gamma$ is understood as a disjunction of *true* formulas (having their position in the succedent Γ) and negated formulas $\neg A_1, \ldots, \neg A_n$ such that $A_1, \ldots, A_n$ are *not true* (having their position in the antecedent Δ). Then for every logical operation there are rules for introducing the operation into the position of formulas that are true and the position of formulas that are not true.

It is well-known that this perspective on sequents can be generalized to n-valued logics. A sequent now is a structure $\Delta_1 \mid \ldots \mid \Delta_n$ with the reading "some formula in Δ_1 has the value 1 or ... or some formula in Δ_n has the value n". A sequent calculus for an n-valued logic then has rules for introducing operations into each of the n positions of a sequent $\Delta_1 \mid \ldots \mid \Delta_n$. It has been known for a long time that there exists a systematic way to define such many-sided sequent systems for any finitely-valued logic defined by a logical matrix together with a set of designated values, see, for example, [16, 17, 125, 218, 290] and references given there. Truth entailment and falsity entailment in *SIXTEEN*$_3$ introduced by Definition 4.1 are, however, not defined in this way. Nevertheless, a kind of many-sided sequents can be used to define sequent calculi for the trilattice logics L^t and L^{t*}. It is not necessary, however, to introduce 16-place sequents $\Delta_1 \mid \ldots \mid \Delta_{16}$. As we have seen, every value from **16** ($=\mathscr{P}(\mathbf{4})$) is determined by a set of *four* characteristic functions. A (generalized) sequent may therefore be defined as a structure $s =$

$$\Delta_1 \circ \Gamma_1 \mid \ldots \mid \Delta_4 \circ \Gamma_4,$$

where Δ_i and $\Gamma_i (i \in \{1,2,3,4\})$ are finite sets of formulas. The intuitive reading of the sequent s is: "(The semantic value of some formula in Δ_1 does not contain **N** or the semantic value of some formula in Γ_1 contains **N**) and (the semantic value of some formula in Δ_2 does not contain **F** or the semantic value of some formula in Γ_2 contains **F**) and (the semantic value of some formula in Δ_3 does not contain **T** or the semantic value of some formula in Γ_3 contains **T**) and (the semantic value of some formula in Δ_4 does not contain **B** or the semantic value of some formula in Γ_4 contains **B**.)"

In the notation for sequents we usually just write A if a set Δ_i or Γ_i is a singleton set $\{A\}$. We write A, Δ_i or Δ_i, A instead of $\{A\} \cup \Delta_i$ (and similarly for Γ_i), and we just leave an empty space if a set Δ_i or Γ_i is empty. Axiomatic sequents are sequents $\Delta_1 \circ \Gamma_1 \mid \Delta_2 \circ \Gamma_2 \mid \Delta_3 \circ \Gamma_3 \mid \Delta_4 \circ \Gamma_4$ where at least one $\Delta_i \circ \Gamma_i$ $(1 \le i \le 4)$ has the shape $\Delta, p \circ \Gamma, p$ for some atomic formula p. A formula A in Δ_i (Γ_i) is said to occur in position Δ_i (Γ_i), and the sequent $\circ \mid \circ \mid \circ \mid \circ$ is called the empty sequent.

The idea is to define a calculus such that the (generalized) sequents

$$A\circ \mid \circ \mid \circ \mid \circ, \circ \mid A\circ \mid \circ \mid \circ, \circ \mid \circ \mid \circ A \mid \circ, \circ \mid \circ \mid \circ \mid \circ A$$

are provable just in case the formula A receives the greatest element of the truth order under every valuation, i.e., iff it is interpreted as $\mathbf{TB} = \{\mathbf{T}, \mathbf{B}\}$ under every valuation. An expression $\Delta_i : A, s$ or $\Gamma_i : A, s$ for each $i \in \{1,2,3,4\}$ is used to denote the place-wise union of s and $\{A\}$ at position Δ_i or Γ_i, respectively; for example, if $s = \Delta_1 \circ \Gamma_1 \mid \Delta_2 \circ \Gamma_2 \mid \Delta_3 \circ \Gamma_3 \mid \Delta_4 \circ \Gamma_4$, then $\Delta_3 : A, s$ means $\Delta_1 \circ \Gamma_1 \mid \Delta_2 \circ \Gamma_2 \mid A, \Delta_3 \circ \Gamma_3 \mid \Delta_4 \circ \Gamma_4$. By s, s' we shall denote the place-wise union of two sequents s and s'. Thus, if $s = \Delta_1 \circ \Gamma_1 \mid \ldots \mid \Delta_4 \circ \Gamma_4$ and $s' = \Delta'_1 \circ \Gamma'_1 \mid \ldots \mid \Delta'_4 \circ \Gamma'_4$, then $s, s' = \Delta_1 \cup \Delta'_1 \circ \Gamma_1 \cup \Gamma'_1 \mid \ldots \mid \Delta_4 \cup \Delta'_4 \circ \Gamma_4 \cup \Gamma'_4$. The introduction rules take their pattern from the introduction rules of the **G3c** sequent system for classical logic, see [252].

Definition 6.9 The system of generalized sequents GL* is defined as the set consisting of all axiomatic sequents together with the cut and introduction rules listed in Figs. 6.1, 6.2, 6.3, 6.4, and 6.5. Provability of sequents is defined in the usual way. We say that a formula A is provable in GL* iff the four sequents $A\circ \mid \circ \mid \circ \mid \circ, \circ \mid A\circ \mid \circ \mid \circ, \circ \mid \circ \mid \circ A \mid \circ$, and $\circ \mid \circ \mid \circ \mid \circ A$ are provable in GL*

The rules for the defined connectives $\neg, \leftrightarrow_t$, and $\leftrightarrow_f$ are easily obtainable. Nevertheless it might be useful to list them because these connectives occur in Odintsov's axioms. For $\neg$ we have the rules

$$\frac{\Gamma_i : A, s}{\Delta_i : \neg A, s}(\neg\Delta_i), \quad \frac{\Delta_i : A, s}{\Gamma_i : \neg A, s}(\neg\Gamma_i).$$

For the biconditionals $\leftrightarrow_t$ and $\leftrightarrow_f$ we have the rules listed in Fig. 6.6.

$$\frac{\Delta_1 : A, s \quad \Gamma_1 : A, s'}{s, s'} \qquad \frac{\Delta_2 : A, s \quad \Gamma_2 : A, s'}{s, s'}$$

$$\frac{\Delta_3 : A, s \quad \Gamma_3 : A, s'}{s, s'} \qquad \frac{\Delta_4 : A, s \quad \Gamma_4 : A, s'}{s, s'}$$

Fig. 6.1 Cut rules $(C_1), \ldots, (C_4)$.

$$\frac{\Delta_3 : A, s}{\Delta_1 : \sim_t A, s} \quad \frac{\Gamma_3 : A, s}{\Gamma_1 : \sim_t A, s} \quad \frac{\Delta_4 : A, s}{\Delta_2 : \sim_t A, s} \quad \frac{\Gamma_4 : A, s}{\Gamma_2 : \sim_t A, s}$$

$$\frac{\Delta_1 : A, s}{\Delta_3 : \sim_t A, s} \quad \frac{\Gamma_1 : A, s}{\Gamma_3 : \sim_t A, s} \quad \frac{\Delta_2 : A, s}{\Delta_4 : \sim_t A, s} \quad \frac{\Gamma_2 : A, s}{\Gamma_4 : \sim_t A, s}$$

$$\frac{\Delta_2 : A, s}{\Delta_1 : \sim_f A, s} \quad \frac{\Gamma_2 : A, s}{\Gamma_1 : \sim_f A, s} \quad \frac{\Delta_1 : A, s}{\Delta_2 : \sim_f A, s} \quad \frac{\Gamma_1 : A, s}{\Gamma_2 : \sim_f A, s}$$

$$\frac{\Delta_4 : A, s}{\Delta_3 : \sim_f A, s} \quad \frac{\Gamma_4 : A, s}{\Gamma_3 : \sim_f A, s} \quad \frac{\Delta_3 : A, s}{\Delta_4 : \sim_f A, s} \quad \frac{\Gamma_3 : A, s}{\Gamma_4 : \sim_f A, s}$$

Fig. 6.2 Introduction rules $(\sim_t\Delta_1), \ldots, (\sim_t\Gamma_4)$ and $(\sim_f\Delta_1), \ldots, (\sim_f\Gamma_4)$.

$$\frac{\Delta_1 : A, s \quad \Delta_1 : B, s}{\Delta_1 : A \wedge_t B, s} \quad \frac{\Gamma_1 : A, \Gamma_1 : B, s}{\Gamma_1 : A \wedge_t B, s} \quad \frac{\Delta_2 : A, s \quad \Delta_2 : B, s}{\Delta_2 : A \wedge_t B, s} \quad \frac{\Gamma_2 : A, \Gamma_2 : B, s}{\Gamma_2 : A \wedge_t B, s}$$

$$\frac{\Delta_3 : A, \Delta_3 : B, s}{\Delta_3 : A \wedge_t B, s} \quad \frac{\Gamma_3 : A, s \quad \Gamma_3 : B, s}{\Gamma_3 : A \wedge_t B, s} \quad \frac{\Delta_4 : A, \Delta_4 : B, s}{\Delta_4 : A \wedge_t B, s} \quad \frac{\Gamma_4 : A, s \quad \Gamma_4 : B, s}{\Gamma_4 : A \wedge_t B, s}$$

$$\frac{\Delta_1 : A, \Delta_1 : B, s}{\Delta_1 : A \vee_t B, s} \quad \frac{\Gamma_1 : A, s \quad \Gamma_1 : B, s}{\Gamma_1 : A \vee_t B, s} \quad \frac{\Delta_2 : A, \Delta_2 : B, s}{\Delta_2 : A \vee_t B, s} \quad \frac{\Gamma_2 : A, s \quad \Gamma_2 : B, s}{\Gamma_2 : A \vee_t B, s}$$

$$\frac{\Delta_3 : A, s \quad \Delta_3 : B, s}{\Delta_3 : A \vee_t B, s} \quad \frac{\Gamma_3 : A, \Gamma_3 : B, s}{\Gamma_3 : A \vee_t B, s} \quad \frac{\Delta_4 : A, s \quad \Delta_4 : B, s}{\Delta_4 : A \vee_t B, s} \quad \frac{\Gamma_4 : A, \Gamma_4 : B, s}{\Gamma_4 : A \vee_t B, s}$$

Fig. 6.3 Introduction rules $(\wedge_t\Delta_1), \ldots, (\wedge_t\Gamma_4)$ and $(\vee_t\Delta_1), \ldots, (\vee_t\Gamma_4)$.

$$\frac{\Delta_1 : A, \Delta_1 : B, s}{\Delta_1 : A \wedge_f B, s} \quad \frac{\Gamma_1 : A, s \quad \Gamma_1 : B, s}{\Gamma_1 : A \wedge_f B, s} \quad \frac{\Delta_2 : A, s \quad \Delta_2 : B, s}{\Delta_2 : A \wedge_f B, s} \quad \frac{\Gamma_2 : A, \Gamma_2 : B, s}{\Gamma_2 : A \wedge_f B, s}$$

$$\frac{\Delta_3 : A, \Delta_3 : B, s}{\Delta_3 : A \wedge_f B, s} \quad \frac{\Gamma_3 : A, s \quad \Gamma_3 : B, s}{\Gamma_3 : A \wedge_f B, s} \quad \frac{\Delta_4 : A, s \quad \Delta_4 : B, s}{\Delta_4 : A \wedge_f B, s} \quad \frac{\Gamma_4 : A, \Gamma_4 : B, s}{\Gamma_4 : A \wedge_f B, s}$$

$$\frac{\Delta_1 : A, s \quad \Delta_1 : B, s}{\Delta_1 : A \vee_f B, s} \quad \frac{\Gamma_1 : A, \Gamma_1 : B, s}{\Gamma_1 : A \vee_f B, s} \quad \frac{\Delta_2 : A, \Delta_2 : B, s}{\Delta_2 : A \vee_f B, s} \quad \frac{\Gamma_2 : A, s \quad \Gamma_2 : B, s}{\Gamma_2 : A \vee_f B, s}$$

$$\frac{\Delta_3 : A, s \quad \Delta_3 : B, s}{\Delta_3 : A \vee_f B, s} \quad \frac{\Gamma_3 : A, \Gamma_3 : B, s}{\Gamma_3 : A \vee_f B, s} \quad \frac{\Delta_4 : A, \Delta_4 : B, s}{\Delta_4 : A \vee_f B, s} \quad \frac{\Gamma_4 : A, s \quad \Gamma_4 : B, s}{\Gamma_4 : A \vee_f B, s}$$

Fig. 6.4 Introduction rules $(\wedge_f \Delta_1), \ldots, (\wedge_f \Gamma_4)$ and $(\vee_f \Delta_1), \ldots, (\vee_f \Gamma_4)$.

$$\frac{\Delta_1 : B, \Gamma_1 : A, s}{\Delta_1 : A \rightarrow_t B, s} \quad \frac{\Delta_1 : A, s \quad \Gamma_1 : B, s}{\Gamma_1 : A \rightarrow_t B, s} \quad \frac{\Delta_2 : B, \Gamma_2 : A, s}{\Delta_2 : A \rightarrow_t B, s} \quad \frac{\Delta_2 : A, s \quad \Gamma_2 : B, s}{\Gamma_2 : A \rightarrow_t B, s}$$

$$\frac{\Delta_3 : B, s \quad \Gamma_3 : A, s}{\Delta_3 : A \rightarrow_t B, s} \quad \frac{\Delta_3 : A, \Gamma_3 : B, s}{\Gamma_3 : A \rightarrow_t B, s} \quad \frac{\Delta_4 : B, s \quad \Gamma_4 : A, s}{\Delta_4 : A \rightarrow_t B, s} \quad \frac{\Delta_4 : A, \Gamma_4 : B, s}{\Gamma_4 : A \rightarrow_t B, s}$$

$$\frac{\Gamma_1 : A, s \quad \Delta_1 : B, s}{\Delta_1 : A \rightarrow_f B, s} \quad \frac{\Delta_1 : A, \Gamma_1 : B, s}{\Gamma_1 : A \rightarrow_f B, s} \quad \frac{\Delta_2 : B, \Gamma_2 : A, s}{\Delta_2 : A \rightarrow_f B, s} \quad \frac{\Delta_2 : A, s \quad \Gamma_2 : B, s}{\Gamma_2 : A \rightarrow_f B, s}$$

$$\frac{\Gamma_3 : A, s \quad \Delta_3 : B, s}{\Delta_3 : A \rightarrow_f B, s} \quad \frac{\Delta_3 : A, \Gamma_3 : B, s}{\Gamma_3 : A \rightarrow_f B, s} \quad \frac{\Delta_4 : B, \Gamma_4 : A, s}{\Delta_4 : A \rightarrow_f B, s} \quad \frac{\Delta_4 : A, s \quad \Gamma_4 : B, s}{\Gamma_4 : A \rightarrow_f B, s}$$

Fig. 6.5 Introduction rules $(\rightarrow_t \Delta_1), \ldots, (\rightarrow_t \Gamma_4)$ and $(\rightarrow_f \Delta_1), \ldots, (\rightarrow_f \Gamma_4)$.

Observation 1 *For every formula* $A \in \mathscr{L}^*_{tf}$, *every sequent* $\Delta_1 \circ \Gamma_2 \mid \Delta_2 \circ \Gamma_2 \mid \Delta_3 \circ \Gamma_3 \mid \Delta_4 \circ \Gamma_4$ *in which at least one* $\Delta_i \circ \Gamma_i (1 \leq i \leq 4)$ *has the shape* $\Delta, A \circ \Gamma, A$ *is provable in* GL*.

Proof By induction on A. By way of example, we here consider just one case:

$$\cfrac{\Delta_1, A \circ \Gamma_1, B, A \mid \Delta_2 \circ \Gamma_2 \mid \Delta_3 \circ \Gamma_3 \mid \Delta_4 \circ \Gamma_4 \quad \Delta_1, A, B \circ \Gamma_1, B \mid \Delta_2 \circ \Gamma_2 \mid \Delta_3 \circ \Gamma_3 \mid \Delta_4 \circ \Gamma_4}{\cfrac{\Delta_1, A, A \rightarrow_f B \circ \Gamma_1, B \mid \Delta_2 \circ \Gamma_2 \mid \Delta_3 \circ \Gamma_3 \mid \Delta_4 \circ \Gamma_4}{\Delta_1, A \rightarrow_f B \circ \Gamma_1, A \rightarrow_f B \mid \Delta_2 \circ \Gamma_2 \mid \Delta_3 \circ \Gamma_3 \mid \Delta_4 \circ \Gamma_4}}$$

□

Thus, bottom-up proof search may stop at sequents $\Delta_1 \circ \Gamma_2 \mid \Delta_2 \circ \Gamma_2 \mid \Delta_3 \circ \Gamma_3 \mid \Delta_4 \circ \Gamma_4$, where at least one $\Delta_i \circ \Gamma_i (1 \leq i \leq 4)$ has the shape $\Delta, A \circ \Gamma, A$.

In Sect. 6.3 we observed that cut is an *admissible* rule of the sequent calculi G_B and GL$_B$. We assume the familiar notions of a derivation and a proof(tree) in sequent calculi, depth (alias height) of a tree, admissibility, and depth-preserving admissibility and invertibility of inference rules, see [252, p. 9, p. 76 f.]. The depth

$$\frac{\Delta_1 : B, \Gamma_1 : A, s \quad \Delta_1 : A, \Gamma_1 : B, s}{\Delta_1 : A \leftrightarrow_t B, s} \qquad \frac{\Delta_2 : B, \Gamma_2 : A, s \quad \Delta_2 : A, \Gamma_2 : B, s}{\Delta_2 : A \leftrightarrow_t B, s}$$

$$\frac{\Delta_1 : A, \Delta_1 : B, s \quad \Delta_1 : A, \Gamma_1 : A, s \quad \Gamma_1 : B, \Delta_1 : B, s \quad \Gamma_1 : B, \Gamma_1 : A, s}{\Gamma_1 : A \leftrightarrow_t B, s}$$

$$\frac{\Delta_2 : A, \Delta_2 : B, s \quad \Delta_2 : A, \Gamma_2 : A, s \quad \Gamma_2 : B, \Delta_2 : B, s \quad \Gamma_2 : B, \Gamma_2 : A, s}{\Gamma_2 : A \leftrightarrow_t B, s}$$

$$\frac{\Delta_3 : B, \Gamma_3 : B, s \quad \Delta_3 : B, \Delta_3 : A, s \quad \Gamma_3 : A, \Gamma_3 : B, s \quad \Gamma_3 : A, \Delta_3 : A, s}{\Delta_3 : A \leftrightarrow_t B, s}$$

$$\frac{\Delta_4 : B, \Delta_4 : A, s \quad \Delta_4 : B, \Gamma_4 : A, s \quad \Gamma_4 : A, \Delta_4 : B, s \quad \Gamma_4 : A, \Gamma_4 : Bs}{\Delta_4 : A \leftrightarrow_t B, s}$$

$$\frac{\Delta_3 : A, \Gamma_3 : B, s \quad \Delta_3 : B, \Gamma_3 : A, s}{\Gamma_3 : A \leftrightarrow_t B, s} \qquad \frac{\Delta_4 : A, \Gamma_4 : B, s \quad \Delta_4 : B, \Gamma_4 : A, s}{\Gamma_4 : A \leftrightarrow_t B, s}$$

$$\frac{\Gamma_1 : A, s \quad \Gamma_1 : B, s \quad \Gamma_1 : B, s \quad \Delta_1 : A, s}{\Delta_1 : A \leftrightarrow_f B, s} \qquad \frac{\Delta_1 : A, \Gamma_1 : B, \Delta_1 : B, \Gamma_1 : A, s}{\Gamma_1 : A \leftrightarrow_f B, s}$$

$$\frac{\Delta_2 : B, \Gamma_2 : A, \Delta_2 : A, \Gamma_2 : B, s}{\Delta_2 : A \leftrightarrow_f B, s} \qquad \frac{\Delta_2 : A, s \quad \Gamma_2 : B, s \quad \Delta_2 : B, s \quad \Gamma_2 : A, s}{\Gamma_2 : A \leftrightarrow_f B, s}$$

$$\frac{\Gamma_3 : A, s \quad \Delta_3 : B, s \quad \Gamma_3 : B, s \quad \Delta_3 : A, s}{\Delta_3 : A \leftrightarrow_f B, s} \qquad \frac{\Delta_3 : A, \Gamma_3 : B, \Delta_3 : B, \Gamma_3 : A, s}{\Gamma_3 : A \leftrightarrow_f B, s}$$

$$\frac{\Delta_4 : B, \Gamma_4 : A, \Delta_4 : A, \Gamma_4 : B, s}{\Delta_4 : A \leftrightarrow_f B, s} \qquad \frac{\Delta_4 : A, s \quad \Gamma_4 : B, s \quad \Delta_4 : B, s \quad \Gamma_4 : A, s}{\Gamma_4 : A \leftrightarrow_f B, s}$$

Fig. 6.6 Introduction rules $(\leftrightarrow_t \Delta_1), \ldots, (\leftrightarrow_t \Gamma_4)$ and $(\leftrightarrow_f \Delta_1), \ldots, (\leftrightarrow_f \Gamma_4)$.

of a derivation is thus the maximum length of the branches in the derivation, and the length of a branch is the number of its nodes minus 1. A rule is admissible in a sequent calculus L iff (for every instance of the rule) the provability of all the premise sequents in L guarantees the provability of the conclusion sequent in L. The rule is depth-preserving admissible in L iff it holds that the conclusion sequent is provable with a prooftree of depth at most n in L if all the premise sequents are provable with a prooftree of depth at most n in L. An introduction rule is depth-preserving invertible in a sequent calculus L iff it holds that each of the premise sequents of the rule has a proof-tree of length at most n if the conclusion sequent has a proof-tree of length at most n in L. Quite obviously, weakening (monotonicity) is depth-preserving admissible in GL*.

Observation 2 *The weakening rule*

$$\frac{\Delta_1 \circ \Gamma_1 \mid \Delta_2 \circ \Gamma_2 \mid \Delta_3 \circ \Gamma_3 \mid \Delta_4 \circ \Gamma_4}{\Delta_1, \Delta_1' \circ \Gamma_1, \Gamma_1' \mid \Delta_2, \Delta_2' \circ \Gamma_2, \Gamma_2' \mid \Delta_3, \Delta_3' \circ \Gamma_3, \Gamma_3' \mid \Delta_4, \Delta_4' \circ \Gamma_4, \Gamma_4'}$$

is depth-preserving admissible in GL*.

Observation 3 *All introduction rules of* GL* *are depth-preserving invertible in* GL*.

Proof By induction on the depth of prooftrees in GL*. □

If we used finite multisets of formulas instead of finite sets as components of sequents, we could use Observation 3 to prove the depth-preserving admissibility of the following contraction rules for $1 \leq i \leq 4$:

$$\frac{\Delta_i : A, \Delta_i : A, s}{\Delta_i : A, s} \qquad \frac{\Gamma_i : A, \Gamma_i : A, s}{\Gamma_i : A, s}$$

Observation 4 *The cut rules* (C_1)–(C_8) *are admissible in cut-free* GL*.

Proof First we can note that the cut rules (C_1)–(C_4) can be replaced by their context-sharing versions $(C_1)'$–$(C_4)'$:

$$\frac{\Delta_1 : A, s \quad \Gamma_1 : A, s}{s} \quad \frac{\Delta_2 : A, s \quad \Gamma_2 : A, s}{s} \quad \frac{\Delta_3 : A, s \quad \Gamma_3 : A, s}{s} \quad \frac{\Delta_4 : A, s \quad \Gamma_4 : A, s}{s}$$

Thus, it is enough to show that $(C_1)'$–$(C_4)'$ are admissible in cut-free GL*. The level of (an application of) $(C_i)'(i \in \{1, 2, 3, 4\})$ is the sum of the depths of the derivations of the premise sequents. The rank of (an application of) $(C_i)'$ with cut-formula A is the maximum length of a branch in A's construction tree plus one. The cut-rank of a derivation π in GL* is the maximum of the ranks of the cut-formulas in π. The only difference to the standard proof of cut-elimination for **G3c** (as presented in [252]) is that the main induction on the cut-rank and the subinduction on the cut-level now is *simultaneous* for the four rules $(C_1)'$–$(C_4)'$. We consider just one principal case of the cut-elimination procedure, thus revealing the need for simultaneous induction. The derivation

$$\dfrac{\dfrac{\dfrac{\pi}{\Delta_1 : A, s}}{\Delta_3 : \sim_t A, s} \quad \dfrac{\dfrac{\pi'}{\Gamma'_1 : A, s'}}{\Gamma'_3 : \sim_t A, s'}}{s, s'} \text{ is replaced by } \dfrac{\dfrac{\pi}{\Delta_1 : A, s} \quad \dfrac{\pi'}{\Gamma'_1 : A, s'}}{s, s'}$$

□

Observation 5 *If* $\vdash \Delta_1 \circ \Gamma_1 \mid \ldots \mid \Delta_4 \circ \Gamma_4$ *in* GL*, *then for every valuation* v: *there exists* $A \in \Delta_1$ *with* $\mathbf{N} \notin v(A)$ *or there exits* $A \in \Gamma_1$ *with* $\mathbf{N} \in v(A)$ *or there exists* $A \in \Delta_2$ *with* $\mathbf{F} \notin v(A)$ *or there exits* $A \in \Gamma_2$ *with* $\mathbf{F} \in v(A)$ *or there exists* $A \in \Delta_3$ *with* $\mathbf{T} \notin v(A)$ *or there exits* $A \in \Gamma_3$ *with* $\mathbf{T} \in v(A)$ *or there exists* $A \in \Delta_4$ *with* $\mathbf{B} \notin v(A)$ *or there exits* $A \in \Gamma_4$ *with* $\mathbf{B} \in v(A)$.

Proof By induction on proofs in GL*. For axioms, the claim is obvious since for every propositional variable p and every $x \in \{\mathbf{N}, \mathbf{F}, \mathbf{T}, \mathbf{B}\}, x \in v(p)$ or $x \notin v(p)$. With regard to the introduction rules for $\sim_t$ and $\sim_f$, the result follows by Proposition 3.4 and by Odintsov's matrix characterization of $-_t$ and $-_f$. For rule

$(\sim_t\Delta_1)$, for example, note that $\mathbf{N} \in v(\sim_t A)$ iff $\mathbf{T} \in v(A)$. With regard to the rules for truth/falsity conjunction and truth/falsity disjunction, the result follows from Proposition 3.4 in [232] and from Odintsov's matrix characterization of $\sqcap_t, \sqcup_t, \sqcap_f$, and $\sqcup_f$. For rule $(\vee_f\Gamma_4)$, for example, note that $\mathbf{B} \in v(A \vee_f B)$ iff ($\mathbf{B} \in v(A)$ and $\mathbf{B} \in v(B)$). With regard to the rules for the two implications, we may use Odintsov's matrix characterizations of $\sqsupset_t$ and $\sqsupset_f$. □

Proposition 6.12 *If A is provable in* GL*, *then* $A \in L^{t*}$.

Proof By contraposition. If it is not the case that $A \in L^{t*}$, then there is a valuation v such that $v(A) \neq \begin{vmatrix} 0 & 0 \\ 1 & 1 \end{vmatrix} = \mathbf{TB}$. If each of the four sequents $A\circ \mid \circ \mid \circ \mid \circ$, $\circ \mid A\circ \mid \circ \mid \circ, \circ \mid \circ \mid \circ A \mid \circ$, and $\circ \mid \circ \mid \circ \mid \circ A$ is provable in GL*, then for every valuation v, $v(A) = \mathbf{TB}$, by Observation 4. Therefore not all of these four sequents are provable in GL*. □

Proposition 6.13 *If* $A \in L^{t*}$, *then A is provable in* GL*.

Proof By induction on proofs in HL^{t*}, i.e., it has to be shown that (i) *modus ponens* preserves provability in GL* and (ii) the 38 axioms of HL^{t*} are provable in GL*. As to (i), assume that the following sequents are provable in GL*:

$$\begin{array}{ll} s_1 = A\circ \mid \circ \mid \circ \mid \circ, & s_2 = \circ \mid A\circ \mid \circ \mid \circ, \\ s_3 = \circ \mid \circ \mid \circ A \mid \circ, & s_4 = \circ \mid \circ \mid \circ \mid \circ A, \\ s_1' = (A \to_t B)\circ \mid \circ \mid \circ \mid \circ, & s_2' = \circ \mid (A \to_t B)\circ \mid \circ \mid \circ, \\ s_3' = \circ \mid \circ \mid \circ(A \to_t B) \mid \circ, & s_4' = \circ \mid \circ \mid \circ \mid \circ(A \to_t B). \end{array}$$

Then the formula B can easily be proved in GL* using the four cut rules. Here is one example:

$$\dfrac{s_1' \qquad \dfrac{s_1 \qquad \dfrac{A, B\circ A, B \mid \circ \mid \circ \mid \circ}{B\circ A, (A \to_t B) \mid \circ \mid \circ \mid \circ}\,(\to_t \Gamma_1)}{B\circ(A \to_t B) \mid \circ \mid \circ \mid \circ}\,(C_1)}{B\circ \mid \circ \mid \circ \mid \circ}\,(C_1)$$

Proving the axioms of HL^{t*} in GL* is a tedious task, and we shall not present all the cases. Consider, for instance, Axiom 17. We have to derive Axiom 17 in the positions $\Delta_1, \Delta_2, \Gamma_3$ and Γ_4 of otherwise empty sequents. Consider here position Γ_4:

$$\dfrac{\dfrac{\dfrac{\dfrac{\dfrac{\circ \mid A\circ A \mid \circ \mid \circ}{\circ \sim_f A \mid A\circ \mid \circ \mid \circ}}{\circ \mid A\circ \mid \circ \sim_t \sim_f A \mid \circ}}{\circ \mid A\circ \mid \circ \mid \circ \sim_f \sim_t \sim_f A}}{\circ \mid \circ \mid \circ \mid \sim_t A\circ \sim_f \sim_t \sim_f A} \qquad \dfrac{\dfrac{\dfrac{\dfrac{\circ \mid A\circ A \mid \circ \mid \circ}{\sim_f A\circ \mid \circ A \mid \circ \mid \circ}}{\circ \mid \circ A \mid \sim_t \sim_f A\circ \mid \circ}}{\circ \mid \circ A \mid \circ \mid \sim_f \sim_t \sim_f A\circ}}{\circ \mid \circ \mid \circ \sim_f \sim_t \sim_f A\circ \sim_t A}}{\circ \mid \circ \mid \circ \mid \circ \sim_t A \leftrightarrow_t \sim_f \sim_t \sim_f A}$$

The most remarkable axiom is Axiom 38, which we will call Odintsov's axiom, *Od*. To show the provability of *Od* in GL*, we will introduce some notation:

$$OdT = ((A \wedge_t B) \leftrightarrow_t (A \wedge_f B)) \wedge_t ((A \vee_t B) \leftrightarrow_t (A \vee_f B)) \wedge_t ((A \rightarrow_f B) \leftrightarrow_t (A \rightarrow_t B))$$
$$OdF = ((C \wedge_t D) \leftrightarrow_t (C \vee_f D)) \wedge_t ((C \vee_t D) \leftrightarrow_t (C \wedge_f D)) \wedge_t ((C \rightarrow_f D) \leftrightarrow_t (\neg C \wedge_t D))$$
$$OdT_1 = ((A \wedge_t B) \leftrightarrow_t (A \wedge_f B))$$
$$OdT_2 = ((A \vee_t B) \leftrightarrow_t (A \vee_f B))$$
$$OdT_3 = ((A \rightarrow_f B) \leftrightarrow_t (A \rightarrow_t B))$$
$$OdF_1 = ((C \wedge_t D) \leftrightarrow_t (C \vee_f D))$$
$$OdF_2 = ((C \vee_t D) \leftrightarrow_t (C \wedge_f D))$$
$$OdF_3 = ((C \rightarrow_f D) \leftrightarrow_t (\neg C \wedge_t D))$$

Thus, $Od = OdT \vee_t OdF$, and OdT_i (OdF_i) with $1 \leq i \leq 3$ are the three $\wedge_t$-conjuncts of $OdT(OdF)$. We here consider *Od* in position Γ_3, split the proof into several parts, and annotate some sequents by names for these sequents.

$$\dfrac{\dfrac{s_1 = \circ \mid \circ \mid \circ OdT_1, OdF \mid \circ \qquad \dfrac{s_2 = \circ \mid \circ \mid \circ OdT_2, OdF \mid \circ \quad s_3 = \circ \mid \circ \mid \circ OdT_3, OdF \mid \circ}{\circ \mid \circ \mid \circ OdT_2 \wedge_t OdT_3, OdF \mid \circ}}{\circ \mid \circ \mid \circ OdT, OdF \mid \circ}}{\circ \mid \circ \mid \circ Od \mid \circ}$$

Simple proofs of s_1–s_3 are available.

$$\dfrac{\dfrac{\dfrac{\circ \mid \circ \mid A,B \circ A, OdF \mid \circ \quad \circ \mid \circ \mid A,B \circ B, OdF \mid \circ}{\circ \mid \circ \mid A,B \circ A \wedge_f B, OdF \mid \circ}}{\circ \mid \circ \mid A \wedge_t B \circ A \wedge_f B, OdF \mid \circ} \qquad \dfrac{\dfrac{\circ \mid \circ \mid A,B \circ A, OdF \mid \circ \quad \circ \mid \circ \mid A,B \circ B, OdF \mid \circ}{\circ \mid \circ \mid A,B \circ A \wedge_t B, OdF \mid \circ}}{\circ \mid \circ \mid A \wedge_f B \circ A \wedge_t B, OdF \mid \circ}}{\circ \mid \circ \mid \circ OdT_1, OdF \mid \circ}$$

$$\dfrac{\dfrac{\dfrac{\circ \mid \circ \mid A \circ A,B, OdF \mid \circ \quad \circ \mid \circ \mid B \circ A,B, OdF \mid \circ}{\circ \mid \circ \mid A \vee_t B \circ A,B, OdF \mid \circ}}{\circ \mid \circ \mid A \vee_t B \circ A \vee_f B, OdF \mid \circ} \qquad \dfrac{\dfrac{\circ \mid \circ \mid A \circ A,B, OdF \mid \circ \quad \circ \mid \circ \mid B \circ A,B, OdF \mid \circ}{\circ \mid \circ \mid A \vee_f B \circ A,B, OdF \mid \circ}}{\circ \mid \circ \mid A \vee_f B \circ A \vee_t B, OdF \mid \circ}}{\circ \mid \circ \mid \circ OdT_2, OdF \mid \circ}$$

$$\dfrac{\dfrac{\dfrac{\circ \mid \circ \mid B,A \circ B, OdF \mid \circ \quad \circ \mid \circ \mid A \circ A,B, OdF \mid \circ}{\circ \mid \circ \mid A \rightarrow_t B, A \circ B, OdF \mid \circ}}{\circ \mid \circ \mid A \rightarrow_t B \circ A \rightarrow_f B, OdF \mid \circ} \qquad \dfrac{\dfrac{\circ \mid \circ \mid A \circ A,B, OdF \mid \circ \quad \circ \mid \circ \mid A,B \circ B, OdF \mid \circ}{\circ \mid \circ \mid A, A \rightarrow_f B \circ B, OdF \mid \circ}}{\circ \mid \circ \mid A \rightarrow_f B \circ A \rightarrow_t B, OdF \mid \circ}}{\circ \mid \circ \mid \circ OdT_3, OdF \mid \circ}$$

Note that this proof does not make use of *OdF* at all. In Δ_1 position, we make use of *OdF* instead of *OdT*.

$$\dfrac{s_1' = OdT, OdF_1 \circ \mid \circ \mid \circ \mid \circ \quad \dfrac{s_2' = OdT, OdF_2 \circ \mid \circ \mid \circ \mid \circ \quad s_3' = OdT, OdF_3 \circ \mid \circ \mid \circ \mid \circ}{OdT, OdF_2 \wedge_t OdF_3 \circ \mid \circ \mid \circ \mid \circ}}{\dfrac{OdT, OdF \circ \mid \circ \mid \circ \mid \circ}{Od \circ \mid \circ \mid \circ \mid \circ}}$$

Simple proofs of s_1'–s_3' are available.

$$\dfrac{\dfrac{\dfrac{OdT, C \circ C, D \mid \circ \mid \circ \mid \circ \quad OdT, D \circ C, D \mid \circ \mid \circ \mid \circ}{OdT, C \wedge_t D \circ C, D \mid \circ \mid \circ \mid \circ}}{OdT, C \wedge_t D \circ C \vee_f D \mid \circ \mid \circ \mid \circ} \quad \dfrac{\dfrac{\circ \mid OdT, C \circ C, D \mid \circ \mid \circ \quad OdT, D \circ C, D \mid \circ \mid \circ \mid \circ}{OdT, C \vee_f D \circ C, D \mid \circ \mid \circ \mid \circ}}{OdT, C \vee_f D \circ C \wedge_t D \mid \circ \mid \circ \mid \circ}}{OdT, OdF_1 \circ \mid \circ \mid \circ \mid \circ}$$

$$\dfrac{\dfrac{\dfrac{OdT, C, D \circ C \mid \circ \mid \circ \mid \circ \quad OdT, C, D \circ D \mid \circ \mid \circ \mid \circ}{OdT, C, D \circ C \wedge_f D \mid \circ \mid \circ \mid \circ}}{OdT, C \vee_t D \circ C \wedge_f \mid \circ \mid \circ \mid \circ} \quad \dfrac{\dfrac{\circ \mid OdT, C, D \circ C \mid \circ \mid \circ \quad OdT, C, D \circ D \mid \circ \mid \circ \mid \circ}{OdT, C, D \circ C \vee_t D \mid \circ \mid \circ \mid \circ}}{OdT, C \wedge_f D \circ C \vee_t D \mid \circ \mid \circ \mid \circ}}{OdT, OdF_2 \circ \mid \circ \mid \circ \mid \circ}$$

We divide the proof of s_3' into two parts.

$$\dfrac{s_{31}' = OdT, C \rightarrow_f D \circ \neg C \wedge_t D \mid \circ \mid \circ \mid \circ \quad s_{32}' = OdT, \neg C \wedge_t D \circ C \rightarrow_f D \mid \circ \mid \circ \mid \circ}{OdT, OdF_3 \circ \mid \circ \mid \circ \mid \circ}$$

$$\dfrac{\dfrac{OdT, D \circ \neg C, D \mid \circ \mid \circ \mid \circ}{OdT, D \circ \neg C \wedge_t D \mid \circ \mid \circ \mid \circ} \quad \dfrac{\dfrac{OdT, C \circ C, D \mid \circ \mid \circ \mid \circ}{OdT \circ C, \neg C, D \mid \circ \mid \circ \mid \circ}}{OdT \circ C, \neg C \wedge_t D \mid \circ \mid \circ \mid \circ}}{s_{31}' = OdT, C \rightarrow_f D \circ \neg C \wedge_t D \mid \circ \mid \circ \mid \circ}$$

$$\dfrac{\dfrac{\dfrac{OdT, C \circ C, D \mid \circ \mid \circ \mid \circ}{OdT, C, \neg C \circ D \mid \circ \mid \circ \mid \circ} \quad OdT, C, D \circ D \mid \circ \mid \circ \mid \circ}{OdT, C, \neg C \wedge_t D \circ D \mid \circ \mid \circ \mid \circ}}{OdT, \neg C \wedge_t D \circ C \rightarrow_f D \mid \circ \mid \circ \mid \circ}$$

Similarly, in position Δ_2 the formula *OdT* is decomposed, and in position Γ_4 the formula *OdF* is decomposed. □

The proof of *Od* reflects the indeterministic nature of *Od* and shows how this is taken into account in GL*. Combining Propositions 6.12 and 6.13, we obtain:

Theorem 6.11 *For every formula* $A \in \mathscr{L}^*_{tf}$, $\vdash A$ *in* HL^{t*} *iff* A *is provable in* GL^*.

This characterization also gives a new decidability proof for HL^{t*}.

Proposition 6.14 *Provability of formulas in* HL^{t*} *is decidable.*

Proof Bottom-up proof search in cut-free GL* is obviously terminating. □

The following proposition follows from Observation 4.

Proposition 6.15 *Let* X *be a non-empty subset of* $\{\sim_t, \wedge_t, \vee_t, \rightarrow_t, \sim_f, \wedge_f, \vee_f, \rightarrow_f\}$, *and let* $\mathrm{GL}^* \setminus X$ *be the result of dropping the introduction rules for the connectives in the complement of* X *from* GL*. *Then a sequent* s *in the language based on* X *is provable in* GL* *iff it is provable in* $\mathrm{GL}^* \setminus X$.

We conclude this chapter by remarking that GL* provides a *framework* for capturing all the other semantically defined trilattice logics. In view of Proposition 6.15, in order to deal with the logics L^t and $L^{f'}$ defined with respect to truth entailment, we may just drop the introduction rules for $\rightarrow_f$ and $\rightarrow_t$, respectively. To deal with the logic $\mathscr{L}^{f*}$ defined with respect to falsity entailment, we postulate that a formula A in the full language $\mathscr{L}^*_{tf}$ is provable iff the four sequents

$$\circ A \mid \circ \mid \circ \mid \circ, \circ \mid A\circ \mid \circ \mid \circ, \circ \mid \circ \mid \circ A \mid \circ, \circ \mid \circ \mid \circ \mid A\circ$$

are provable. Sequent systems for fragments of $\mathscr{L}^{f*}$ are obtained by omitting the introduction rules for the connectives in question.

Chapter 7
Intuitionistic Trilattice Logics

Abstract We will take up a suggestion by Odintsov (Studia Logica 93:407–428, 2009) and define intuitionistic variants of certain logics arising from the trilattice *SIXTEEN*$_3$. In a first step, a logic I_{16} is presented as a Gentzen-type sequent calculus for an intuitionistic version of Odintsov's Hilbert-style axiom system L_T from Chap. 5. The cut-elimination theorem for I_{16} is proved using an embedding of I_{16} into Gentzen's sequent system *LJ* for intuitionistic logic. The completeness theorem with respect to a Kripke-style semantics is also proved for I_{16}. The framework of I_{16} is regarded as plausible and natural for the following reasons: (i) the properties of constructible falsity and paraconsistency with respect to some negation connectives hold for I_{16}, and (ii) sequent calculi for Belnap and Dunn's four-valued logic of first-degree entailment and for Nelson's constructive paraconsistent logic **N4** are included as natural subsystems of I_{16}. In a second step, a logic IT_{16} is introduced as a tableau calculus. The tableau system IT_{16} is an intuitionistic counterpart of Odintsov's axiom system for truth entailment $\models_t$ in *SIXTEEN*$_3$ and of the sequent calculus for $\models_t$ presented in Chap. 6. The tableau calculus is also shown to be sound and complete with respect to a Kripke-style semantics. A tableau calculus for falsity entailment can be obtained by suitably modifying the notion of provability.

7.1 Introduction

In Chap. 6, different kinds of cut-free, sound and complete sequent calculi for the first-degree axiom systems $\vdash_T$ and $\vdash_B$ as well as the axiom systems L_T, and L_B have been defined. Cut-free sound and complete sequent systems for the first-degree axiom systems $\vdash^f_T, \vdash^f_B$ and for the axiom systems $L^f_T, L^f_B, L'_T, L'_B, L^{f'}_T$, and $L^{f'}_B$ listed in Table 5.1 from Chap. 5 can be obtained in analogy to the definition of the sequent systems for $\vdash_T, \vdash_B, L_T$, and L_B. It has been remarked that cut-free sequent calculi for the basic first-degree calculus $\vdash_{\text{base}}$ are not known and that constructing

DOI: 10.1007/978-94-007-0907-2_7, © Springer Science+Business Media B.V. 2011

a cut-free sequent calculus for $\vdash_{\text{base}}$ might turn out to be difficult because the method used to show that cut is an admissible rule of the sequent calculi for $\vdash_T, \vdash_B, L_T$, and L_B does not apply to $\vdash_{\text{base}}$. Moreover, it has been pointed out that the axiomatic systems $L_T, L_B, L_T^f, L_B^f, L_T', L_B', L_T^{f'}$, and $L_B^{f'}$ are all auxiliary in the sense that they do not capture truth entailment $\models_t$ or falsity entailment $\models_f$ in the respective languages. The main axiom systems are those capturing the semantically defined logics $L^t, L^f, L', L^{f'}, L^{t*}$, and L^{f*}. Cut-free sound and complete calculi of generalized sequents for all these axiom systems have been defined in Chap. 6. In the present chapter we will define sequent calculi and tableau systems for *intuitionistic* sequent calculi.[1]

An important step in obtaining the axiom systems for truth and falsity entailment in $SIXTEEN_3$ is Odintsov's idea to represent the lattice operations of the truth and the falsity order as so-called twist-operations over the two-element Boolean algebra. Odintsov makes the following comment:

> It turns out that the matrix of $SIXTEEN_3$ can be represented as a four-component twist-structure over the two-element Boolean algebra, which suggests also a way of defining an intuitionistic version of the consequence relations $\models_t$ and $\models_f$ via replacing in this twist-structure the two-element Boolean algebra by an arbitrary Heyting algebra.

This remark provides the motivation for the present chapter. The implication $\rightarrow_t$ is Boolean and hence it satisfies the Deduction Theorem. Intuitionistic implicational logic is known to be "the minimal pure implicational calculus having *modus ponens* as its sole rule and satisfying the Deduction Theorem" [77, p. 133]. It is therefore of theoretical interest to investigate trilattice logics with implications weaker than Boolean implication. We will take up Odintsov's suggestion first from a purely proof-theoretical point of view. The idea is to impose on existing sequent calculi for certain trilattice logics the well-known restriction leading from Gentzen's sequent calculus *LK* for classical logic to the sequent system *LJ* for intuitionistic logic, namely the restriction to at most one formula in the succedent of a sequent. We consider the sequent calculi GL_T and GL_B introduced in Chap. 6. These sequent systems capture Odintsov's axiomatic trilattice logics L_T and L_B. The completeness proof uses a kind of Kripke semantics. Since the cases of L_T and L_B are fully analogous, we will focus on a Gentzen-type sequent calculus I_{16} to obtain an intuitionistic counterpart of L_T. The system I_{16} is plausible and natural for the following reasons: (i) the properties of constructible falsity and paraconsistency with respect to some negation connectives hold for I_{16}, and (ii) sequent calculi for Belnap and Dunn's four-valued logic and for Nelson's constructive paraconsistent logic **N4** are included as natural subsystems of I_{16}.

In Chap. 3 it has been pointed out that the first trilattice structure on a 16-valued set of truth values had been introduced in [231]. In this trilattice $SIXTEEN_3^c$, in addition to an information order and a truth order, there is a constructiveness order. The constructivity order is motivated by considerations of constructive truth and

[1] We here use the material from [274].

constructive falsity in intuitionistic logic and constructive logic with strong negation, and the constructivity of truth and falsity is taken as a dimension of comparison between notions of truth and falsity. In this chapter, another perspective on the constructiveness of reasoning is explored. In a first step we will apply a familiar proof-theoretic strategy for obtaining intuitionistic variants of *SIXTEEN*$_3$-related logics. In a second step we will replace the classically defined representation of the algebraic operations in *SIXTEEN*$_3$ from Chap. 5 with an intuitionistically defined representation in order to obtain intuitionistic trilattice logics. It is these two moves and not an appeal to a notion of intuitionistic truth values that explain the title of the present chapter.

As pointed out in Chap. 6, although the sequent calculi GL_T and GL_B are sequent systems for the trilattice logics L_T and L_B, and both are *related to* truth entailment in *SIXTEEN*$_3$, they do not *characterize* the relation $\models_t$. We presented calculi of generalized sequents which characterize $\models_t$ and $\models_f$. These generalized sequents consist of four components corresponding to the constituents of Odintsov's four-component twist-structures, and the matrix representation of the algebraic operations of *SIXTEEN*$_3$ is transformed into introduction rules for the corresponding logical connectives. The strategy for obtaining intuitionistic variants in this case is to recast the matrix representations by intuitionistic tableau rules, cf. [198]. In addition to tableau rules for the intuitionistic connectives, we also use tableau rules for the co-implication connective from Heyting–Brouwer logic. The completeness proof with respect to a Kripke-style semantics uses familiar tableau methods.

First, we will define a sequent calculus I_{16} which is an intuitionistic counterpart of L_T and prove cut-elimination for I_{16}. The sequent system I_{16} is shown to be complete with respect to a Kripke-style semantics in Sect. 7.3. Finally, we turn to intuitionistic versions of the sequent calculi for truth and falsity entailment in *SIXTEEN*$_3$. In Sect. 7.4 we introduce the tableau rules for the intuitionistic trilattice connectives and define a tableau calculus as an intuitionistic counterpart of the sequent calculus GL^* from Chap. 6. This tableau calculus IT_{16} for intuitionistic truth entailment is shown to be complete with respect to another intuitionistic Kripke-style semantics in Sect. 7.5.

7.2 Sequent Calculus I_{16}

In this section, we introduce an intuitionistic version of the sequent calculus GL_T for the trilattice logic L_T. We will consider the language $\mathscr{L}_{tf}^{\rightarrow_t}$ together with the constants $\top$ and $\bot$. Since we have only one implication, we will omit the subscript and simply write $\rightarrow$. The connective $\rightarrow$ is just the intuitionistic implication. The symbol $\sim_b$ is used to denote $\sim_t\sim_f$, or $\sim_f\sim_t$, the symbol $\sim_d$ is used to denote $\sim_t\sim_t, \sim_f\sim_f$ or $\sim_b\sim_b$, and the symbol $\sim_e$ is now used to denote $\sim_t$ or $\sim_b$. The symbol $\triangleq$ is used to denote the identity of (sets of) symbols. Letters $A, B, \ldots$ are used to denote formulas, and Greek capital letters $\Gamma, \Delta, \ldots$ are used to represent finite (possibly empty) sets of formulas. An expression of the form $\Gamma \Rightarrow \Delta$ where Δ is either empty or a single formula is called a *sequent*. Again, the expression $L \vdash S$ (or $\vdash S$) is used to denote the fact that a sequent S is provable in a sequent calculus L.

Definition 7.1 For any propositional variable p,[2] the initial sequents of I_{16} are of the form:

$$p \Rightarrow p \quad \sim_e p \Rightarrow \sim_e p \quad \sim_f p \Rightarrow \sim_f p$$

$$\sim_f \sim_t p \Rightarrow \sim_t \sim_f p \quad \sim_t \sim_f p \Rightarrow \sim_f \sim_t p$$

$$\bot \Rightarrow \quad \Rightarrow \top \quad \Rightarrow \sim_e \bot \quad \sim_e \top \Rightarrow \quad \sim_f \bot \Rightarrow \quad \Rightarrow \sim_f \top.$$

The structural inference rules of I_{16} are of the form:

$$\frac{\Gamma \Rightarrow A \quad A, \Sigma \Rightarrow \Pi}{\Gamma, \Sigma \Rightarrow \Pi}(\mathrm{cut}) \quad \frac{\Gamma \Rightarrow \Delta}{A, \Gamma \Rightarrow \Delta}(\mathrm{w-l}) \quad \frac{\Gamma \Rightarrow}{\Gamma \Rightarrow A}(\mathrm{w-r}).$$

The logical inference rules of I_{16} are of the form:

$$\frac{\Gamma \Rightarrow A \quad B, \Delta \Rightarrow \Pi}{A \rightarrow B, \Gamma, \Delta \Rightarrow \Pi}(\rightarrow \mathrm{l}) \quad \frac{A, \Gamma \Rightarrow B}{\Gamma \Rightarrow A \rightarrow B}(\rightarrow \mathrm{r})$$

$$\frac{A, B, \Gamma \Rightarrow \Delta}{A \wedge_t B, \Gamma \Rightarrow \Delta}(\wedge_t \mathrm{l}) \quad \frac{\Gamma \Rightarrow A \quad \Gamma \Rightarrow B}{\Gamma \Rightarrow A \wedge_t B}(\wedge_t \mathrm{r})$$

$$\frac{A, \Gamma \Rightarrow \Delta \quad B, \Gamma \Rightarrow \Delta}{A \vee_t B, \Gamma \Rightarrow \Delta}(\vee_t \mathrm{l}) \quad \frac{\Gamma \Rightarrow A}{\Gamma \Rightarrow A \vee_t B}(\vee_t \mathrm{r1}) \quad \frac{\Gamma \Rightarrow B}{\Gamma \Rightarrow A \vee_t B}(\vee_t \mathrm{r2})$$

$$\frac{A, B, \Gamma \Rightarrow \Delta}{A \wedge_f B, \Gamma \Rightarrow \Delta}(\wedge_f \mathrm{l}) \quad \frac{\Gamma \Rightarrow A \quad \Gamma \Rightarrow B}{\Gamma \Rightarrow A \wedge_f B}(\wedge_f \mathrm{r})$$

$$\frac{A, \Gamma \Rightarrow \Delta \quad B, \Gamma \Rightarrow \Delta}{A \vee_f B, \Gamma \Rightarrow \Delta}(\vee_f \mathrm{l}) \quad \frac{\Gamma \Rightarrow A}{\Gamma \Rightarrow A \vee_f B}(\vee_f \mathrm{r1}) \quad \frac{\Gamma \Rightarrow B}{\Gamma \Rightarrow A \vee_f B}(\vee_f \mathrm{r2})$$

$$\frac{A, \Gamma \Rightarrow \Delta}{\sim_d A, \Gamma \Rightarrow \Delta}(\sim_d \mathrm{l}) \quad \frac{\Gamma \Rightarrow A}{\Gamma \Rightarrow \sim_d A}(\sim_d \mathrm{r})$$

$$\frac{\sim_f A, \Gamma \Rightarrow \Delta}{\sim_t \sim_f \sim_t A, \Gamma \Rightarrow \Delta}(\sim_t \sim_f \sim_t \mathrm{l}) \quad \frac{\Gamma \Rightarrow \sim_f A}{\Gamma \Rightarrow \sim_t \sim_f \sim_t A}(\sim_t \sim_f \sim_t \mathrm{r})$$

$$\frac{\sim_t A, \Gamma \Rightarrow \Delta}{\sim_f \sim_t \sim_f A, \Gamma \Rightarrow \Delta}(\sim_f \sim_t \sim_f \mathrm{l}) \quad \frac{\Gamma \Rightarrow \sim_t A}{\Gamma \Rightarrow \sim_f \sim_t \sim_f A}(\sim_f \sim_t \sim_f \mathrm{r})$$

[2] Again, the setting of the propositional-variable-based initial sequents will be needed for proving an embedding theorem, cf. Chap. 6. We will show later that the (general) formula-based initial sequents are also provable in I_{16}.

$$\frac{{\sim_f}A, \Gamma \Rightarrow \Delta}{{\sim_f}{\sim_t}{\sim_t}A, \Gamma \Rightarrow \Delta}({\sim_f}{\sim_t}{\sim_t}\mathrm{l}) \quad \frac{\Gamma \Rightarrow {\sim_f}A}{\Gamma \Rightarrow {\sim_f}{\sim_t}{\sim_t}A}({\sim_f}{\sim_t}{\sim_t}\mathrm{r})$$

$$\frac{{\sim_t}A, \Gamma \Rightarrow \Delta}{{\sim_t}{\sim_f}{\sim_f}A, \Gamma \Rightarrow \Delta}({\sim_t}{\sim_f}{\sim_f}\mathrm{l}) \quad \frac{\Gamma \Rightarrow {\sim_t}A}{\Gamma \Rightarrow {\sim_t}{\sim_f}{\sim_f}A}({\sim_t}{\sim_f}{\sim_f}\mathrm{r})$$

$$\frac{A, {\sim_e}B, \Gamma \Rightarrow \Delta}{{\sim_e}(A \to B), \Gamma \Rightarrow \Delta}({\sim_e} \to \mathrm{l}) \quad \frac{\Gamma \Rightarrow A \quad \Gamma \Rightarrow {\sim_e}B}{\Gamma \Rightarrow {\sim_e}(A \to B)}({\sim_e} \to \mathrm{r})$$

$$\frac{{\sim_e}A, \Gamma \Rightarrow \Delta \quad {\sim_e}B, \Gamma \Rightarrow \Delta}{{\sim_e}(A \wedge_t B), \Gamma \Rightarrow \Delta}({\sim_e}\wedge_t\mathrm{l})$$

$$\frac{\Gamma \Rightarrow {\sim_e}A}{\Gamma \Rightarrow {\sim_e}(A \wedge_t B)}({\sim_e}\wedge_t\mathrm{r1}) \quad \frac{\Gamma \Rightarrow {\sim_e}B}{\Gamma \Rightarrow {\sim_e}(A \wedge_t B)}({\sim_e}\wedge_t\mathrm{r2})$$

$$\frac{{\sim_e}A, {\sim_e}B, \Gamma \Rightarrow \Delta}{{\sim_e}(A \vee_t B), \Gamma \Rightarrow \Delta}({\sim_e}\vee_t\mathrm{l}) \quad \frac{\Gamma \Rightarrow {\sim_e}A \quad \Gamma \Rightarrow {\sim_e}B}{\Gamma \Rightarrow {\sim_e}(A \vee_t B)}({\sim_e}\vee_t\mathrm{r})$$

$$\frac{{\sim_e}A, {\sim_e}B, \Gamma \Rightarrow \Delta}{{\sim_e}(A \wedge_f B), \Gamma \Rightarrow \Delta}({\sim_e}\wedge_f\mathrm{l}) \quad \frac{\Gamma \Rightarrow {\sim_e}A \quad \Gamma \Rightarrow {\sim_e}B}{\Gamma \Rightarrow {\sim_e}(A \wedge_f B)}({\sim_e}\wedge_f\mathrm{r})$$

$$\frac{{\sim_e}A, \Gamma \Rightarrow \Delta \quad {\sim_e}B, \Gamma \Rightarrow \Delta}{{\sim_e}(A \vee_f B), \Gamma \Rightarrow \Delta}({\sim_e}\vee_f\mathrm{l})$$

$$\frac{\Gamma \Rightarrow {\sim_e}A}{\Gamma \Rightarrow {\sim_e}(A \vee_f B)}({\sim_e}\vee_f\mathrm{r1}) \quad \frac{\Gamma \Rightarrow {\sim_e}B}{\Gamma \Rightarrow {\sim_e}(A \vee_f B)}({\sim_e}\vee_f\mathrm{r2})$$

$$\frac{\Gamma \Rightarrow {\sim_f}A \quad {\sim_f}B, \Delta \Rightarrow \Pi}{{\sim_f}(A \to B), \Gamma, \Delta \Rightarrow \Pi}({\sim_f} \to \mathrm{l}) \quad \frac{{\sim_f}A, \Gamma \Rightarrow {\sim_f}B}{\Gamma \Rightarrow {\sim_f}(A \to B)}({\sim_f} \to \mathrm{r})$$

$$\frac{{\sim_f}A, {\sim_f}B, \Gamma \Rightarrow \Delta}{{\sim_f}(A \wedge_t B), \Gamma \Rightarrow \Delta}({\sim_f}\wedge_t\mathrm{l}) \quad \frac{\Gamma \Rightarrow {\sim_f}A \quad \Gamma \Rightarrow {\sim_f}B}{\Gamma \Rightarrow {\sim_f}(A \wedge_t B)}({\sim_f}\wedge_t\mathrm{r})$$

$$\frac{{\sim_f}A, \Gamma \Rightarrow \Delta \quad {\sim_f}B, \Gamma \Rightarrow \Delta}{{\sim_f}(A \vee_t B), \Gamma \Rightarrow \Delta}({\sim_f}\vee_t\mathrm{l})$$

$$\frac{\Gamma \Rightarrow {\sim_f}A}{\Gamma \Rightarrow {\sim_f}(A \vee_t B)}({\sim_f}\vee_t\mathrm{r1}) \quad \frac{\Gamma \Rightarrow {\sim_f}B}{\Gamma \Rightarrow {\sim_f}(A \vee_t B)}({\sim_f}\vee_t\mathrm{r2})$$

$$\frac{{\sim_f}A, \Gamma \Rightarrow \Delta \quad {\sim_f}B, \Gamma \Rightarrow \Delta}{{\sim_f}(A \wedge_f B), \Gamma \Rightarrow \Delta}({\sim_f}\wedge_f\mathrm{l})$$

$$\frac{\Gamma \Rightarrow {\sim_f}A}{\Gamma \Rightarrow {\sim_f}(A \wedge_f B)}({\sim_f}\wedge_f\mathrm{r1}) \quad \frac{\Gamma \Rightarrow {\sim_f}B}{\Gamma \Rightarrow {\sim_f}(A \wedge_f B)}({\sim_f}\wedge_f\mathrm{r2})$$

$$\frac{\sim_f A, \sim_f B, \Gamma \Rightarrow \Delta}{\sim_f (A \vee_f B), \Gamma \Rightarrow \Delta}(\sim_f \vee_f \mathrm{l}) \quad \frac{\Gamma \Rightarrow \sim_f A \quad \Gamma \Rightarrow \sim_f B}{\Gamma \Rightarrow \sim_f (A \vee_f B)}(\sim_f \vee_f \mathrm{r}).$$

Definition 7.2 The $\{\bot, \top, \wedge_t, \vee_t, \rightarrow\}$-fragment of I_{16} gives a familiar definition of the sequent calculus *LJ* for intuitionistic logic.

Note that the $\{\wedge_t, \vee_t, \sim_t\}$-fragment of I_{16} is a sequent calculus for Belnap and Dunn's four-valued logic [3, 22, 23, 75], and the $\{\rightarrow, \wedge_t, \vee_t, \sim_t\}$-fragment of I_{16} is a sequent calculus for Nelson's constructive paraconsistent logic [2]. Thus, I_{16} may be regarded as a natural extension and generalization of Belnap and Dunn's logic and Nelson's logic.

Note that a sequent calculus I_{16}^* for the intuitionistic version of Odintsov's L_B is obtained from I_{16} by replacing the inference rules $\{(\wedge_f \mathrm{l}), (\wedge_f \mathrm{r}), (\vee_f \mathrm{l}), (\vee_f \mathrm{r})\}$ with the inference rules of the form:

$$\frac{A, \Gamma \Rightarrow \Delta \quad B, \Gamma \Rightarrow \Delta}{A \wedge_f B, \Gamma \Rightarrow \Delta}(\wedge_f \mathrm{l}^*) \quad \frac{\Gamma \Rightarrow A}{\Gamma \Rightarrow A \wedge_f B}(\wedge_f \mathrm{r1}^*) \quad \frac{\Gamma \Rightarrow B}{\Gamma \Rightarrow A \wedge_f B}(\wedge_f \mathrm{r2}^*)$$

$$\frac{A, B, \Gamma \Rightarrow \Delta}{A \vee_f B, \Gamma \Rightarrow \Delta}(\vee_f \mathrm{l}^*) \quad \frac{\Gamma \Rightarrow A \quad \Gamma \Rightarrow B}{\Gamma \Rightarrow A \vee_f B}(\vee_f \mathrm{r}^*).$$

Since we can treat both I_{16} and I_{16}^* in a similar way, we will deal only with I_{16} in the following.

Proposition 7.1 *The following rules are admissible in cut-free* I_{16}:

$$\frac{\sim_t \sim_f A, \Gamma \Rightarrow \Delta}{\sim_f \sim_t A, \Gamma \Rightarrow \Delta}(\sim_f \sim_t \mathrm{l}) \quad \frac{\Gamma \Rightarrow \sim_t \sim_f A}{\Gamma \Rightarrow \sim_f \sim_t A}(\sim_f \sim_t \mathrm{r})$$

$$\frac{\sim_f \sim_t A, \Gamma \Rightarrow \Delta}{\sim_t \sim_f A, \Gamma \Rightarrow \Delta}(\sim_t \sim_f \mathrm{l}) \quad \frac{\Gamma \Rightarrow \sim_f \sim_t A}{\Gamma \Rightarrow \sim_t \sim_f A}(\sim_t \sim_f \mathrm{r}).$$

Proof We show the claim only for the case $(\sim_f \sim_t \mathrm{l})$.

- $(\sim_f \sim_t \mathrm{l})$: We show that the rule $(\sim_f \sim_t \mathrm{l})$ is admissible in cut-free I_{16}, i.e., we show that $\mathrm{I}_{16} - (\mathrm{cut}) \vdash \sim_t \sim_f A, \Gamma \Rightarrow \Delta$ implies $\mathrm{I}_{16} - (\mathrm{cut}) \vdash \sim_f \sim_t A, \Gamma \Rightarrow \Delta$. This is proved by induction on a cut-free proof π of $\sim_t \sim_f A,\ \Gamma \Rightarrow \Delta$ in I_{16}. We distinguish the cases according to the last inference of π. We show some cases.

  Case $(\sim_t \sim_f p \Rightarrow \sim_t \sim_f p)$: The proof π is of the form: $\sim_t \sim_f p \Rightarrow \sim_t \sim_f p$ where p is a propositional variable. Then, $\sim_f \sim_t p \Rightarrow \sim_t \sim_f p$ is also an initial sequent of I_{16}.

  Case $(\sim_t \sim_f \sim_t \mathrm{l})$: The last inference rule of π is of the form:

$$\frac{\sim_f B, \Gamma \Rightarrow \Delta}{\sim_t \sim_f \sim_t B, \Gamma \Rightarrow \Delta}(\sim_t \sim_f \sim_t \mathrm{l})$$

where $A \mathrel{\mathring{=}} {\sim_t} B$. By the induction hypothesis, we have $\mathrm{I}_{16} - (\text{cut}) \vdash {\sim_f} B, \Gamma \Rightarrow \Delta$, and then we obtain:

$$\dfrac{\begin{array}{c}\vdots\\ {\sim_f} B, \Gamma \Rightarrow \Delta\end{array}}{{\sim_f}{\sim_t}{\sim_t} B, \Gamma \Rightarrow \Delta}\,({\sim_f}{\sim_t}{\sim_t}\text{l}).$$

Case $({\sim_b}\wedge_t\text{l})$: The last inference rule of π is of the form:

$$\dfrac{{\sim_t}{\sim_f} A', \Gamma \Rightarrow \Delta \quad {\sim_t}{\sim_f} A'', \Gamma \Rightarrow \Delta}{{\sim_t}{\sim_f}(A' \wedge_t A''), \Gamma \Rightarrow \Delta}\,({\sim_b}\wedge_t\text{l})$$

where $A \mathrel{\mathring{=}} A' \wedge_t A''$. By the induction hypothesis, we have $\mathrm{I}_{16} - (\text{cut}) \vdash {\sim_f}{\sim_t} A', \Gamma \Rightarrow \Delta$ and $\mathrm{I}_{16} - (\text{cut}) \vdash {\sim_f}{\sim_t} A'', \Gamma \Rightarrow \Delta$. Thus, we obtain:

$$\dfrac{\begin{array}{c}\vdots\\ {\sim_f}{\sim_t} A', \Gamma \Rightarrow \Delta\end{array} \quad \begin{array}{c}\vdots\\ {\sim_f}{\sim_t} A'', \Gamma \Rightarrow \Delta\end{array}}{{\sim_f}{\sim_t}(A' \wedge_t A''), \Gamma \Rightarrow \Delta}\,({\sim_b}\wedge_t\text{l}).$$

□

Proposition 7.2 *The following rules are admissible in cut-free* I_{16}:

$$\dfrac{{\sim_d} A, \Gamma \Rightarrow \Delta}{A, \Gamma \Rightarrow \Delta}\,({\sim_d}\text{l}^{-1}) \qquad \dfrac{\Gamma \Rightarrow {\sim_d} A}{\Gamma \Rightarrow A}\,({\sim_d}\text{r}^{-1})$$

$$\dfrac{{\sim_t}{\sim_f}{\sim_t} A, \Gamma \Rightarrow \Delta}{{\sim_f} A, \Gamma \Rightarrow \Delta}\,({\sim_t}{\sim_f}{\sim_t}\text{l}^{-1}) \qquad \dfrac{\Gamma \Rightarrow {\sim_t}{\sim_f}{\sim_t} A}{\Gamma \Rightarrow {\sim_f} A}\,({\sim_t}{\sim_f}{\sim_t}\text{r}^{-1})$$

$$\dfrac{{\sim_f}{\sim_t}{\sim_f} A, \Gamma \Rightarrow \Delta}{{\sim_t} A, \Gamma \Rightarrow \Delta}\,({\sim_f}{\sim_t}{\sim_f}\text{l}^{-1}) \qquad \dfrac{\Gamma \Rightarrow {\sim_f}{\sim_t}{\sim_f} A}{\Gamma \Rightarrow {\sim_t} A}\,({\sim_f}{\sim_t}{\sim_f}\text{r}^{-1})$$

$$\dfrac{{\sim_f}{\sim_t}{\sim_t} A, \Gamma \Rightarrow \Delta}{{\sim_f} A, \Gamma \Rightarrow \Delta}\,({\sim_f}{\sim_t}{\sim_t}\text{l}^{-1}) \qquad \dfrac{\Gamma \Rightarrow {\sim_f}{\sim_t}{\sim_t} A}{\Gamma \Rightarrow {\sim_f} A}\,({\sim_f}{\sim_t}{\sim_t}\text{r}^{-1})$$

$$\dfrac{{\sim_t}{\sim_f}{\sim_f} A, \Gamma \Rightarrow \Delta}{{\sim_t} A, \Gamma \Rightarrow \Delta}\,({\sim_t}{\sim_f}{\sim_f}\text{l}^{-1}) \qquad \dfrac{\Gamma \Rightarrow {\sim_t}{\sim_f}{\sim_f} A}{\Gamma \Rightarrow {\sim_t} A}\,({\sim_t}{\sim_f}{\sim_f}\text{r}^{-1}).$$

Proof Straightforward. □

Proposition 7.3 *The following sequents are provable in cut-free* I_{16}, *for any formula* A:

1. ${\sim_t} A \Rightarrow {\sim_t} A$,
2. ${\sim_f} A \Rightarrow {\sim_f} A$,
3. ${\sim_b} A \Rightarrow {\sim_b} A$,
4. ${\sim_d} A \Rightarrow {\sim_d} A$.

Proof By simultaneous induction on the structure of A.

(1) We show some cases.

(Case $A \stackrel{\circ}{=} {\sim_f} B$): By the induction hypothesis for (3), we obtain the required fact $\mathrm{I}_{16} - (\mathrm{cut}) \vdash {\sim_t}{\sim_f} B \Rightarrow {\sim_t}{\sim_f} B$.

(Case $A \stackrel{\circ}{=} {\sim_t} B$): By the induction hypothesis for (4), we obtain the required fact $\mathrm{I}_{16} - (\mathrm{cut}) \vdash {\sim_t}{\sim_t} B \Rightarrow {\sim_t}{\sim_t} B$.

(Case $A \stackrel{\circ}{=} B \to C$): We show $\mathrm{I}_{16} - (\mathrm{cut}) \vdash {\sim_t}(B \to C) \Rightarrow {\sim_t}(B \to C)$. By the induction hypothesis for (4), we have $\mathrm{I}_{16} - (\mathrm{cut}) \vdash {\sim_t}{\sim_t} B \Rightarrow {\sim_t}{\sim_t} B$, and by the induction hypothesis for (1), we have $\mathrm{I}_{16} - (\mathrm{cut}) \vdash {\sim_t} C \Rightarrow {\sim_t} C$. Then, we obtain the required fact:

$$\dfrac{\dfrac{\dfrac{\dfrac{\dfrac{\vdots}{{\sim_t}{\sim_t}B \Rightarrow {\sim_t}{\sim_t}B}}{B \Rightarrow {\sim_t}{\sim_t}B}\,({\sim_t}{\sim_t}\mathrm{l}^{-1})}{B \Rightarrow B}\,({\sim_t}{\sim_t}\mathrm{r}^{-1})}{B, {\sim_t}C \Rightarrow B}\,(\text{w-l})}{{\sim_t}(B \to C) \Rightarrow B}\,({\sim_t} \to \mathrm{l}) \quad \dfrac{\dfrac{\dfrac{\vdots}{{\sim_t}C \Rightarrow {\sim_t}C}}{B, {\sim_t}C \Rightarrow {\sim_t}C}\,(\text{w-l})}{{\sim_t}(B \to C) \Rightarrow {\sim_t}C}\,({\sim_t} \to \mathrm{l})$$
$$\overline{\qquad {\sim_t}(B \to C) \Rightarrow {\sim_t}(B \to C) \qquad}\,({\sim_t} \to \mathrm{r})$$

where $({\sim_t}{\sim_t}\mathrm{l}^{-1})$ and $({\sim_t}{\sim_t}\mathrm{r}^{-1})$ are admissible in cut-free I_{16} by Proposition 7.2.

(2) Similar to (1).

(3) We show only the case for ${\sim_b} = {\sim_f}{\sim_t}$ since the case for ${\sim_b} = {\sim_t}{\sim_f}$ can be obtained similarly. We show some cases.

(Case $A \stackrel{\circ}{=} {\sim_t} B$): We show $\mathrm{I}_{16} - (\mathrm{cut}) \vdash {\sim_f}{\sim_t}{\sim_t} B \Rightarrow {\sim_f}{\sim_t}{\sim_t} B$. By the induction hypothesis for (2), we have $\mathrm{I}_{16} - (\mathrm{cut}) \vdash {\sim_f} B \Rightarrow {\sim_f} B$. Then, we obtain the required fact:

$$\dfrac{\dfrac{\dfrac{\vdots}{{\sim_f}B \Rightarrow {\sim_f}B}}{{\sim_f}{\sim_t}{\sim_t}B \Rightarrow {\sim_f}B}\,({\sim_f}{\sim_t}{\sim_t}\mathrm{l})}{{\sim_f}{\sim_t}{\sim_t}B \Rightarrow {\sim_f}{\sim_t}{\sim_t}B}\,({\sim_f}{\sim_t}{\sim_t}\mathrm{r}).$$

(Case $A \stackrel{\circ}{=} B \wedge_t C$): We show $\mathrm{I}_{16} - (\mathrm{cut}) \vdash {\sim_f}{\sim_t}(B \wedge_t C) \Rightarrow {\sim_f}{\sim_t}(B \wedge_t C)$. By the induction hypothesis for (3), we have $\mathrm{I}_{16} - (\mathrm{cut}) \vdash {\sim_f}{\sim_t} B \Rightarrow {\sim_f}{\sim_t} B$ and $\mathrm{I}_{16} - (\mathrm{cut}) \vdash {\sim_f}{\sim_t} C \Rightarrow {\sim_f}{\sim_t} C$. Then, we obtain the required fact:

$$\dfrac{\dfrac{\dfrac{\vdots}{{\sim_f}{\sim_t}B \Rightarrow {\sim_f}{\sim_t}B}}{{\sim_f}{\sim_t}B \Rightarrow {\sim_f}{\sim_t}(B \wedge_t C)}\,({\sim_f}{\sim_t}\mathrm{r1}) \quad \dfrac{\dfrac{\vdots}{{\sim_f}{\sim_t}C \Rightarrow {\sim_f}{\sim_t}C}}{{\sim_f}{\sim_t}C \Rightarrow {\sim_f}{\sim_t}(B \wedge_t C)}\,({\sim_f}{\sim_t}\mathrm{r2})}{{\sim_f}{\sim_t}(B \wedge_t C) \Rightarrow {\sim_f}{\sim_t}(B \wedge_t C)}\,({\sim_f}{\sim_t}\wedge_t\mathrm{l}).$$

(4) We show some cases.

(Case $A \doteq p$ for a propositional variable p): We can obtain $\mathrm{I}_{16} - (\mathrm{cut}) \vdash \sim_d p \Rightarrow \sim_d p$ from the initial sequent $p \Rightarrow p$ by applying $(\sim_d \mathrm{l})$ and $(\sim_d \mathrm{r})$.

(Case $A \doteq B \to C$): We show $\mathrm{I}_{16} - (\mathrm{cut}) \vdash \sim_d(B \to C) \Rightarrow \sim_d(B \to C)$. By the induction hypothesis for (4), we have $\mathrm{I}_{16} - (\mathrm{cut}) \vdash \sim_d B \Rightarrow \sim_d B$ and $\mathrm{I}_{16} - (\mathrm{cut}) \vdash \sim_d C \Rightarrow \sim_d C$. Then, we obtain the required fact:

$$
\dfrac{\dfrac{\dfrac{\dfrac{\dfrac{\vdots}{\sim_d B \Rightarrow \sim_d B}}{B \Rightarrow \sim_d B}\,(\sim_d \mathrm{l}^{-1})}{B \Rightarrow B}\,(\sim_d \mathrm{r}^{-1}) \quad \dfrac{\dfrac{\dfrac{\vdots}{\sim_d C \Rightarrow \sim_d C}}{C \Rightarrow \sim_d C}\,(\sim_d \mathrm{l}^{-1})}{C \Rightarrow C}\,(\sim_d \mathrm{r}^{-1})}{\dfrac{\dfrac{\dfrac{B, B \to C \Rightarrow C}{B \to C \Rightarrow B \to C}\,(\to \mathrm{r})}{\sim_d(B \to C) \Rightarrow B \to C}\,(\sim_d \mathrm{l})}{\sim_d(B \to C) \Rightarrow \sim_d(B \to C)}\,(\sim_d \mathrm{r})}\,(\to \mathrm{l})
$$

where $(\sim_d \mathrm{l}^{-1})$ and $(\sim_d \mathrm{r}^{-1})$ are admissible in cut-free I_{16} by Proposition 7.2. □

Proposition 7.4 *The following sequents are provable in cut-free* I_{16}, *for any formula* A:

1. $\sim_f \sim_t A \Rightarrow \sim_t \sim_f A$,
2. $\sim_t \sim_f A \Rightarrow \sim_f \sim_t A$.

Proof By induction on A. We use Propositions 7.3 (3) and 7.1. □

Proposition 7.5 *For any formula* A, $\mathrm{I}_{16} - (\mathrm{cut}) \vdash A \Rightarrow A$.

Proof By induction on A. We use Proposition 7.3 (1) (2) for the cases $A \doteq \sim_t B$ and $A \doteq \sim_f B$. □

An expression $A \Leftrightarrow B$ means the sequents $A \Rightarrow B$ and $B \Rightarrow A$. The following proposition shows that the sequents which correspond to Odintsov's axioms are provable in I_{16}.

Proposition 7.6 *The following sequents are provable in cut-free* I_{16} *for any formulas* A *and* B:

1. $A \Leftrightarrow \sim_d A$,
2. $\sim_t \sim_f A \Leftrightarrow \sim_f \sim_t A$,
3. $\sim_f A \Leftrightarrow \sim_t \sim_f \sim_t A$,
4. $\sim_t A \Leftrightarrow \sim_f \sim_t \sim_f A$,
5. $A \wedge_t B \Leftrightarrow A \wedge_f B$,
6. $A \vee_t B \Leftrightarrow A \vee_f B$,
7. $\sim_e(A \wedge_t B) \Leftrightarrow \sim_e A \vee_t \sim_e B$,
8. $\sim_e(A \vee_t B) \Leftrightarrow \sim_e A \wedge_t \sim_e B$,

9. $\sim_e(A \wedge_f B) \Leftrightarrow \sim_e A \wedge_f \sim_e B$,
10. $\sim_e(A \vee_f B) \Leftrightarrow \sim_e A \vee_f \sim_e B$,
11. $\sim_f(A \wedge_f B) \Leftrightarrow \sim_f A \vee_f \sim_f B$,
12. $\sim_f(A \vee_f B) \Leftrightarrow \sim_f A \wedge_f \sim_f B$,
13. $\sim_f(A \wedge_t B) \Leftrightarrow \sim_f A \wedge_t \sim_f B$,
14. $\sim_f(A \vee_t B) \Leftrightarrow \sim_f A \vee_t \sim_f B$.

In order to show the cut-elimination theorem for I_{16}, we introduce an embedding of I_{16} into *LJ*.

Definition 7.3 We fix a set Φ of propositional variables and define three disjoint sets of propositional variables: $\Phi' := \{p' | p \in \Phi\}, \Phi'' := \{p'' | p \in \Phi\}$, and $\Phi''' := \{p''' | p \in \Phi\}$. The language $\mathscr{L}_{I16}$ of I_{16} is defined by using $\Phi, \top, \bot, \rightarrow, \wedge_t, \vee_t, \wedge_f, \vee_f, \sim_t$, and $\sim_f$. The language $\mathscr{L}_{LJ}$ of *LJ* is obtained from $\mathscr{L}_{I16}$ by adding $\{\Phi', \Phi'', \Phi'''\}$ and deleting $\{\wedge_f, \vee_f, \sim_t, \sim_f\}$.

A mapping f from $\mathscr{L}_{I16}$ to $\mathscr{L}_{LJ}$ is defined as follows.

1. $f(p) := p, f(\sim_t p) := p' \in \Phi', f(\sim_f p) := p'' \in \Phi''$, and $f(\sim_t \sim_f p) = f(\sim_f \sim_t p) := p''' \in \Phi'''$ for any $p \in \Phi$.
2. $f(\sharp) := \sharp, f(\sim_e \bot) := \top, f(\sim_e \top) := \bot$ and $f(\sim_f \sharp) := \sharp$ where $\sharp \in \{\bot, \top\}$.
3. $f(A \circ B) := f(A) \circ f(B)$ where $\circ \in \{\rightarrow, \wedge_t, \vee_t\}$.
4. $f(A \wedge_f B) := f(A) \wedge_t f(B)$.
5. $f(A \vee_f B) := f(A) \vee_t f(B)$.
6. $f(\sim_d A) := f(A)$.
7. $f(\sim_t \sim_f \sim_t A) := f(\sim_f A)$.
8. $f(\sim_f \sim_t \sim_f A) := f(\sim_t A)$.
9. $f(\sim_f \sim_t \sim_t A) := f(\sim_f A)$.
10. $f(\sim_t \sim_f \sim_f A) := f(\sim_t A)$.
11. $f(\sim_e(A \rightarrow B)) := f(A) \wedge_t f(\sim_e B)$.
12. $f(\sim_e(A \wedge_t B)) := f(\sim_e A) \vee_t f(\sim_e B)$.
13. $f(\sim_e(A \vee_t B)) := f(\sim_e A) \wedge_t f(\sim_e B)$.
14. $f(\sim_e(A \wedge_f B)) := f(\sim_e A) \wedge_t f(\sim_e B)$.
15. $f(\sim_e(A \vee_f B)) := f(\sim_e A) \vee_t f(\sim_e B)$.
16. $f(\sim_f(A \rightarrow B)) := f(\sim_f A) \rightarrow f(\sim_f B)$.
17. $f(\sim_f(A \wedge_t B)) := f(\sim_f A) \wedge_t f(\sim_f B)$.
18. $f(\sim_f(A \vee_t B)) := f(\sim_f A) \vee_t f(\sim_f B)$.
19. $f(\sim_f(A \wedge_f B)) := f(\sim_f A) \vee_t f(\sim_f B)$.
20. $f(\sim_f(A \vee_f B)) := f(\sim_f A) \wedge_t f(\sim_f B)$.

Let Γ be a set of formulas in $\mathscr{L}_{I16}$. Then, $f(\Gamma)$ denotes the result of replacing every occurrence of a formula A in Γ by an occurrence of $f(A)$.

Theorem 7.1 *Let* Γ and Δ *be sets of formulas in* $\mathscr{L}_{I16}$, *and* f *be the mapping defined in Definition 7.3. Then*

1. *If* $I_{16} \vdash \Gamma \Rightarrow \Delta$, *then* $LJ \vdash f(\Gamma) \Rightarrow f(\Delta)$.
2. *If* $LJ - (\text{cut}) \vdash f(\Gamma) \Rightarrow f(\Delta)$, *then* $I_{16} - (\text{cut}) \vdash \Gamma \Rightarrow \Delta$.

Proof

(1) By induction on a proof π of $\Gamma \Rightarrow \Delta$ in I_{16}. We distinguish the cases according to the last inference of π. We show some cases.

Case $(\sim_f\sim_t p \Rightarrow \sim_t\sim_f p)$: π is of the form $\sim_f\sim_t p \Rightarrow \sim_t\sim_f p$ where p is a propositional variable. By the definition of f, we have $f(\sim_f\sim_t p) = f(\sim_t\sim_f p) = p''' \in \Phi'''$, and hence $LJ \vdash p''' \Rightarrow p'''$.

Case $(\sim_t \rightarrow \text{l})$: The last inference rule of π is of the form:

$$\frac{A, \sim_t B, \Sigma \Rightarrow \Delta}{\sim_t(A \rightarrow B), \Sigma \Rightarrow \Delta}(\sim_t \rightarrow \text{l}).$$

By the induction hypothesis, we have $LJ \vdash f(A), f(\sim_t B), f(\Sigma) \Rightarrow f(\Delta)$. Then, we obtain the required fact:

$$\frac{\begin{array}{c}\vdots\\ f(A), f(\sim_t B), f(\Sigma) \Rightarrow f(\Delta)\end{array}}{f(A) \wedge_t f(\sim_t B), f(\Sigma) \Rightarrow f(\Delta)}(\wedge_t \text{l})$$

where $f(A) \wedge_t f(\sim_t B) = f(\sim_t(A \rightarrow B))$ by the definition of f.

Case $(\sim_t \rightarrow \text{r})$: The last inference rule of π is of the form:

$$\frac{\Sigma \Rightarrow A \quad \Sigma \Rightarrow \sim_t B}{\Sigma \Rightarrow \sim_t(A \rightarrow B)}(\sim_t \rightarrow \text{r}).$$

By the induction hypothesis, we have $LJ \vdash f(\Sigma) \Rightarrow f(A)$ and $LJ \vdash f(\Sigma) \Rightarrow f(\sim_t B)$. Then, we obtain the required fact:

$$\frac{\begin{array}{c}\vdots\\ f(\Sigma) \Rightarrow f(A)\end{array} \quad \begin{array}{c}\vdots\\ f(\Sigma) \Rightarrow f(\sim_t B)\end{array}}{f(\Sigma) \Rightarrow f(A) \wedge_t f(\sim_t B)}(\wedge_t \text{l})$$

where $f(A) \wedge_t f(\sim_t B) = f(\sim_t(A \rightarrow B))$ by the definition of f.

Case $(\sim_t\sim_f\sim_t \text{l})$: The last inference rule of π is of the form:

$$\frac{\sim_f A, \Sigma \Rightarrow \Delta}{\sim_t\sim_f\sim_t A, \Sigma \Rightarrow \Delta}(\sim_t\sim_f\sim_t \text{l}).$$

By the induction hypothesis, we have $LJ \vdash f(\sim_f A), f(\Sigma) \Rightarrow f(\Delta)$, and hence $LJ \vdash f(\sim_t\sim_f\sim_t A), f(\Sigma) \Rightarrow f(\Delta)$ by the definition of f.

Case $(\sim_f\sim_t\wedge_t\mathrm{l})$: The last inference rule of π is of the form:

$$\frac{\sim_f\sim_t A, \Sigma \Rightarrow \Delta \quad \sim_f\sim_t B, \Sigma \Rightarrow \Delta}{\sim_f\sim_t(A\wedge_t B), \Sigma \Rightarrow \Delta}\,(\sim_f\sim_t\wedge_t\mathrm{l}).$$

By the induction hypothesis, we have $LJ \vdash f(\sim_f\sim_t A), f(\Sigma) \Rightarrow f(\Delta)$ and $LJ \vdash f(\sim_f\sim_t B), f(\Sigma) \Rightarrow f(\Delta)$. Thus, we obtain

$$\frac{\begin{matrix}\vdots \\ f(\sim_f\sim_t A), f(\Sigma) \Rightarrow f(\Delta)\end{matrix} \quad \begin{matrix}\vdots \\ f(\sim_f\sim_t B), f(\Sigma) \Rightarrow f(\Delta)\end{matrix}}{f(\sim_f\sim_t A)\vee_t f(\sim_f\sim_t B), f(\Sigma) \Rightarrow f(\Delta)}\,(\vee_t\mathrm{l}).$$

Therefore we obtain the required fact $LJ \vdash f(\sim_f\sim_t(A\wedge_t B)), f(\Sigma) \Rightarrow f(\Delta)$ by the definition of f.

(2) By induction on a proof π of $f(\Gamma) \Rightarrow f(\Delta)$ in LJ – (cut). We distinguish the cases according to the last inference of π. We only show the following case. The last inference rule of π is of the form:

$$\frac{f(\sim_f\sim_t A), f(\Sigma) \Rightarrow f(\Delta) \quad f(\sim_f\sim_t B), f(\Sigma) \Rightarrow f(\Delta)}{f(\sim_f\sim_t A)\vee_t f(\sim_f\sim_t B), f(\Sigma) \Rightarrow f(\Delta)}\,(\vee_t\mathrm{l})$$

where $f(\sim_f\sim_t A)\vee_t f(\sim_f\sim_t B) = f(\sim_f\sim_t(A\wedge_t B))$ by the definition of f. By the induction hypothesis, we have L_{16} – (cut) $\vdash \sim_f\sim_t A, \Sigma \Rightarrow \Delta$ and L_{16} – (cut) $\vdash \sim_f\sim_t B, \Sigma \Rightarrow \Delta$. Thus, we obtain the required fact:

$$\frac{\begin{matrix}\vdots \\ \sim_f\sim_t A, \Sigma \Rightarrow \Delta\end{matrix} \quad \begin{matrix}\vdots \\ \sim_f\sim_t B, \Sigma \Rightarrow \Delta\end{matrix}}{\sim_f\sim_t(A\wedge_t B), \Sigma \Rightarrow \Delta}\,(\sim_f\sim_t\wedge_t\mathrm{l}).$$

We can now prove the cut-elimination theorem.

Theorem 7.2 *The rule* (cut) *is admissible in cut-free* I_{16}.

Proof Suppose $\mathrm{I}_{16} \vdash \Gamma \Rightarrow \Delta$. Then, we have $LJ \vdash f(\Gamma) \Rightarrow f(\Delta)$ by Theorem 7.1 (1), and hence LJ – (cut) $\vdash f(\Gamma) \Rightarrow f(\Delta)$ by the well-known cut-elimination theorem for LJ. By Theorem 7.1 (2), we obtain the required fact I_{16} – (cut) $\vdash \Gamma \Rightarrow \Delta$. □

Theorem 7.3 I_{16} *is decidable.*

Proof By Theorem 7.1, the provability of sequents in I_{16} can be reduced to the provability of sequents in LJ. Moreover, the translation from a formula of I_{16} into the corresponding formula of LJ can be performed in finitely many steps. Since LJ is decidable, I_{16} is also decidable. □

The following theorem states that I_{16} satisfies the constructible falsity property with respect to the negations $\sim_e$, which may be seen as an indication of the constructiveness of these negations and the system I_{16}. Whereas Nelson's constructive logics with strong negation [2, 180] satisfy the constructible falsity property with respect to strong negation, intuitionistic logic does not satisfy the constructible falsity property with respect to intuitionistic negation.

Theorem 7.4 *If* $I_{16} \vdash\ \Rightarrow \sim_e(A \wedge_t B)$, *then* $I_{16} \vdash\ \Rightarrow \sim_e A$ or $I_{16} \vdash\ \Rightarrow \sim_e B$.

Proof This is a consequence of Theorem 7.2. □

Recall the notions of explosiveness and paraconsistency of a sequent calculus with respect to a negation(-like) connective introduced by Definition 6.4.

Proposition 7.7 *Let* $\sharp$ be $\sim_f$ *or* $\sim_e$. *Then,* I_{16} *is paraconsistent with respect to* $\sharp$.

Proof Consider a sequent $p, \sharp p \Rightarrow q$ where p and q are distinct propositional variables. Then, the unprovability of this sequent is guaranteed by Theorem 7.2. □

7.3 Kripke Completeness for I_{16}

We introduce a possible worlds semantics modeled after Kripke's semantics for **IPL**.

Definition 7.4 A *Kripke frame* is a structure $\langle M, R \rangle$, where M is a non-empty set and R is a reflexive and transitive binary relation on M.

Definition 7.5 The *valuations* $\models_n, \models_t, \models_f$ and $\models_b$ on a Kripke frame $\langle M, R \rangle$ are mappings from the set *Atom* of all propositional variables to the power set 2^M of M such that for any $* \in \{n, t, f, b\}$, any $p \in$ *Atom* and any $x, y \in M$, if $x \in \models_* (p)$ and xRy, then $y \in \models_* (p)$. We will write $x \models_* p$ for $x \in \models_* (p)$. These valuations $\models_*$ ($* \in \{n, t, f, b\}$) are extended to mappings from the set of all formulas to 2^M for any $e \in \{t, b\}$ by stipulating:

1. $x \models_n \top$ holds,
2. $x \models_n \bot$ does not hold,
3. $x \models_n A \rightarrow B$ iff $\forall y \in M\,[xRy$ and $y \models_n A$ imply $y \models_n B]$,
4. $x \models_n A \wedge_t B$ iff $x \models_n A$ and $x \models_n B$,
5. $x \models_n A \vee_t B$ iff $x \models_n A$ or $x \models_n B$,
6. $x \models_n A \wedge_f B$ iff $x \models_n A$ and $x \models_n B$,
7. $x \models_n A \vee_f B$ iff $x \models_n A$ or $x \models_n B$,
8. $x \models_e \top$ does not hold,
9. $x \models_e \bot$ holds,
10. $x \models_e A \rightarrow B$ iff $x \models_n A$ and $x \models_e B$,
11. $x \models_e A \wedge_t B$ iff $x \models_e A$ or $x \models_e B$,
12. $x \models_e A \vee_t B$ iff $x \models_e A$ and $x \models_e B$,
13. $x \models_e A \wedge_f B$ iff $x \models_e A$ and $x \models_e B$,
14. $x \models_e A \vee_f B$ iff $x \models_e A$ or $x \models_e B$,
15. $x \models_f \top$ holds,
16. $x \models_f \bot$ does not hold,
17. $x \models_f A \rightarrow B$ iff $\forall y \in M\,[xRy$ and $y \models_f A$ imply $y \models_f B]$,
18. $x \models_f A \wedge_t B$ iff $x \models_f A$ and $x \models_f B$,

19. $x \models_f A \vee_t B$ iff $x \models_f A$ or $x \models_f B$,
20. $x \models_f A \wedge_f B$ iff $x \models_f A$ or $x \models_f B$,
21. $x \models_f A \wedge_f B$ iff $x \models_f A$ and $x \models_f B$,
22. $x \models_n {\sim_t} A$ iff $x \models_t A$,
23. $x \models_n {\sim_f} A$ iff $x \models_f A$,
24. $x \models_n {\sim_b} A$ iff $x \models_b A$,
25. $x \models_t {\sim_t} A$ iff $x \models_n A$,
26. $x \models_t {\sim_f} A$ iff $x \models_b A$,
27. $x \models_t {\sim_b} A$ iff $x \models_f A$,
28. $x \models_f {\sim_t} A$ iff $x \models_b A$,
29. $x \models_f {\sim_f} A$ iff $x \models_n A$,
30. $x \models_f {\sim_b} A$ iff $x \models_t A$,
31. $x \models_b {\sim_t} A$ iff $x \models_f A$,
32. $x \models_b {\sim_f} A$ iff $x \models_t A$,
33. $x \models_b {\sim_b} A$ iff $x \models_n A$.

Note that the following conditions hold: for any $* \in \{n, t, f, b\}$,

1. $x \models_* A \wedge_t B$ iff $x \models_* A \wedge_f B$,
2. $x \models_* A \vee_t B$ iff $x \models_* A \vee_f B$,
3. $x \models_t {\sim_f} A$ iff $x \models_f {\sim_t} A$,
4. $x \models_t A$ iff $x \models_n {\sim_t} A$ iff $x \models_f {\sim_b} A$ iff $x \models_b {\sim_f} A$,
5. $x \models_f A$ iff $x \models_n {\sim_f} A$ iff $x \models_t {\sim_b} A$ iff $x \models_b {\sim_t} A$,
6. $x \models_b A$ iff $x \models_n {\sim_b} A$ iff $x \models_t {\sim_f} A$ iff $x \models_f {\sim_t} A$.

Proposition 7.8 *For any $* \in \{n, t, f, b\}$, any formula A and any $x, y \in M$, if $x \models_* A$ and xRy, then $y \models_* A$.*

Proof By (simultaneous) induction on the complexity of A. □

The expression $\Gamma^\wedge$ stands for $C_1 \wedge_t C_2 \wedge_t \cdots \wedge_t C_n$ or $\top$ if $\Gamma \doteq \{C_1, C_2, \ldots, C_n\}$ or $\varnothing$, respectively. The expression Δ^* stands for A or $\perp$ if $\Delta \doteq \{A\}$ or $\varnothing$, respectively. The expression $(\Gamma \Rightarrow \Delta)^*$ abbreviates $\Gamma^\wedge \to \Delta^*$.

Definition 7.6 A *Kripke model* is a structure $\langle M, R, \models_n, \models_t, \models_f, \models_b \rangle$ such that (1) $\langle M, R \rangle$ is a Kripke frame and (2) $\models_*$ $(* \in \{n, t, f, b\})$ are valuations on $\langle M, R \rangle$.

A formula A is *true* in a Kripke model $\langle M, R, \models_n, \models_t, \models_f, \models_b \rangle$ if $x \models_n A$ for any $x \in M$, and it is *valid* in a Kripke frame $\langle M, R \rangle$ if it is true for any valuations $\models_*$ $(* \in \{n, t, f, b\})$ on the Kripke frame.

A sequent $\Gamma \Rightarrow \Delta$ is true in a Kripke model $\langle M, R, \models_n, \models_t, \models_f, \models_b \rangle$ if the formula $(\Gamma \Rightarrow \Delta)^*$ is true in the Kripke model, and it is valid in a Kripke frame $\langle M, R \rangle$ if it is true for any valuations $\models_*$ $(* \in \{n, t, f, b\})$ on the Kripke frame.

The soundness of I_{16} with respect to the above semantics can be shown straightforwardly.

Theorem 7.5 *Let C be the class of all Kripke frames, $L := \{\Gamma \Rightarrow \Delta \mid \mathrm{I}_{16} \vdash \Gamma \Rightarrow \Delta\}$ and $L(C) := \{\Gamma \Rightarrow \Delta \mid \Gamma \Rightarrow \Delta$ is valid in all frames of $C\}$. Then, $L \subseteq L(C)$.*

We now start to prove the completeness theorem.

Definition 7.7 Let x and y be sets of formulas. The pair (x,y) is *consistent* iff for any $A_1,\ldots,A_m \in x$ and any $B_1,\ldots,B_n \in y$ with $(m,n \geq 0)$, the sequent

$$A_1,\ldots,A_m \Rightarrow B_1 \vee_t \cdots \vee_t B_n$$

is not provable in I_{16}. The pair (x,y) is *maximal consistent* iff it is consistent and for every formula A, $A \in x$ or $A \in y$.

The following extension lemma can be proved using (cut).

Lemma 7.1 *Let x and y be sets of formulas. If the pair (x,y) is consistent, then there is a maximal consistent pair (x',y') such that $x \subseteq x'$ and $y \subseteq y'$.*

Proof Let $C_1, C_2, \ldots$ be an enumeration of all formulas of I_{16}. Define a sequence of pairs (x_n, y_n) $(n = 0, 1, \ldots)$ inductively by $(x_0, y_0) := (x, y)$, and $(x_{m+1}, y_{m+1}) := (x_m, y_m \cup \{C_{m+1}\})$ if $(x_m, y_m \cup \{C_{m+1}\})$ is consistent and $(x_{m+1}, y_{m+1}) := (x_m \cup \{C_{m+1}\}, y_m)$ otherwise. We can determine the fact that if (x_m, y_m) is consistent, then so is (x_{m+1}, y_{m+1}). To verify this, suppose (x_{m+1}, y_{m+1}) is not consistent. Then, there are formulas $A_1,\ldots,A_i,A'_1,\ldots,A'_j \in x_m$ and $B_1,\ldots,B_k,B'_1,\ldots,B'_l \in y_m$ such that $\mathrm{I}_{16} \vdash A_1,\ldots,A_i \Rightarrow B_1 \vee_t \cdots \vee_t B_k \vee_t C_{m+1}$ and $\mathrm{I}_{16} \vdash A'_1,\ldots,A'_j, C_{m+1} \Rightarrow B'_1 \vee_t \cdots \vee_t B'_l$. By using (cut) and some other rules, we can obtain $\mathrm{I}_{16} \vdash A_1,\ldots,A_i, A'_1,\ldots,A'_j \Rightarrow B_1 \vee_t \cdots \vee_t B_k \vee_t B'_1 \vee_t \cdots \vee_t B'_l$. This contradicts the consistency of (x_m, y_m). Hence, a produced pair (x_k, y_k) is consistent for any k. We thus obtain a maximal consistent pair $(\bigcup_{n=0}^{\infty} x_n, \bigcup_{n=0}^{\infty} y_n)$. □

Definition 7.8 Let M_L be the set of all maximal consistent pairs. A binary relation R_L on M_L is defined by $(x,w)R_L(y,z)$ iff $x \subseteq y$. Valuations $\models_N (p), \models_T (p), \models_F (p)$ and $\models_B (p)$ for any propositional variable p are defined by

1. $\models_N (p) := \{(x,w) \in M_L \mid p \in x\}$,
2. $\models_T (p) := \{(x,w) \in M_L \mid \sim_t p \in x\}$,
3. $\models_F (p) := \{(x,w) \in M_L \mid \sim_f p \in x\}$,
4. $\models_B (p) := \{(x,w) \in M_L \mid \sim_b p \in x\}$.

Lemma 7.2 *The structure $\langle M_L, R_L, \models_N, \models_T, \models_F, \models_B \rangle$ defined in Definition 7.8 is a Kripke model such that for any formula A and any $(x,w) \in M_L$,*

1. $A \in x$ iff $(x,w) \models_N A$,
2. $\sim_t A \in x$ iff $(x,w) \models_T A$,
3. $\sim_f A \in x$ iff $(x,w) \models_F A$,
4. $\sim_b A \in x$ iff $(x,w) \models_B A$.

Proof It can be shown that (1) M_L is a non-empty set, by Lemma 7.1, (2) R_L is a reflexive and transitive relation on M_L, and (3) for any $* \in \{N,T,F,B\}$, any atomic formula p and any $(x,w),(y,z) \in M_L$, if $(x,w)R_L(y,z)$ and $(x,w) \models_* (p)$, then $(y,z) \models_* (p)$. Thus, the structure $\langle M_L, R_L, \models_N, \models_T, \models_F, \models_B\rangle$ is in fact a Kripke model. □

It remains to be shown that in this model, for any formula A and any $(x,w) \in M_L$,

1. $A \in x$ iff $(x,w) \models_N A$,
2. $\sim_t A \in x$ iff $(x,w) \models_T A$,
3. $\sim_f A \in x$ iff $(x,w) \models_F A$,
4. $\sim_b A \in x$ iff $(x,w) \models_B A$.

This is shown by (simultaneous) induction on the complexity of A. The base step is obvious by Definition 7.8. We now consider the induction step below.

(1) The proof is almost the same as that of *LJ*. We show only the cases which are different from those of *LJ*.

(Case $A \stackrel{\circ}{=} \sim_t B$): $\sim_t B \in x$ iff $(x,w) \models_T B$ (by the induction hypothesis for (2)) iff $(x,w) \models_N \sim_t B$.

(Cases $A \stackrel{\circ}{=} \sim_f B$ and $A \stackrel{\circ}{=} \sim_b B$): Similar to the above case.

(2) We show some cases.

(Case $A \stackrel{\circ}{=} \top$): By the consistency of (x,w), $\sim_t \top \in x$ does not hold.

(Case $A \stackrel{\circ}{=} C \to D$): Suppose $\sim_t(C \to D) \in x$. Since $\mathrm{I}_{16} \vdash \sim_t(C \to D) \Rightarrow C$, the fact $C \in w$ contradicts the consistency of (x,w), and hence $C \in x$. Similarly, we obtain $\sim_t D \in x$. By the induction hypotheses for (1) and (2), we obtain $(x,w) \models_N C$ and $(x,w) \models_T D$, and hence $(x,w) \models_T C \to D$. Conversely, suppose $(x,w) \models_T C \to D$, i.e., $(x,w) \models_N C$ and $(x,w) \models_T D$. Then, we obtain $C \in x$ and $\sim_t D \in x$ by the induction hypotheses for (1) and (2). Since $\mathrm{I}_{16} \vdash C, \sim_t D \Rightarrow \sim_t(C \to D)$, the fact $\sim_t(C \to D) \in w$ contradicts the consistency of (x,w), and hence $\sim_t(C \to D) \notin w$. By the maximality of (x,w), we obtain $\sim_t(C \to D) \in x$.

(Case $A \stackrel{\circ}{=} C \wedge_t D$): Suppose $\sim_t(C \wedge D) \in x$. Since $\mathrm{I}_{16} \vdash \sim_t(C \wedge_t D) \Rightarrow \sim_t C \vee_t \sim_t D$, the fact $\sim_t C, \sim_t D \in w$ contradicts the consistency of (x,w), and hence $\sim_t C \notin w$ or $\sim_t D \notin w$. Thus, we obtain $\sim_t C \in x$ or $\sim_t D \in x$ by the maximality of (x,w). By the induction hypothesis for (2), we obtain $(x,w) \models_T C$ or $(x,w) \models_T D$, and hence $(x,w) \models_T C \wedge_t D$. Conversely, suppose $(x,w) \models_T C \wedge_t D$, i.e., $(x,w) \models_T C$ or $(x,w) \models_T D$. By the induction hypothesis for (1), we obtain $\sim_t C \in x$ or $\sim_t D \in x$. Since $\mathrm{I}_{16} \vdash \sim_t C \Rightarrow \sim_t(C \wedge_t D)$ and $\mathrm{I}_{16} \vdash \sim_t D \Rightarrow \sim_t(C \wedge_t D)$, the fact $\sim_t(C \wedge_t D) \in w$ contradicts the consistency of (x,w), and hence $\sim_t(C \wedge_t D) \notin w$. By the maximality of (x,w), we obtain $\sim_t(C \wedge_t D) \in x$.

(3) We show some cases.

(Case $A \stackrel{\circ}{=} C \to D$): Suppose $\sim_f(C \to D) \in x$. We will show $(x,w) \models_F C \to D$, i.e., $\forall (y,z) \in M_L[(x,w)\, R_L\, (y,z)$ and $(y,z); \models_F C$ imply $(y,z) \models_F C]$. Suppose $(x,w)R_L(y,z)$ and $(y,z) \models_F C$. Then, we have (*): $\sim_f(C \to D) \in y$ by the definition of R_L, and obtain (**): $\sim_f C \in y$ by the induction hypothesis for (3).

Since (*), (**) and $I_{16} \vdash {\sim_f}(C \to D), {\sim_f}C \Rightarrow {\sim_f}D$, the fact ${\sim_f}D \in z$ contradicts the consistency of (y,z), and hence ${\sim_f}D \notin z$. By the maximality of (y,z), we obtain ${\sim_f}D \in y$. By the induction hypothesis for (3), we obtain the required fact $(y,z) \models_F D$. Conversely, suppose ${\sim_f}(C \to D) \notin x$. Then, ${\sim_f}(C \to D) \in w$ by the maximality of (x,w). Then, the pair $(x \cup \{{\sim_f}C\}, \{{\sim_f}D\})$ is consistent because of the following reason. If it is not consistent, $I_{16} \vdash \Gamma, {\sim_f}C \Rightarrow {\sim_f}D$ for some Γ consisting of formulas in x, and hence $I_{16} \vdash \Gamma \Rightarrow {\sim_f}(C \to D)$. This fact contradicts the consistency of (x,w). By Lemma 7.1, there is a maximal consistent pair (y,z) such that $x \cup \{{\sim_f}C\} \subseteq y$ and $\{{\sim_f}D\} \subseteq z$ (thus, we have ${\sim_f}D \notin y$ by the consistency of (y,z)). As a result, we have (x,w) $R_L(y,z)$, $(y,z) \models_F C$ and not-$[(y,z) \models_F D]$ by the induction hypothesis for (3). Therefore $(x,w) \models_F C \to D$ does not hold.

(Case $A \doteq C \wedge_t D$): Suppose ${\sim_f}(C \wedge_t D) \in x$. Since $I_{16} \vdash {\sim_f}(C \wedge_t D) \Rightarrow {\sim_f}C$, the fact ${\sim_f}C \in w$ contradicts the consistency of (x,w), and hence ${\sim_f}C \in x$. Similarly, we obtain ${\sim_f}D \in x$. By the induction hypothesis for (3), we obtain $(x,w) \models_F C$ and $(x,w) \models_F D$, and hence $(x,w) \models_F C \wedge_t D$. Conversely, suppose $(x,w) \models_F C \wedge_t D$, i.e., $(x,w) \models_F C$ and $(x,w) \models_F D$. Then, we obtain ${\sim_f}C \in x$ and ${\sim_f}D \in x$ by the induction hypothesis for (3). Since $I_{16} \vdash {\sim_f}C, {\sim_f}D \Rightarrow {\sim_f}(C \wedge_t D)$, the fact ${\sim_f}(C \wedge_t D) \in w$ contradicts the consistency of (x,w), and hence ${\sim_f}(C \wedge_t D) \notin w$. By the maximality of (x,w), we obtain ${\sim_f}(C \wedge_t D) \in x$.

(4) Similar to (2).

Theorem 7.6 (*Completeness*) *Let C be the class of all Kripke frames, $L := \{\Gamma \Rightarrow \Delta \mid I_{16} \vdash \Gamma \Rightarrow \Delta\}$ and $L(C) := \{\Gamma \Rightarrow \Delta \mid \Gamma \Rightarrow \Delta$ is valid in all frames of $C\}$. Then, $L(C) \subseteq L$.*

Proof It is sufficient to show that for any sequent $\Gamma \Rightarrow \Delta$, if $\Gamma \Rightarrow \Delta$ is valid in an arbitrary frame in C, then it is provable in I_{16}. To show this, we show that if $\Gamma \Rightarrow \Delta$ is not provable in I_{16}, then there is a frame $F = \langle M_L, R_L \rangle \in C$ such that $\Gamma \Rightarrow \Delta$ is not valid in F, i.e., there is a Kripke model $\langle M_L, R_L, \models_N, \models_T, \models_F, \models_B \rangle$ such that $\Gamma \Rightarrow \Delta$ is not true in it.

Suppose that $\Gamma \Rightarrow \Delta$ is not provable in I_{16}. Then, the pair (Γ, Δ) is consistent. By Lemma 7.1, there is a maximal consistent pair (u,v) such that $\Gamma \subseteq u$ and $\Delta \subseteq v$. It is noted that if $\Delta \doteq \{A\}$, then $A \notin u$ by the consistency of (u,v).

Then, our goal is to show that $(u,v) \models_N \Gamma \Rightarrow \Delta$ does not hold in the constructed model. Here we consider only the case $\Gamma \neq \emptyset$. We show that $(u,v) \models_N \Gamma^\wedge \to \Delta^*$ does not hold, $\exists (x,z) \in M_L[[(u,v)R_L(x,z)$ and $(x,z) \models_N \Gamma^\wedge]$ and $[(x,z) \models_N \Delta^*$ does not hold]]. Taking (u,v) for (x,z), we can verify that there is $(u,v) \in M_L$ such that $[(u,v)R_L(u,v)$ and $(u,v) \models_N \Gamma^\wedge]$ and $[(u,v) \models_N \Delta^*$ does not hold]. The first argument is obvious because of the reflexivity of R_L and the fact $\Gamma \subseteq u$. The second argument is shown below. The case $\Delta \doteq \emptyset$ is obvious because $(u,v) \models_N \bot$ does not hold. The case $\Delta \doteq \{A\}$ can be proved by using Lemma 7.2 and the fact $A \notin u$ because we have the fact $A \notin u$ iff $[(u,v) \models_N A$ does not hold] by Lemma 7.2. □

7.4 Tableau Calculus IT_{16}

In order to obtain further intuitionistic trilattice logics, we will now proceed in a less strictly proof-theoretical way. The logic L^{t*} is defined as a set of formulas, and the generalized sequent calculus GL^* for L^{t*} in [272] is a proof system for proving single formulas. Restricting the generalized sequents of GL^* to at most single formulas in the various positions of a sequent would block the provability of intuitionistically unproblematic formulas such as $A \rightarrow_t (A \vee_t B)$. We therefore replace the metalogical connectives used in Odintsov's matrix presentation with their intuitionistic counterparts. Note that in the matrix definition of $\sqsupset_f$, for example, $\neg b \wedge b'$ could also be expressed using a symbol for classical co-implication. If we want to have constructive counterparts of the matrix definitions of $\rightarrow_t$ and $\rightarrow_f$, then in addition to intuitionistic implication, we have to consider the constructive co-implication from Heyting–Brouwer logic, also known as bi-intuitionistic logic. Thus, each of the compound formulas that are used in the matrix presentations may be seen as a formula from bi-intuitionistic logic, and for each of these four formulas, we state its corresponding pair of tableau rules.

We will introduce a tableau calculus IT_{16} in the style of the tableau systems presented in [198]. A tableau is a rooted tree. The construction of a tableau starts with an expression $A, 0, \Theta$, where A is a formula from $\mathscr{L}^*_{tf}$, $\Theta \in \{\Delta_i, \Gamma_i\}$, and $1 \leq i \leq 4$. Intuitively, $A, 0, \Delta_i$ can be read as "at state 0, the value of formula A does not contain the element from **4** encoded by i", and $A, 0, \Gamma_i$ can be read as "at state 0, the value of formula A contains the element from **4** encoded by i".

The symbols Δ_i and Γ_i are distinct formal signs generalizing the signs T and F used in tableaux for intuitionistic logic. If j names a state, then A, j, Δ_i and A, j, Γ_i are a pair of contradictory expressions, the appearance of which closes a tableau branch.

To an expression $A, 0, \Theta$ decomposition rules and structural tableau rules may (or may not) be applied to obtain a tableau. Some decomposition rules introduce expressions of the form $j\mathsf{r}k$. Intuitively, $j\mathsf{r}k$ provides semantical information about an intuitionistic Kripke frame. It indicates that the information state k is a possible expansion of the information state j.

The indices j, k, l are (names of) elements from $\mathbb{N}$, and a new index is the (name of the) smallest natural number not already used in the tableau. If an expression is placed below an arrow, it is a rule output; otherwise it is a rule input. Note that the structural rule ref for which no input is stated may be applied to any node to introduce an expression $j\mathsf{r}j$ if j occurs on the tableau, that some decomposition rules introduce a new information state, and that if an expression of the form $j\mathsf{r}k$ $(k\mathsf{r}j)$ is a rule input on a tableau branch, the rule has to be re-applied after the introduction of an expression $j\mathsf{r}l$ $(l\mathsf{r}j)$ not already on the branch. If a rule application yields an expression already on the branch, the rule is not applied.

Definition 7.9 A tableau branch is said to be *closed* iff there are expressions of the form A, k, Δ_i and A, k, Γ_i on the branch. A tableau is called *closed* iff all of its branches are closed. If a tableau (tableau branch) is not closed, it is called *open*. A tableau branch is said to be *complete* iff no more rules can be applied to expand it. A tableau is said to be *complete* iff each of its branches is complete.

Definition 7.10 The tableau calculus IT_{16} consists of the structural and persistence rules from Table 7.1 and the following decomposition rules:

$$
\begin{array}{cccc}
\sim_t A,j,\Delta_1 & \sim_t A,j,\Gamma_1 & \sim_t A,j,\Delta_2 & \sim_t A,j,\Gamma_2 \\
\downarrow & \downarrow & \downarrow & \downarrow \\
A,j,\Delta_3 & A,j,\Gamma_3 & A,j,\Delta_4 & A,j,\Gamma_4
\end{array}
$$

$$
\begin{array}{cccc}
\sim_t A,j,\Delta_3 & \sim_t A,j,\Gamma_3 & \sim_t A,j,\Delta_4 & \sim_t A,j,\Gamma_4 \\
\downarrow & \downarrow & \downarrow & \downarrow \\
A,j,\Delta_1 & A,j,\Gamma_1 & A,j,\Delta_2 & A,j,\Gamma_2
\end{array}
$$

$$
\begin{array}{cccc}
A \vee_t B,j,\Delta_1 & A \vee_t B,j,\Gamma_1 & A \vee_t B,j,\Delta_2 & A \vee_t B,j,\Gamma_2 \\
\swarrow\searrow & \downarrow & \swarrow\searrow & \downarrow \\
A,j,\Delta_1 \quad B,j,\Delta_1 & A,j,\Gamma_1\, B,j,\Gamma_1 & A,j,\Delta_2 \quad B,j,\Delta_2 & A,j,\Gamma_2\, B,j,\Gamma_2
\end{array}
$$

$$
\begin{array}{cccc}
A \vee_t B,j,\Delta_3 & A \vee_t B,j,\Gamma_3 & A \vee_t B,j,\Delta_4 & A \vee_t B,j,\Gamma_4 \\
\downarrow & \swarrow\searrow & \downarrow & \swarrow\searrow \\
A,j,\Delta_3\, B,j,\Delta_3 & A,j,\Gamma_3 \quad B,j,\Gamma_3 & A,j,\Delta_4\, B,j,\Delta_4 & A,j,\Gamma_4 \quad B,j,\Gamma_4
\end{array}
$$

$$
\begin{array}{cccc}
A \wedge_t B,j,\Delta_1 & A \wedge_t B,j,\Gamma_1 & A \wedge_t B,j,\Delta_2 & A \wedge_t B,j,\Gamma_2 \\
\downarrow & \swarrow\searrow & \downarrow & \swarrow\searrow \\
A,j,\Delta_1 B,j,\Delta_1 & A,j,\Gamma_1 \quad B,j,\Gamma_1 & A,j,\Delta_2\, B,j,\Delta_2 & A,j,\Gamma_2 \quad B,j,\Gamma_2
\end{array}
$$

$$
\begin{array}{cccc}
A \wedge_t B,j,\Delta_3 & A \wedge_t B,j,\Gamma_3 & A \wedge_t B,j,\Delta_4 & A \wedge_t B,j,\Gamma_4 \\
\swarrow\searrow & \downarrow & \swarrow\searrow & \downarrow \\
A,j,\Delta_3 \quad B,j,\Delta_3 & A,j,\Gamma_3\, B,j,\Gamma_3 & A,j,\Delta_4 \quad B,j,\Delta_4 & A,j,\Gamma_4\, B,j,\Gamma_4
\end{array}
$$

$$
\begin{array}{cccc}
A \rightarrow_t B,j,\Delta_1 & A \rightarrow_t B,j,\Gamma_1 & A \rightarrow_t B,j,\Delta_2 & A \rightarrow_t B,j,\Gamma_2 \\
krj & \downarrow & krj & \downarrow \\
\swarrow\searrow & krj & \swarrow\searrow & krj \\
A,k,\Gamma_1 \quad B,k,\Delta_1 & A,k,\Delta_1\, B,k,\Gamma_1 & A,k,\Gamma_2 \quad B,k,\Delta_2 & A,k,\Delta_2\, B,k,\Gamma_2
\end{array}
$$

$$
\begin{array}{cccc}
A \rightarrow_t B,j,\Delta_3 & A \rightarrow_t B,j,\Gamma_3 & A \rightarrow_t B,j,\Delta_4 & A \rightarrow_t B,j,\Gamma_4 \\
\downarrow & jrk & \downarrow & jrk \\
jrk & \swarrow\searrow & jrk & \swarrow\searrow \\
A,k,\Gamma_3\, B,k,\Delta_3 & A,k,\Delta_3 \quad B,k,\Gamma_3 & A,k,\Gamma_4\, B,k,\Delta_4 & A,k,\Delta_4 \quad B,k,\Gamma_4
\end{array}
$$

$$\begin{array}{cccc}
\sim_f A,j,\Delta_1 & \sim_f A,j,\Gamma_1 & \sim_f A,j,\Delta_2 & \sim_f A,j,\Gamma_2 \\
\downarrow & \downarrow & \downarrow & \downarrow \\
A,j,\Delta_2 & A,j,\Gamma_2 & A,j,\Delta_1 & A,j,\Gamma_1
\end{array}$$

$$\begin{array}{cccc}
\sim_f A,j,\Delta_3 & \sim_f A,j,\Gamma_3 & \sim_f A,j,\Delta_4 & \sim_f A,j,\Gamma_4 \\
\downarrow & \downarrow & \downarrow & \downarrow \\
A,j,\Delta_4 & A,j,\Gamma_4 & A,j,\Delta_3 & A,j,\Gamma_3
\end{array}$$

$$\begin{array}{cccc}
A \vee_f B,j,\Delta_1 & A \vee_f B,j,\Gamma_1 & A \vee_f B,j,\Delta_2 & A \vee_f B,j,\Gamma_2 \\
\downarrow & \swarrow\searrow & \swarrow\searrow & \downarrow \\
A,j,\Delta_1\; B,j,\Delta_1 & A,j,\Gamma_1 \quad B,j,\Gamma_1 & A,j,\Delta_2 \quad B,j,\Delta_2 & A,j,\Gamma_2\; B,j,\Gamma_2
\end{array}$$

$$\begin{array}{cccc}
A \vee_f B,j,\Delta_3 & A \vee_f B,j,\Gamma_3 & A \vee_f B,j,\Delta_4 & A \vee_f B,j,\Gamma_4 \\
\downarrow & \swarrow\searrow & \swarrow\searrow & \downarrow \\
A,j,\Delta_3\; B,j,\Delta_3 & A,j,\Gamma_3 \quad B,j,\Gamma_3 & A,j,\Delta_4 \quad B,j,\Delta_4 & A,j,\Gamma_4\; B,j,\Gamma_4
\end{array}$$

$$\begin{array}{cccc}
A \wedge_f B,j,\Delta_1 & A \wedge_f B,j,\Gamma_1 & A \wedge_f B,j,\Delta_2 & A \wedge_f B,j,\Gamma_2 \\
\swarrow\searrow & \downarrow & \downarrow & \swarrow\searrow \\
A,j,\Delta_1 \quad B,j,\Delta_1 & A,j,\Gamma_1\; B,j,\Gamma_1 & A,j,\Delta_2\; B,j,\Delta_2 & A,j,\Gamma_2 \quad B,j,\Gamma_2
\end{array}$$

$$\begin{array}{cccc}
A \wedge_f B,j,\Delta_3 & A \wedge_f B,j,\Gamma_3 & A \wedge_f B,j,\Delta_4 & A \wedge_f B,j,\Gamma_4 \\
\swarrow\searrow & \downarrow & \downarrow & \swarrow\searrow \\
A,j,\Delta_3 \quad B,j,\Delta_3 & A,j,\Gamma_3\; B,j,\Gamma_3 & A,j,\Delta_4\; B,j,\Delta_4 & A,j,\Gamma_4 \quad B,j,\Gamma_4
\end{array}$$

$$\begin{array}{cccc}
A \rightarrow_f B,j,\Delta_1 & A \rightarrow_f B,j,\Gamma_1 & A \rightarrow_f B,j,\Delta_2 & A \rightarrow_f B,j,\Gamma_2 \\
\downarrow & j\,r\,k & k\,r\,j & \downarrow \\
j\,r\,k & \swarrow\searrow & \swarrow\searrow & k\,r\,j \\
A,k,\Gamma_1\; B,k,\Delta_1 & A,k,\Delta_1 \quad B,k,\Gamma_1 & A,k,\Gamma_2 \quad B,k,\Delta_2 & A,k,\Delta_2\; B,k,\Gamma_2
\end{array}$$

$$\begin{array}{cccc}
A \rightarrow_f B,j,\Delta_3 & A \rightarrow_f B,j,\Gamma_3 & A \rightarrow_f B,j,\Delta_4 & A \rightarrow_f B,j,\Gamma_4 \\
\downarrow & j\,r\,k & k\,r\,j & \downarrow \\
j\,r\,k & \swarrow\searrow & \swarrow\searrow & k\,r\,j \\
A,k,\Gamma_3\; B,k,\Delta_3 & A,k,\Delta_3 \quad B,k,\Gamma_3 & A,k,\Gamma_4 \quad B,k,\Delta_4 & A,k,\Delta_4\; B,k,\Gamma_4
\end{array}$$

If k occurs below an arrow, then it is a *new* index.

Table 7.1 Structural tableau rules and persistence rules

ref	tran	pers
$\cdot$	jrk	p, j, Γ_i
$\downarrow$	krl	jrk
krk	$\downarrow$	$\downarrow$
	jrl	p, k, Γ_i
for all k on the tableau		$1 \leq i \leq 4$

The decomposition rule for a connective $\sharp$ in position Δ_i or Γ_i for $i \in \{1,2,3,4\}$ is denoted by $(\sharp\Delta_i)$ or $(\sharp\Gamma_i)$, respectively. Note that a tableau of the tableau calculus IT_{16} may contain infinite branches. In such a case, the tableau is said to be infinite.

Example 7.1 The following tableau, where p is an atomic formula, has an infinite branch indicated by vertical dots:

$$((p \to_f (p \to_f p)) \to_f (p \to_f p)) \to_f p, 0, \Delta_1$$
$$\downarrow$$
$$0r0$$
$$0r1$$
$$(p \to_f (p \to_f p)) \to_f (p \to_f p), 1, \Gamma_1$$
$$p, 1, \Delta_1$$
$$\downarrow$$
$$1r1$$
$$\swarrow \qquad \searrow$$
$$(p \to_f (p \to_f p)), 1, \Delta_1 \quad \ldots$$
$$\downarrow$$
$$1r2$$
$$p, 2, \Gamma_1$$
$$p \to_f p, 2, \Delta_1$$
$$\swarrow \qquad \searrow$$
$$(p \to_f (p \to_f p)), 2, \Delta_1 \quad \ldots$$
$$\downarrow$$
$$2r3$$
$$p, 3, \Gamma_1$$
$$p \to_f p, 3, \Delta_1$$
$$\vdots$$

Definition 7.11 An $\mathscr{L}^*_{tf}$-formula A is *provable in position* Δ_i $(1 \leq i \leq 4)$ iff there is a closed complete tableau for $A, 0, \Gamma_i$. The formula A is provable in position Γ_i $(1 \leq i \leq 4)$ iff there is a closed complete tableau for $A, 0, \Delta_i$.

Definition 7.12 An $\mathscr{L}^*_{tf}$-formula A is *provable in the tableau calculus* IT_{16} iff A is provable in positions $\Delta_1, \Delta_2, \Gamma_3$, and Γ_4.

Thus, a formula A is not provable iff A is not provable in position Δ_1 or it is not provable in position Δ_2 or it is not provable in position Γ_3 or it is not provable in Γ_4. This is the case iff there is an open branch in every complete tableau for $A, 0, \Gamma_1$ or an open branch in every complete tableau for $A, 0, \Gamma_2$ or an open branch in every complete tableau for $A, 0, \Delta_3$ or an open branch in every complete tableau for $A, 0, \Delta_4$. If we want to prove a formula in IT_{16}, we need not, of course, expand a closed branch.

Example 7.2 A proof of a formula consists of four closed tableau. Let p and q be any propositional variables and consider $p \rightarrow_t (q \rightarrow_t p)$. The provability of this formula in positions Δ_1 and Δ_2 can be shown by completely analogous closed tableaux and so can the provability in positions Γ_3 and Γ_4. We show only one tableau for provability in position Δ_1 and one for provability in position Γ_3. In both tableaux the final rule applied is pers.

$p \rightarrow_t (q \rightarrow_t p), 0, \Gamma_1$	$p \rightarrow_t (q \rightarrow_t p), 0, \Delta_3$
$\downarrow$	$\downarrow$
1r0	0r1
$p, 1, \Delta_1$	$p, 1, \Gamma_1$
$q \rightarrow_t p, 1, \Gamma_1$	$q \rightarrow_t p, 1, \Delta_1$
$\downarrow$	$\downarrow$
2r1	1r2
$q, 2, \Delta_1$	$q, 2, \Gamma_1$
$p, 2, \Gamma_1$	$p, 2, \Delta_1$
$\downarrow$	$\downarrow$
$p, 1, \Gamma_1$	$p, 2, \Gamma_1$

7.5 Kripke Completeness for IT_{16}

We again define a relational semantics with four kinds of evaluation relations.

Definition 7.13 A *Kripke model* is a structure $\mathscr{M} = \langle \mathscr{F}, \models_1, \models_2, \models_3, \models_4, v \rangle$, where $\mathscr{F} = \langle M, R \rangle$ is a Kripke frame as introduced in Definition 7.4 and v is a function from $Atom \times M$ into **16** satisfying the following persistence condition: for all

$x, y \in M$ and $\sharp \in \mathbf{4}$, if xRy and $\sharp \in v(p,x)$, then $\sharp \in v(p,y)$. For $\mathscr{M} = \langle\langle M, R\rangle, \models_1, \models_2, \models_3, \models_4, v\rangle$, the relations $\models_i \subseteq (M \times \mathscr{L}^*_{tf})(1 \le i \le 4)$ are defined as follows:

$$x \models_1 p \text{ iff } \mathbf{N} \in v(p,x),$$
$$x \models_2 p \text{ iff } \mathbf{F} \in v(p,x),$$
$$x \models_3 p \text{ iff } \mathbf{T} \in v(p,x),$$
$$x \models_4 p \text{ iff } \mathbf{B} \in v(p,x),$$

$$x \models_1 \sim_t A \text{ iff } x \models_3 A,$$
$$x \models_2 \sim_t A \text{ iff } x \models_4 A,$$
$$x \models_3 \sim_t A \text{ iff } x \models_1 A,$$
$$x \models_4 \sim_t A \text{ iff } x \models_2 A,$$

$$x \models_1 A \wedge_t B \text{ iff } x \models_1 A \text{ or } x \models_1 B,$$
$$x \models_2 A \wedge_t B \text{ iff } x \models_2 A \text{ or } x \models_2 B,$$
$$x \models_3 A \wedge_t B \text{ iff } x \models_3 A \text{ and } x \models_3 B,$$
$$x \models_4 A \wedge_t B \text{ iff } x \models_4 A \text{ and } x \models_4 B.$$

$$x \models_1 A \vee_t B \text{ iff } x \models_1 A \text{ and } x \models_1 B,$$
$$x \models_2 A \vee_t B \text{ iff } x \models_2 A \text{ and } x \models_2 B,$$
$$x \models_3 A \vee_t B \text{ iff } x \models_3 A \text{ or } x \models_3 B,$$
$$x \models_4 A \vee_t B \text{ iff } x \models_4 A \text{ or } x \models_4 B,$$

$$x \models_1 A \rightarrow_t B \text{ iff } (\exists y \in M)yRx \text{ and } y \not\models_1 A \text{ and } y \models_1 B,$$
$$x \models_2 A \rightarrow_t B \text{ iff } (\exists y \in M)yRx \text{ and } y \not\models_2 A \text{ and } y \models_2 B,$$
$$x \models_3 A \rightarrow_t B \text{ iff } (\forall y \in M)xRy \text{ implies } (y \not\models_3 A \textit{ or } y \models_3 B),$$
$$x \models_4 A \rightarrow_t B \text{ iff } (\forall y \in M)xRy \text{ implies } (y \not\models_4 A \text{ or } y \models_4 B),$$

$$x \models_1 \sim_f A \text{ iff } x \models_2 A,$$
$$x \models_2 \sim_f A \text{ iff } x \models_1 A,$$
$$x \models_3 \sim_f A \text{ iff } x \models_4 A,$$
$$x \models_4 \sim_f A \text{ iff } x \models_3 A,$$

$$x \models_1 A \wedge_f B \text{ iff } x \models_1 A \text{ and } x \models_1 B,$$
$$x \models_2 A \wedge_f B \text{ iff } x \models_2 A \text{ or } x \models_2 B,$$
$$x \models_3 A \wedge_f B \text{ iff } x \models_3 A \text{ and } x \models_3 B,$$
$$x \models_4 A \wedge_f B \text{ iff } x \models_4 A \text{ or } x \models_4 B,$$

$$x \models_1 A \vee_f B \text{ iff } x \models_1 A \text{ or } x \models_1 B,$$
$$x \models_2 A \vee_f B \text{ iff } x \models_2 A \text{ and } x \models_2 B,$$
$$x \models_3 A \vee_f B \text{ iff } x \models_3 A \text{ or } x \models_3 B,$$
$$x \models_4 A \vee_f B \text{ iff } x \models_4 A \text{ and } x \models_4 B,$$

$$x \models_1 A \rightarrow_f B \text{ iff } (\forall y \in M) xRy \text{ implies } (y \not\models_1 A \text{ or } y \models_1 B),$$
$$x \models_2 A \rightarrow_f B \text{ iff } (\exists y \in M) yRx \text{ and } y \not\models_2 A \text{ and } y \models_2 B,$$
$$x \models_3 A \rightarrow_f B \text{ iff } (\forall y \in M) xRy \text{ implies } (y \not\models_3 A \text{ or } y \models_3 B),$$
$$x \models_4 A \rightarrow_f B \text{ iff } (\exists y \in M) yRx \text{ and } y \not\models_4 A \text{ and } y \models_4 B.$$

The following proposition shows how the relations $\models_i$ ($i \in \{1, 2, 3, 4, \}$) are related to the valuations $\models_*$ ($* \in \{n, t, f, b\}$).

Proposition 7.9 *For every (implication-free) $\mathcal{L}_{tf}$-formula A:*

1. $(\forall \mathcal{M}(\forall x \in M)\forall \models_n, \models_f, \models_t, \models_b \colon x \models_n A)$ iff
   $(\forall \mathcal{M}(\forall x \in M)\forall \models_1, \models_2, \models_3, \models_4 \colon x \models_4 A)$,
2. $(\forall \mathcal{M}(\forall x \in M)\forall \models_n, \models_f, \models_t, \models_b \colon x \models_f A)$ iff
   $(\forall \mathcal{M}(\forall x \in M)\forall \models_1, \models_2, \models_3, \models_4 \colon x \models_3 A)$,
3. $(\forall \mathcal{M}(\forall x \in M)\forall \models_n, \models_f, \models_t, \models_b \colon x \models_t A)$ iff
   $(\forall \mathcal{M}(\forall x \in M)\forall \models_1, \models_2, \models_3, \models_4 \colon x \models_2 A)$, and
4. $(\forall \mathcal{M}(\forall x \in M)\forall \models_n, \models_f, \models_t, \models_b \colon x \models_b A)$ iff
   $(\forall \mathcal{M}(\forall x \in M)\forall \models_1, \models_2, \models_3, \models_4 \colon x \models_1 A)$.

Proof By simultaneous induction on A. For atoms and negated atoms the claims hold trivially. □

Proposition 7.10 *Let $\mathcal{M} = \langle \mathcal{F}, \models_1, \models_2, \models_3, \models_4, v\rangle$ be a Kripke model and A be an $\mathcal{L}^*_{tf}$-formula. Then for all $x, y \in M$ and $i \in \{1, 2, 3, 4\}$, if xRy and $x \models_i A$, then $y \models_i A$.*

Proof By (simultaneous) induction on the complexity of A. □

Definition 7.14 Suppose that $\mathcal{M} = \langle\langle M, R\rangle, \models_1, \models_2, \models_3, \models_4, v\rangle$ is a Kripke model. An $\mathcal{L}^*_{tf}$-formula A is IT$_{16}$-*valid in* $\mathcal{M}$ iff for every $x \in M : x \not\models_1 A, x \not\models_2 A, x \models_3 A, x \models_4 A$. An $\mathcal{L}^*_{tf}$-formula A is IT$_{16}$-*valid* iff it is IT$_{16}$-valid in every Kripke model.

Example 7.3 Here is a simple example of a complete and closed tableau.

$$p \vee_t (p \rightarrow_t q), 0, \Delta_4$$
$$\downarrow$$
$$p, 0, \Delta_4$$
$$p \rightarrow_t q, 0, \Delta_4$$
$$\downarrow$$
$$0\mathsf{r}1$$
$$p, 1, \Gamma_4$$
$$q, 1, \Delta_4$$
$$\downarrow$$
$$0\mathsf{r}0$$
$$1\mathsf{r}1$$

From this tableau one can easily read off a model based on the set of indices $\{0, 1\}$ with $0 \not\models_4 p, 0 \not\models_4 q, 1 \models_4 p$, and $1 \not\models_4 q$.Thus, $0 \not\models_4 p \vee_t (p \rightarrow_t q)$ and therefore $p \vee_t (p \rightarrow_t q)$ fails to be IT_{16}-valid.

Definition 7.15 Let $\mathcal{M} = \langle\langle M, R\rangle, \models_1, \models_2, \models_3, \models_4, v\rangle$ be a Kripke model and *br* be a tableau branch. The model $\mathcal{M}$ is *faithful to br* iff there exists a function *f* from $\mathbb{N}$ into *M* such that:

1. for every expression A, j, Γ_i on $br, f(j) \models_i A$;
2. for every expression A, j, Δ_i on $br, f(j) \not\models_i A$;
3. for every expression $j\mathsf{r}k$ on $br, f(j)Rf(k)$.

The function *f* is said to *show that $\mathcal{M}$ is faithful to branch br*.

Lemma 7.3 *Let $\mathcal{M} = \langle\langle M, R\rangle, \models_1, \models_2, \models_3, \models_4, v\rangle$ be a Kripke model and br be a tableau branch. If $\mathcal{M}$ is faithful to br and a tableau rule is applied to br, then the application produces at least one extension br' of br such that $\mathcal{M}$ is faithful to br'.*

Proof Assume that *f* is a function that shows $\mathcal{M}$ to be faithful to *br*. We have to consider all tableau rules. The cases of the tableau rules encoding intuitionistic connectives (including the decomposition rules for $\wedge_t$, $\wedge_f$, $\vee_t$, and $\vee_f$, and part of the rules for $\rightarrow_t$ and $\rightarrow_f$) and the rules **ref**, **tran**, and **pers** are familiar from the tableau calculi for intuitionistic logic, cf. [198, Sect. 6.4]. Consider one case of a decomposition rule for $\sim_t$, the cases of the other decomposition rules for $\sim_t$ and the decomposition rules for $\sim_f$ are similar.

($\sim_t\Delta_1$): Suppose that $\sim_t A, j, \Delta_1$ is on branch *br* and that $(\sim_t\Delta_1)$ has been applied to obtain A, j, Δ_3. Since *f* shows that $\mathcal{M}$ is faithful to $br, f(j) \not\models_1 A$. But then $f(k) \not\models_3 A$ and $\mathcal{M}$ is faithful to *br*.

Finally, consider $(\rightarrow_f \Delta_4)$ and $(\rightarrow_f \Gamma_4)$. That is, consider two of the four decomposition rules for $\rightarrow_f$ that encode the backward-looking co-implication connective from Heyting–Brouwer logic.

$(\rightarrow_f \Delta_4)$: Suppose that $A \rightarrow_f B, j, \Delta_4$ and $k\mathfrak{r}j$ are on br and that $(\rightarrow_f \Delta_4)$ has been applied to get one branch br_1 with A, k, Γ_4 on it and another branch br_2 with B, k, Δ_4 on it. By faithfulness, $f(j) \not\models_4 A \rightarrow_f B$. Therefore, for every $y \in M, yRf(j)$ implies $(y \models_4 A$ or $y \not\models_4 B)$. Since, by faithfulness, $f(k)Rf(j)$, it follows that $f(k) \models_4 A$ or $f(k) \not\models_4 B$. Thus, $\mathcal{M}$ is faithful to br_1 or to br_2.

$(\rightarrow_f \Gamma_4)$: Let $A \rightarrow_f B, j, \Gamma_4$ be on branch br. We apply $(\rightarrow_f \Gamma_4)$ and obtain a new branch br' with $k\mathfrak{r}j, A, k, \Delta_4$, and B, k, Γ_4 on it, where k is new to the tableau. Then $f(j) \models_4 A \rightarrow_f B$ and thus there exists $y \in M$ with $yRf(j)$, $y \not\models_4 A$ and $y \models_4 B$. Define f' to be the same function as f, except that $f'(k) = y$, Then f' shows $\mathcal{M}$ to be faithful to br'. □

Theorem 7.7 *If an $\mathcal{L}^*_{tf}$-formula A is s provable in* IT_{16}, *then A is* IT_{16}*-valid.*

Proof Suppose A is not IT_{16}-valid. Then there is a model $\langle\langle M, R\rangle, \models_1, \models_2, \models_3, \models_4, v\rangle$ and $x \in M$ such that $x \models_1 A, x \models_2 A, x \not\models_3 A$, or $x \not\models_4 A$. Assume now, for reductio, that A is provable in IT_{16}. There are complete tableaux $\tau_1, \tau_2, \tau_3, \tau_4$ for $A, 0, \Gamma_1; A, 0, \Gamma_2; A, 0, \Delta_3; A, 0, \Delta_4$; respectively. The model $\mathcal{M}$ is faithful to the four branches consisting of the initial nodes $A, 0, \Gamma_1$; $A, 0, \Gamma_2; A, 0, \Delta_3; A, 0, \Delta_4$. By Lemma 7.3, the tableaux $\tau_1, \tau_2, \tau_3, \tau_4$ contain branches such that $\mathcal{M}$ is faithful to every initial segment of these branches. If one of these branches, say br, is closed, there are expressions B, n, Δ_i and B, n, Γ_i $(1 \leq i \leq 4)$ on some initial segment of br. Since $\mathcal{M}$ is faithful to br, this is impossible. Thus, A is not provable in IT_{16}. □

Corollary 7.1 *The rule* PERS

$$\begin{array}{c} A, j, \Gamma_i \\ j\mathfrak{r}k \\ \downarrow \\ A, k, \Gamma_i \\ 1 \leq i \leq 4 \end{array}$$

is admissible for arbitrary $\mathcal{L}^*_{tf}$-formulas A.

Proof By Proposition 7.10, applications of PERS preserve the faithfulness of a Kripke model to a tableau branch. □

We now prove completeness of the tableau calculus IT_{16} with respect to the class of all Kripke models.

Definition 7.16 Let br be an open branch of a complete tableau. The model $\mathcal{M}_{br} = \langle\langle M_{br}, R_{br}\rangle, \models_1, \models_2, \models_3, \models_4, v_{br}\rangle$ *induced by* br is defined as follows:

1. $M_{br} := \{w_j \mid j$ occurs on $br\}$,
2. $w_j R_{br} w_k$ iff $j\mathfrak{r}k$ occurs on br,

3. $\mathbf{N} \in v_{br}(p, w_j)$ iff p, j, Γ_1 occurs on br,
4. $\mathbf{F} \in v_{br}(p, w_j)$ iff p, j, Γ_2 occurs on br,
5. $\mathbf{T} \in v_{br}(p, w_j)$ iff p, j, Γ_3 occurs on br,
6. $\mathbf{B} \in v_{br}(p, w_j)$ iff p, j, Γ_4 occurs on br.

The model $\mathcal{M}_{br}$ is well-defined, because br is open. There exists no $w_j \in M_{br}$, no atomic formula p, and no $\sharp \in \{\mathbf{N}, \mathbf{F}, \mathbf{T}, \mathbf{B}\}$ such that $\sharp \in v_{br}(p, w_j)$ and $\sharp \notin v_{br}(p, w_j)$. Moreover, the relation R_{br} is reflexive and transitive and the persistence condition is satisfied, because br is complete and hence the rules ref, tran, and pers have been applied.

Lemma 7.4 *Suppose that br is an open branch of a complete tableau, and let* $\mathcal{M}_{br} = \langle\langle M_{br}, R_{br}\rangle, \models_1, \models_2, \models_3, \models_4, v_{br}\rangle$ *be the model induced by br. Then for every* $i \in \{1, 2, 3, 4\}$:

1. if A, j, Γ_i occurs on br, then $w_j \models_i A$,
2. if A, j, Δ_i occurs on br, then $w_j \not\models_i A$.

Proof By straightforward induction on A. If A is atomic, Claim 1 holds by definition of $\mathcal{M}_{br}$. Claim 2 holds by definition of $\mathcal{M}_{br}$ and the fact that br is open. We here consider one case where A is an implication $B \rightarrow_f C$.

1. Suppose $B \rightarrow_f C, j, \Gamma_2$ occurs on br. Since br is complete, there is a k with $k\mathsf{r}j, B, k, \Delta_2$, and C, k, Γ_2 on br. By definition of $\mathcal{M}_{br}$ and the induction hypothesis, there exists w_k with $w_k R_{br} w_j, w_k \models_2 C$, and $w_k \not\models_2 B$. Therefore, $w_j \models_2 B \rightarrow_f C$.
2. Suppose $B \rightarrow_f C, j, \Delta_2$ occurs on br. For every k such that $k\mathsf{r}j$ occurs on br, B, k, Γ_2 is on br or C, k, Δ_2 is on br. By definition of $\mathcal{M}_{br}$ and the induction hypothesis, for every k such that $w_k R_{br} w_j, w_k \models_2 B$ or $w_k \not\models_2 C$. Thus, $w_j \not\models_2 B \rightarrow_f C$. □

Theorem 7.8 *Let A be an* $\mathcal{L}^*_{tf}$*-formula. If A is* IT_{16}*-valid, then A is provable in* IT_{16}.

Proof Suppose that A is not provable in IT_{16}. Then there is an open branch br_1 of a complete tableau for $A, 0, \Gamma_1$ or an open branch br_2 of a complete tableau for $A, 0, \Gamma_2$ or an open branch br_3 of a complete tableau for $A, 0, \Delta_3$ or an open branch br_4 of a complete tableau for $A, 0, \Delta_4$. By Lemma 7.4, $w_0 \models_1 A$ or $w_0 \models_2 A$ or $w_0 \not\models_3 A$ or $w_0 \not\models_4 A$. In any case, A is not valid in $\mathcal{M}_{br_i}$ for some $i \in \{1, 2, 3, 4\}$, and hence A is not IT_{16}-valid. □

Corollary 7.2 *If one complete tableau for* $A, 0, \Gamma_1$ *is closed, one complete tableau for* $A, 0, \Gamma_2$ *is closed, one complete tableau for* $A, 0, \Delta_3$ *is closed, and one complete tableau for* $A, 0, \Delta_4$ *is closed, then every complete tableau for* $A, 0, \Gamma_1$ *is closed, every complete tableau for* $A, 0, \Gamma_2$ *is closed, every complete tableau for* $A, 0, \Delta_3$ *is closed, and every complete tableau for* $A, 0, \Delta_4$ *is closed.*

Proof Suppose (i) that there are complete tableaux for $A, 0, \Gamma_1, A, 0, \Gamma_2, A, 0, \Delta_3$, and $A, 0, \Delta_4$ that are closed, but (ii) that there is a complete and open tableau for $A, 0, \Gamma_1, A, 0, \Gamma_2, A, 0, \Delta_3$, or $A, 0, \Delta_4$. By (i) and Theorem 7.7, the formula A is IT_{16}-valid. By (ii) and Lemma 7.4, A is not IT_{16}-valid, a contradiction. □

The calculus IT_{16} is an intuitionistic tableau-counterpart of the sequent calculus GL^* for truth entailment in $SIXTEEN_3$. A sequent calculus for falsity entailment in $SIXTEEN_3$ can be obtained form GL^* by suitably modifying the notion of provability of an $\mathscr{L}^*_{tf}$-formula in the sequent calculus. In a similar way, the notion of provability in IT_{16} can be modified.

Definition 7.17 The tableau calculus IF_{16} is the same as IT_{16}. An $\mathscr{L}^*_{tf}$-formula A is *provable in* IF_{16} iff A is provable in positions $\Gamma_1, \Delta_2, \Gamma_3$, and Δ_4.

Chapter 8
Generalized Truth Values and Many-Valued Logics: Harmonious Many-Valued Logics

Abstract In this chapter, we reconsider the notion of an n-valued propositional logic. In many-valued logic, sometimes a distinction is made not only between designated and undesignated (not designated) truth values, but also between designated and antidesignated truth values. Even if the set of truth values is, in fact, tripartitioned, usually only a single semantic consequence relation is defined that preserves the possession of a designated value from the premises to the conclusions of an inference. We argue that if in the set of semantical values the sets of designated and antidesignated truth values are not complements of each other, it is natural to define at least *two* entailment relations, a "positive" one that preserves the possession of a designated value from the premises to the conclusions of an inference, and a "negative" one that preserves the possession of an antidesignated value from the conclusions to the premises. Once this distinction has been drawn, it is quite natural to reflect it in the logical object language and to contemplate many-valued logics Λ whose language is split into a positive and a matching negative logical vocabulary. If the positive and the negative entailment relations do not coincide, if the interpretations of matching pairs of connectives are distinct, and if the positive entailment relation restricted to the positive vocabulary is nevertheless isomorphic to the negative entailment relation restricted to the negative vocabulary, then we say that Λ is a *harmonious* many-valued logic. We reconstruct some of the logical systems considered in this book as harmonious, finitely-valued logics. At the end of the chapter, we outline some possible ways of generalizing the notion of a harmonious n-valued propositional logic.

8.1 Many-Valued Propositional Logics Generalized

Let us hark back to the notion of a valuation system considered in Sect. 1.6. Recall that one of the most important components of any valuation system $\mathbf{V} = \langle \mathscr{V}, \mathscr{D}, \mathscr{F} \rangle$ is the set $\mathscr{D}$ of *designated truth values*. This set is essential,

Y. Shramko and H. Wansing, *Truth and Falsehood*, Trends in Logic 36, DOI: 10.1007/978-94-007-0907-2_8, © Springer Science+Business Media B.V. 2011

particularly from a logical point of view because it determines a relation of semantical consequence (entailment) for the given valuation system, as can be defined, e.g., by Definition 1.3. This definition introduces a relation between a set of formulas (premises) and a single formula (conclusion). One can easily extend this definition to the multiple-conclusions case as follows:

$$\Delta \models_V \Gamma \text{ iff } \forall a(\forall A \in \Delta : v_a(A) \in \mathcal{D} \Rightarrow \exists B \in \Gamma : v_a(B) \in \mathcal{D}). \qquad (8.1)$$

A set of formulas Δ *entails* a set of formulas Γ ($\Delta \models \Gamma$) iff for every valuation function v_a the following holds true: if for every $A \in \Delta, v_a(A) \in \mathcal{D}$, then $v_a(B) \in \mathcal{D}$ for some $B \in \Gamma$. An n-valued tautology then is a formula A such that $\varnothing \models A$.

In some writings, the definition of an n-valued propositional logic and the terminology is slightly different. First, the elements of $\mathcal{V}$ are sometimes referred to as *quasi truth values*. Gottwald [124, p. 2] explains that one reason for using the term 'quasi truth value' is that there is no convincing and uniform interpretation[1] of the truth values that in many-valued logic are taken in addition to the classical truth values *the True* and *the False*. According to Gottwald, this understanding associates the additional values with the naive understanding of being true, respectively the naive understanding of degrees of being true. In later publications, Gottwald has changed his terminology and states that "to [a]void any confusion with the case of classical logic one prefers in many-valued logic to speak of *truth degrees* and to use the word "truth value" only for classical logic" [125, p. 4], cf. also Sect. 1.7. In what follows, we do not adopt this terminology and continue to speak of the elements of n-valued valuation systems as truth values. Also, we will use the term 'semantical value' (or just 'value') which seems to be non-committal.

What is perhaps more important than these differences in terminology is that in part of the literature, for example in [124, 125, 165, 207], an explicit distinction is drawn between a set $\mathcal{D}^+$ of *designated* values and a set $\mathcal{D}^-$ of *antidesignated* values, where the latter need not coincide with the complement of $\mathcal{D}^+$.[2] Usually, this distinction is, however, not fully exploited in many-valued logic: The notion of entailment is defined with respect to the designated values, and no independent additional entailment relation is defined with respect to the antidesignated values.

Gottwald is quite aware of the distinction between designated and antidesignated values. When he discusses the notion of a contradiction (or logical falsity), for example, he explains that there are two ways of generalizing this notion from classical logic to many-valued logic, [125, p. 32, notation adjusted]:

---

[1] At least there was no such interpretation at the time of the writing of [124].

[2] Incidentally, this distinction is relevant for an assessment of Suszko's Thesis, see [246], the claim that "there are but two logical values, true and false" [43, p. 169], which is given formal contents by the so-called Suszko Reduction, the proof that every Tarskian n-valued propositional logic is also characterized by a bivalent semantics. For a recent treatment and references to the literature, see [43]. A critical discussion of Suszko's Thesis will be presented in the next chapter.

1. In the case that the given system S of propositional many-valued logic has a suitable negation connective $\sim$, one can take as logical falsities all those wffs H for which $\sim H$ is a logical truth.
2. In the case that the given system S of propositional many-valued logic has antidesignated truth degrees, one can take as logical falsities all those wffs H which assume only antidesignated truth degrees...

Gottwald assumes that $\mathscr{D}^+ \cap \mathscr{D}^- = \varnothing$ and remarks that if the designated and antidesignated values exhaust all truth degrees and, moreover, the negation operation $\sim$ satisfies the following standard condition (notation adjusted):

$$f_{\sim}(x) \in \mathscr{D}^+ \text{ iff } x \notin \mathscr{D}^+,$$

then the two notions of a contradiction coincide [125, p. 32]. Since $\mathscr{D}^-$ may differ from the complement of $\mathscr{D}^+$, another standard condition for negation $\sim$ is:

$$f_{\sim}(x) \in \mathscr{D}^+ \text{ iff } x \in \mathscr{D}^- \quad \text{and} \quad f_{\sim}(x) \in \mathscr{D}^- \text{ iff } x \in \mathscr{D}^+.$$

If the latter condition is satisfied, the two ways of defining the notion of a contradiction are equivalent also if $\mathscr{V} \setminus \mathscr{D}^+ \neq \mathscr{D}^-$.

Rescher [207, p. 68], who does not consider semantic consequence but only tautologies and contradictions, explains that "there may be good reason for *letting one and the same truth-value be both designated and antidesignated.*" However, he also warns that "we would not want it to happen that there is some truth-value v which is both designated and antidesignated when it is also the case that there is some formula which uniformly assumes this truth-value, for then this formula would be both a tautology and a contradiction" [207, p. 67].

Although Gottwald recognizes that $\mathscr{V} \setminus \mathscr{D}^+$ may be distinct from $\mathscr{D}^-$,[3] he nevertheless follows the tradition in defining only a single semantic consequence relation in terms of $\mathscr{D}^+$. If this privileged treatment of $\mathscr{D}^+$ is given up, a more general definition of a symmetric valuation system and a symmetric n-valued propositional logic emerges.

---

[3] On p. 30 of [125] he explains (notation adjusted):

> Even in the case that $\mathscr{D}^+ \neq \varnothing$ and $\mathscr{D}^- \neq \varnothing$ it is, however, not necessarily $\mathscr{D}^+ \cup \mathscr{D}^- = \mathscr{V}$, which means that together with designated and antidesignated truth degrees also *undesignated* truth degrees may exist. This possibility indicates two essentially different positions regarding the designation of truth degrees. The first one assumes only a binary division of the set of truth degrees and can proceed by simply marking a set of designated truth degrees, treating the undesignated ones like antidesignated ones. The second position assumes a tripartition and marks some truth degrees as designated, some others as antidesignated, and has besides these both types, also some undesignated truth degrees...

Definition 8.1 Let $\mathscr{L}$ be a language in a denumerable set of sentence letters and a finite non-empty set of finitary connectives $\mathscr{C}$. A *symmetric* n-valued valuation system is a structure

$$\langle \mathscr{V}, \mathscr{D}^+, \mathscr{D}^-, \{f_c : c \in \mathscr{C}\}\rangle,$$

where $\mathscr{V}$ is a non-empty set containing n elements $(2 \leq n)$, $\mathscr{D}^+$ and $\mathscr{D}^-$ are distinct non-empty proper subsets of $\mathscr{V}$, and every f_c is a function on $\mathscr{V}$ with the same arity as c.

A symmetric valuation system again can be equipped with an assignment function a which maps the set of atomic sentences into $\mathscr{V}$. Each assignment can be inductively extended to a function v_a from the set of *all* $\mathscr{L}$-formulas into $\mathscr{V}$ in accordance with Conditions (1.1) and (1.2) from Chap. 1. For all sets of $\mathscr{L}$-formulas Δ, Γ, semantic consequence (entailment) relations $\models^+$ and $\models^-$ are defined then as follows:

$$\Delta \models^+ \Gamma \text{ iff } \forall v_a(\forall A \in \Delta, v_a(A) \in \mathscr{D}^+ \Rightarrow \exists B \in \Gamma v_a(B) \in \mathscr{D}^+); \tag{8.2}$$

$$\Delta \models^- \Gamma \text{ iff } \forall v_a(\forall A \in \Gamma, v_a(A) \in \mathscr{D}^- \Rightarrow \exists B \in \Delta v_a(B) \in \mathscr{D}^-). \tag{8.3}$$

An n-valued tautology is a formula A such that $\varnothing \models^+ A$, and an n-valued contradiction is a formula A such that $A \models^- \varnothing$. A symmetric n-valued valuation system together with entailment relations defined by (8.2) and (8.3) constitutes a symmetric n-valued propositional logic. In what follows, we take for granted that any valuation system is canonically equipped with entailment relations $\models^+$ and $\models^-$, and thus we frequently speak of valuation systems *as* logics.

It is not difficult to see that *any* n-valued propositional logic can be "symmetrized", i.e., represented as a symmetric logic by setting just $\mathscr{D}^- = \mathscr{V} \setminus \mathscr{D}^+$. Therefore, we will be especially interested in logical systems which are symmetric in a non-trivial way.

Definition 8.2 Let $\Lambda = \langle \mathscr{V}, \mathscr{D}^+, \mathscr{D}^-, \{f_c : c \in \mathscr{C}\}, \models^+, \models^-\rangle$ be a symmetric n-valued propositional logic. Λ is called a *separated* n-valued propositional logic if $\mathscr{V} \setminus \mathscr{D}^+ \neq \mathscr{D}^-$ (that is, if $\mathscr{V}$ is not partitioned into the non-empty sets $\mathscr{D}^+$ and $\mathscr{D}^-$), and Λ is said to be a *refined* n-valued propositional logic, if it is separated and $\models^+ \neq \models^-$.

Clearly, if a symmetric n-valued logic is not separated, the two entailment relations $\models^+$ and $\models^-$ coincide. In a refined n-valued propositional logic, however, neither "positive" entailment $\models^+$ nor "negative" entailment $\models^-$ need enjoy a privileged status in comparison to each other. Moreover, an entailment relation is usually defined with respect to a given formal language. For the separate entailment relations $\models^+$ and $\models^-$ one might, therefore, expect that they come with their own, in a sense "dual", languages, $\mathscr{L}^+$ and $\mathscr{L}^-$ (see the next section for further explanations). Then one might be interested in n-valued logics in which these languages have the same signature.

Definition 8.3 Let $\mathscr{C}$ be a finite non-empty set of finitary connectives, let $\mathscr{L}^+$ be the language based on $\mathscr{C}^+ = \{c^+ \mid c \in \mathscr{C}\}$, and let $\mathscr{L}^-$ be the language based on $\mathscr{C}^- = \{c^- \mid c \in \mathscr{C}\}$. If A is an $\mathscr{L}^+$-formula, let A^- be the result of replacing every connective c^+ in A by c^-. If Δ is a set of $\mathscr{L}^+$ formulas, let $\Delta^- = \{A^- \mid A \in \Delta\}$. If the language $\mathscr{L}$ of a refined n-valued logic Λ is based on $\mathscr{C}^+ \cup \mathscr{C}^-$, then Λ is said to be *harmonious* iff (i) for all sets of $\mathscr{L}^+$-formulas Δ, Γ : $\Delta \models^+ \Gamma$ iff $\Delta^- \models^- \Gamma^-$, and (ii) for every $c \in \mathscr{C}, f_{c^+} \neq f_{c^-}$.

In the present chapter, we shall first of all argue that the distinction between designated and antidesignated values (and therefore also the distinction between positive entailment $\models^+$ and negative entailment $\models^-$) is an important distinction (Sect. 8.2). In Sect. 8.3, we shall consider some separated n-valued propositional logics, and in Sects. 8.4 and 8.5, we shall present natural examples of harmonious finitely-valued logics. Finally, we shall make some brief remarks on generalizing harmony (Sect. 8.6).

8.2 Designated and Antidesignated Values

Why is it important to draw a distinction between designated and antidesignated values? The notion of a set of designated values is often considered as a generalization of the notion of truth. Similarly, the set of antidesignated values can be regarded as representing a generalized concept of falsity. However, logic and its terminology is to a large extent predominated by the notion of truth. The Fregean Bedeutung (reference) of a declarative sentence (or, in the first place, a thought) is a truth value. According to Frege, there are exactly two truth values, *The True* and *The False*, henceforth just *true* (T) and *false* (F). Whereas T and F are referred to as *truth* values, neither F nor T is called a *falsity* value.

Moreover, Frege explicitly characterized logic as “the science of the most general laws of being true”, see Hans Sluga’s translation in [241, p. 86]. This view finds its manifestation in the fact that most of the fundamental logical notions—logical operations, relations, etc.—are usually defined through the category of truth. Thus, valid consequence is usually defined as preserving truth in passing from the premises to the conclusions of an inference. It is required that for every model $\mathfrak{M}$, if all the premises are true in $\mathfrak{M}$, then so is at least one conclusion. By contraposition, in a valid inference, not being true is preserved from the conclusions to at least one of the premises. For every model $\mathfrak{M}$, if every conclusion is not true in $\mathfrak{M}$, then so is at least one of the premises. As to falsity, its role in such definitions frequently remain a subordinated one, if any. When constructing a semantic model, “false” is often understood as a mere abbreviation for “not true”, as for instance when the classical truth-table definition for conjunction is stated as: “A conjunctive sentence is true if both of its conjuncts are true, otherwise it is not true (i.e., false)”.

In general, failing to be true and being false may, however, fall apart conceptually. Although this is the very point of many-valued logic, a distinction between falsity and the absence of truth (or truth and the absence of falsity) is often represented only by the values that $\mathscr{V}$ contains in addition to T and F and not by distinguishing between a set of designated values $\mathscr{D}^+$ and another set of antidesignated values $\mathscr{D}^-$ (and the consequence relations induced by these sets). We are interested in inferences that preserve truth because we are interested in true beliefs. But likewise we are interested in inferences in which false conclusions are bound to depend on at least one false premise because we are interested in avoiding false beliefs. This point has vividly been made by William James in his essay "The Will to Believe":

> Believe truth! Shun error!—These, we see, are two materially different laws; and by choosing between them we may end by coloring differently our whole intellectual life. We may regard the chase for truth as paramount, and the avoidance of error as secondary; or we may, on the other hand, treat the avoidance of error as more imperative, and let truth take its chance [134, p. 18].

It seems then quite natural to modify (or to extend) in a certain respect the Fregean definition of logic by saying that its scope is studying not just the laws of being true, but rather of being true *and* being false. An immediate effect of this modification consists in acknowledging the importance of falsity (and more generally, antidesignated values) for defining logical notions. Every definition formulated in terms of truth should be "counterbalanced" with a "parallel" (dual) definition formulated in terms of falsity.

This observation not only justifies the distinction between positive ($\models^+$) and negative ($\models^-$) entailment relations but also clarifies the idea of the corresponding languages $\mathscr{L}^+$ and $\mathscr{L}^-$ with "positive" and "negative" connectives. As an example, let us take the operation of conjunction in classical logic. The connective $\wedge^+$ can naturally be defined through its truth conditions: $v(A \wedge^+ B) = T$ iff $v(A) = T$ and $v(B) = T$. But we may also wish to consider the connective $\wedge^-$, exhaustively defined by means of the *falsity conditions*: $v(A \wedge^- B) = F$ iff $v(A) = F$ or $v(B) = F$. Now, although in classical logic $\wedge^+$ and $\wedge^-$ are, obviously, coincident, in the general case, for example in various many-valued logics, they may well differ from each other. Thus, if the relation between designated and antidesignated values is not as straightforward as in classical logic, a separate introduction of the positive and negative connectives (in fact: "truth-connectives" and "falsity-connectives") may acquire especial significance.

Let us generalize this point with respect to the propositional connectives of conjunction and disjunction in the context of many-valued logics. There is a view that in a many-valued semantics $f_\wedge$ and $f_\vee$ are just the functions of taking the minimum and the maximum of their arguments. As Richard Dewitt put it:

> In many-valued systems, intuitions concerning the appropriate truth-conditions for disjunction and conjunction are the most widely agreed on. In particular, there is general agreement that a disjunction should take the maximum value of the disjuncts, while a conjunction should take the minimum value of the conjuncts [64, p. 552].

As a result, we get the following definitions:

$$v(A \wedge B) = \min(v(A), v(B));$$
$$v(A \vee B) = \max(v(A), v(B)),$$

which Dewitt refers to as the *standard conditions* for conjunction and disjunction. One may note, however, that these conditions are justified only if the set of semantical values is in some way *linearly ordered* so that any two values are mutually comparable. And although this is frequently indeed the case, e.g., when the semantical values are identified with some points on a numerical segment, there also exist a number of many-valued systems where not all of the values are comparable with each other.[4] In such systems, the standard conditions cannot be employed directly.

The idea of truth *degrees* in many-valued logic naturally implies that semantical values may differ in their truth-content. Moreover, it is assumed that any of the designated values is "more true" (is of a higher degree in its truth-content) than any of the values which is not designated. Thus, taking into account the "minimality–maximality" intuition described above, the conjunction can be more generally regarded as an operation $\wedge^+$ that in a sense *minimizes the truth-content* (or the "designatedness") of the conjuncts, and disjunction can be more generally regarded as an operation $\vee^+$ that *maximizes the truth-content* of the disjuncts.[5] That is, for truth values that are comparable in their truth degrees, $f_{\wedge^+}$ is just the standard min-function, but if two truth values x and y are incomparable, the "less true" relation nevertheless should be such that it determines $f_{\wedge^+}(x, y)$, the outcome of which is *less true* than *both* of the conjuncts. This is similarly true for disjunction. In this way, we should be able to obtain definitions of notions of conjunction $\wedge^+$ and disjunction $\vee^+$ purely in terms of truth:

$$v(A \wedge^+ B) = \min{}^+(v(A), v(B));$$
$$v(A \vee^+ B) = \max{}^+(v(A), v(B)),$$

where $\min^+$ and $\max^+$ are generalized functions of truth-minimizing and truth-maximizing, correspondingly.

Now, if falsity is not the same as non-truth, an independent consideration of propositional connectives from the standpoint of antidesignated values is appropriate. In this sense, conjunction should be regarded as the falsity-maximizer and disjunction as the falsity-minimizer, and thus, we should be able to obtain generalized functions $\max^-$ and $\min^-$ of maximizing and minimizing, respectively, the falsity-content of their arguments, so that operations $\wedge^-$ and $\vee^-$ can be defined purely in terms of falsity:

---

4 Cf., for instance, the values **N** and **B** under the order $\leq_t$ in the four-valued logic B_4 considered in Sect. 8.4.

5 And of course, a conjunction should maximize the non-truth of the conjuncts, while a disjunction should minimize the non-truth of the disjuncts.

$$v(A \wedge^- B) = \max{}^-(v(A), v(B));$$
$$v(A \vee^- B) = \min{}^-(v(A), v(B)).$$

We also briefly observe the difference between the two kinds of negation. Namely, whereas $f_{\sim^+}$ can be viewed as a function that (within every semantical value) turns truth into non-truth and vice versa, $f_{\sim^-}$ can be treated as an operation that interchanges exclusively between falsity and non-falsity.

One might object that the distinction between designated and antidesignated values makes sense *only* for doxastically or epistemically interpreted semantical values. Certainly, if a proposition is not believed (known) to be true, this does not imply that the proposition is believed (known) to be false. The distinction is, however, also sensible for other, non-doxastical and non-epistemical adverbial qualifications of truth and falsity. If a proposition is not necessarily true, for instance, it need not be necessarily false. The designated semantical values are used to define an entailment relation $\models^+$ that preserves the possession of a doxastically wanted value in passing from the premises to the conclusions of an inference. Analogously, the antidesignated values may be used to define an entailment relation $\models^-$ by requiring that if the conclusions are doxastically unwanted, at least one of the assumptions is doxastically unwanted, too. Among the doxastically wanted values, there may be values interpreted, for example, as "true", "neither true nor false", "known to be true", "unknown to be false", "necessarily true", "possibly true", etc. Among the doxastically unwanted values, there may be values interpreted, for example, as "false", "neither true nor false", "known to be false", "unknown to be true", "necessarily false", "possibly false", etc.

Whether "neither true nor false" is doxastically wanted or unwanted, may be a matter of perspective. A proposition evaluated as "neither true nor false" is not falsified and hence possibly wanted, but it is also not verified and therefore possibly unwanted. If one wants to take into account that both perspectives are legitimate, a value read as "neither true nor false" may even sensibly be classified as both wanted and unwanted. That is, in general, it is reasonable to distinguish between designated, antidesignated, and undesignated semantical values, and also not to exclude the possibility of values that are both designated and antidesignated.

These considerations can be generalized, see Sect. 8.6.

8.3 Some Separated Finitely-Valued Logics

We first consider separated versions of the well-known three-valued logics of Kleene and Łukasiewicz.

Definition 8.4 Kleene's strong three-valued logic K_3 and Łukasiewicz's three-valued logic $Ł_3$ are semantically defined by the following symmetric valuation systems:

1. $\mathfrak{K}_3 := \langle\{T, \varnothing, F\}, \{T\}, \{\varnothing, F\}, \{f_c : c \in \{\sim, \wedge, \vee, \supset\}\}\rangle$, where the functions f_c are defined as follows:

$f_\sim$	
T	F
$\varnothing$	$\varnothing$
F	T

$f_\wedge$	T	$\varnothing$	F
T	T	$\varnothing$	F
$\varnothing$	$\varnothing$	$\varnothing$	F
F	F	F	F

$f_\vee$	T	$\varnothing$	F
T	T	T	T
$\varnothing$	T	$\varnothing$	$\varnothing$
F	T	$\varnothing$	F

$f_\supset$	T	$\varnothing$	F
T	T	$\varnothing$	F
$\varnothing$	T	$\varnothing$	$\varnothing$
F	T	T	T

2. $\mathfrak{L}_3 := \langle\{T, \varnothing, F\}, \{T\}, \{\varnothing, F\}, \{f_c : c \in \{\sim, \wedge, \vee, \supset\}\}\rangle$, where the functions f_c are defined as in K_3 except that:

$f_\supset$	T	$\varnothing$	F
T	T	$\varnothing$	F
$\varnothing$	T	T	$\varnothing$
F	T	T	T

Definition 8.5 The separated three-valued propositional logics $K_3{}^*$ and Ł$_3{}^*$ are semantically defined by the following symmetric valuation systems:

$\mathfrak{K}_3{}^* := \langle\{T, \varnothing, F\}, \{T\}, \{F\}, \{f_c : c \in \{\sim, \wedge, \vee, \supset\}\}\rangle$, where the functions f_c are defined as in K_3;

$\mathfrak{L}_3{}^* := \langle\{T, \varnothing, F\}, \{T\}, \{F\}, \{f_c : c \in \{\sim, \wedge, \vee, \supset\}\}\rangle$, where the functions f_c are defined as in Ł$_3$.

Proposition 8.1 *$K_3{}^*$ and Ł$_3{}^*$ are refined, i.e., the relations $\models^+$ and $\models^-$ do not coincide.*

Proof In both logics, formulas of the form $A \wedge (A \supset B)$ have the same truth table:

	A	B	$A \wedge (A \supset B)$
	T	T	T
	T	$\varnothing$	$\varnothing$
	T	F	F
	$\varnothing$	T	$\varnothing$
	$\varnothing$	$\varnothing$	$\varnothing$
*	$\varnothing$	F	$\varnothing$
	F	T	F
	F	$\varnothing$	F
	F	F	F

Whereas $A \wedge (A \supset B) \models^+ B$, the row marked with an asterisk shows that $A \wedge (A \supset B) \not\models^- B$. □

Although it is not surprising, perhaps, that in a separated n-valued logic, the relations $\models^+$ and $\models^-$ need not coincide, there are separated n-valued logics which are not refined. Let again $\mathbf{N} := \varnothing, \mathbf{T} := \{T\}, \mathbf{F} := \{F\}$ and $\mathbf{B} := \{T, F\}$.

Definition 8.6 The useful four-valued logic B_4 of Dunn and Belnap is semantically defined by the symmetric valuation system $\mathfrak{B}_4 = \langle\{\mathbf{N}, \mathbf{T}, \mathbf{F}, \mathbf{B}\}, \{\mathbf{T}, \mathbf{B}\}, \{\mathbf{N}, \mathbf{F}\}, \{f_c : c \in \{\sim, \wedge, \vee\}\}\rangle$, where the functions f_c are defined as follows:

$f_\sim$	
T	**F**
B	**B**
N	**N**
F	**T**

$f_\wedge$	**T**	**B**	**N**	**F**
T	**T**	**B**	**N**	**F**
B	**B**	**B**	**F**	**F**
N	**N**	**F**	**N**	**F**
F	**F**	**F**	**F**	**F**

$f_\vee$	**T**	**B**	**N**	**F**
T	**T**	**T**	**T**	**T**
B	**T**	**B**	**T**	**B**
N	**T**	**T**	**N**	**N**
F	**T**	**B**	**N**	**F**

Definition 8.7 The separated four-valued propositional logic $B_4{}^*$ is semantically defined by the symmetric valuation system $\mathfrak{B}_4{}^* = \langle\{\mathbf{N}, \mathbf{T}, \mathbf{F}, \mathbf{B}\}, \{\mathbf{T}, \mathbf{B}\}, \{\mathbf{F}, \mathbf{B}\}, \{f_c : c \in \{\sim, \wedge, \vee, \}\}\rangle$, where the functions f_c are defined as in B_4.

The set $\{\mathbf{N}, \mathbf{T}, \mathbf{F}, \mathbf{B}\}$ is also referred to as **4**, cf. Chap. 3. Note that in $B_4{}^*$ not only $\mathbf{4} \setminus \mathscr{D}^+ \neq \mathscr{D}^-$, but also $\mathscr{D}^+ \cap \mathscr{D}^- \neq \varnothing$.

Proposition 8.2 *The separated logic $B_4{}^*$ is not refined:* $\models^+ = \models^-$.

Proof A proof (for the case of single premises and conclusions) using Dunn's method of "dual" valuations is given, e.g., in [80, p. 10]. It is observed that the relation $\models^+$ alias $\models^-$ also coincides with the relation $\models$ defined as follows: $\Delta \models \Gamma$ iff ($\Delta \models^+ \Gamma$ and $\Delta \models^- \Gamma$). □

Clearly, K_3 does not have any tautologies because any formula takes the value $\varnothing$ if every propositional variable occurring in it takes the value $\varnothing$. It has been observed in [207] that for the same reason, the three-valued logic which is now known as the Logic of Paradox [196], *LP*, has no contradictions.

Definition 8.8 The Logic of Paradox *LP* is the three-valued propositional logic $\langle\{T, \varnothing, F\}, \{T, \varnothing\}, \{F\}, \{f_c : c \in \{\sim, \wedge, \vee, \supset\}\}\rangle$, where the functions f_c are defined as in K_3.

In B_4 there are neither any tautologies (consider the constant valuation that assigns only **N**) nor any contradictions (consider the constant valuation that assigns only **B**).[6] Obviously, $\models^+$ in B_4 coincides with $\models^+$ and $\models^-$ in $B_4{}^*$. Our main question is: Are there *natural* examples of harmonious finitely-valued logics?

[6] Rescher [207, p. 67] seems to interpret the fact that the logic *LP* has no contradictions as a reason for distinguishing between antidesignated and undesignated values because in *LP* no formula receives an undesignated truth value under any valuation.

8.4 A Harmonious Logic Inspired by the Logic of *SIXTEEN*$_3$

Let us now consider the logic of *SIXTEEN*$_3$ "in a harmonious perspective". Recall that the logic of *SIXTEEN*$_3$ in the language $\mathscr{L}_{tf}$ can be semantically presented as a *bi-consequence system*, namely the structure $(\mathscr{L}_{tf}, \models_t, \models_f)$, where the two entailment relations $\models_t$ and $\models_f$ are defined with respect to the truth order $\leq_t$ and the falsity order $\leq_f$, respectively. These relations have been defined in Chap. 4 by Definitions 4.1 and 4.2 as relations between single formulas. It is not difficult to extend these definitions to obtain the relations $\models_t \subseteq \mathscr{P}(\mathscr{L}_{tf}) \times \mathscr{P}(\mathscr{L}_{tf})$ and $\models_f \subseteq \mathscr{P}(\mathscr{L}_{tf}) \times \mathscr{P}(\mathscr{L}_{tf})$.

Definition 8.9 $\Delta \models_t^{16} \Gamma$ iff $\forall v^{16} \sqcap_t \{v^{16}(A) \mid A \in \Delta\} \leq_t \sqcup_t \{v^{16}(A) \mid A \in \Gamma\}$.

Definition 8.10 $\Delta \models_f^{16} \Gamma$ iff $\forall v^{16} \sqcup_f \{v^{16}(A) \mid A \in \Gamma\} \leq_f \sqcap_f \{v^{16}(A) \mid A \in \Delta\}$.

Till the end of this section, we will omit the superscript and write just $\models_t$ and $\models_f$ instead of $\models_t^{16}$ and $\models_f^{16}$. Let the sets x^t and x^f be defined as they were on page 53. We then define a separated 16-valued logic based on *SIXTEEN*$_3$ in the language $\mathscr{L}_{tf}$.

Definition 8.11 The separated 16-valued logic B_{16} is semantically defined by the symmetric valuation system $\mathfrak{B}_{16} = \langle \mathbf{16}, \{x \in \mathbf{16} \mid x^t \text{ is non-empty}\}, \{x \in \mathbf{16} \mid x^f \text{is non-empty}\}, \{-_t, \sqcap_t, \sqcup_t, -_f, \sqcup_f, \sqcap_f\}\rangle$, where for all sets of $\mathscr{L}_{tf}$-formulas Δ, Γ, semantic consequence relations $\models^+$ and $\models^-$ are defined in accordance with (8.2) and (8.3).

Proposition 8.3 *B_{16} is refined.*

Proof It can easily be seen that in B_{16} the relations $\models^+$ and $\models^-$ are distinct, e.g., in view of the following counterexample: $(A \wedge_f B) \models^- A$ but $(A \wedge_f B) \not\models^+ A$ (since, e.g., $\boldsymbol{F} \sqcup_f \mathbf{FTB} = \mathbf{FB}$). □

Proposition 8.4 *The* 16*-valued propositional logic B_{16} is harmonious.*

Proof Obviously, we may view the language $\mathscr{L}_{tf}$ as being based on a set of "positive connectives" $\mathscr{C}^+ = \{\sim_t, \wedge_t, \vee_t\}$ and a set of "negative connectives" with matching arity $\mathscr{C}^- = \{\sim_f, \wedge_f, \vee_f\}$, i.e., $\mathscr{C} = \{\sim, \wedge, \vee\}$. Moreover, we have already observed that the condition $f_{c_t} \neq f_{c_f}$ is satisfied for every $c \in \mathscr{C}$. If A is an $\mathscr{L}_t$-formula, let A^f be the result of replacing every connective c_t in A by c_f. If Δ is a set of $\mathscr{L}_t$ formulas, let $\Delta^f = \{A^f \mid A \in \Delta\}$. It remains to be shown that in B_{16} for all sets of $\mathscr{L}_t$-formulas Δ, Γ,

$$(\dagger)\ \Delta \models^+ \Gamma \text{ iff } \Delta^f \models^- \Gamma^f.$$

This follows from Lemmas 4.1, 4.2 and Theorems 4.1, 4.2 in Chap. 4 for $\leq_t$ and the analogous versions of these statements for $\leq_f$. Lemma 4.2 says that for every $A, B \in \mathscr{L}_t$: $A \models_t B$ iff $\forall v(\mathbf{T} \in v(A) \Rightarrow \mathbf{T} \in v(B))$. According to Lemma 4.1, within language $\mathscr{L}_t$, the condition $\forall v(\mathbf{T} \in v(A) \Rightarrow \mathbf{T} \in v(B))$ is equivalent to $\forall v(\mathbf{B} \in v(A) \Rightarrow \mathbf{B} \in v(B))$. Thus, for every $A, B \in \mathscr{L}_t$: $A \models_t B$ iff $\forall v(\mathbf{T} \in v(A)$ or $\mathbf{B} \in v(A) \Rightarrow \mathbf{T} \in v(B)$ or $\mathbf{B} \in v(B))$. This means that $\models^+$ restricted to $\mathscr{L}_t$ is the same relation as $\models_t$ restricted to $\mathscr{L}_t$, and it is then axiomatized as first-degree entailment (Theorems 4.1 and 4.2). Since for every $A, B \in \mathscr{L}_f$: $A \models_f B$ iff $\forall v(\mathbf{F} \in v(B) \Rightarrow \mathbf{F} \in v(A))$ iff $\forall v(\mathbf{B} \in v(B) \Rightarrow \mathbf{B} \in v(A))$, and the restriction of $\models^-$ to $\mathscr{L}_f$(= the restriction of $\models_f$ to $\mathscr{L}_f$) is *also* axiomatized as first-degree entailment, condition (†) is satisfied. □

But what is the situation with the whole language $\mathscr{L}_{tf}$ and the corresponding entailment relations? In fact, we have two logics, $(\mathscr{L}_{tf}, \models_t, \models_f)$ and a harmonious many-valued logic $(\mathscr{L}_{tf}, \models^+, \models^-)$, and we may note that the corresponding entailment relations of the logics $(\mathscr{L}_{tf}, \models_t, \models_f)$ and $(\mathscr{L}_{tf}, \models^+, \models^-)$ are distinct: $\models_t \neq \models^+$ and $\models_f \neq \models^-$. We may, for example, note that for every $\mathscr{L}_{tf}$-formula $A, A \models^+ \sim_f A$ and $A \models^- \sim_t A$, whereas there exists no atomic $\mathscr{L}_{tf}$-formula A, such that $A \models_t \sim_f A$ or $A \models_f \sim_t A$.

8.5 Harmony ad Infinitum

Recall that the structure *SIXTEEN*$_3$ is an example of a *Belnap trilattice* introduced in Sect. 4.3. Belnap trilattices are obtained by iterated powerset-formation applied to the set **4** and by generalizing the definitions of a truth order and a falsity order on **16**. In Sect. 4.3, we obtained an infinite collection of sets of generalized semantical values by considering $\mathscr{P}^n(\mathbf{4})$. A Belnap trilattice has been defined there as a structure $\mathscr{M}_3^n := (\mathscr{P}^n(\mathbf{4}), \sqcap_i, \sqcup_i, \sqcap_t, \sqcup_t, \sqcap_f, \sqcup_f)$, where $\sqcap_i$ $(\sqcap_t, \sqcap_f)$ is the lattice meet and $\sqcup_i$ $(\sqcup_t, \sqcup_f)$ is the lattice join with respect to the ordering $\leq_i$ $(\leq_t, \leq_f)$ on $\mathscr{P}^n(\mathbf{4}), n \geq 1$. It has been observed that *SIXTEEN*$_3(=\mathscr{M}_3^1)$ is the smallest Belnap trilattice.

Considering again the languages $\mathscr{L}_t$, $\mathscr{L}_f$, $\mathscr{L}_{tf}$ defined in Sect. 4.1 and n-valuations as defined on page 77, we extend Definitions 4.7 and 4.8 of t-entailment and f-entailment for any n (for arbitrary formulas A, B from $\mathscr{L}_{tf}$) to the case with arbitrary sets of formulas:

Definition 8.12 The relations $\models_t^n \subseteq \mathscr{P}(\mathscr{L}_{tf}) \times \mathscr{P}(\mathscr{L}_{tf})$ and $\models_f^n \subseteq \mathscr{P}(\mathscr{L}_{tf}) \times \mathscr{P}(\mathscr{L}_{tf})$ are defined by the following equivalences:

$$\Delta \models_t^n \Gamma \text{ iff } \forall v^n \ \sqcap_t\{v^n(A) \mid A \in \Delta\} \leq_t \sqcup_t\{v^n(A) \mid A \in \Gamma\};$$
$$\Delta \models_f^n \Gamma \text{ iff } \forall v^n \ \sqcup_f\{v^n(A) \mid A \in \Gamma\} \leq_f \sqcap_f\{v^n(A) \mid A \in \Delta\}.$$

Thus, semantically, the logic of a Belnap trilattice $\mathscr{M}_3^n$ is the bi-consequence system $(\mathscr{L}_{tf}, \models_t^n, \models_f^n)$. We now define an infinite chain of separated finitely-valued logics.

Definition 8.13 Let $\sharp n$ be the cardinality of $\mathscr{P}^n(\mathbf{4})$. The $\sharp n$-valued logic $B_{\sharp n}$ is the structure $\langle \mathscr{P}^n(\mathbf{4}), \mathscr{D}^{n+}, \mathscr{D}^{n-}, \{-_t, \sqcap_t, \sqcup_t, -_f, \sqcup_f, \sqcap_f\}\rangle$, where $\mathscr{D}^{n+} := \{x \in \mathscr{P}^n(\mathbf{4}) \mid$

x is t-positive$\}$ and $\mathscr{D}^{n-} := \{x \in \mathscr{P}^n(\mathbf{4}) \mid x$ is f-positive$\}$. For every logic $B_{\sharp n}$, and for all sets of $\mathscr{L}_{tf}$-formulas Δ, Γ, the semantic consequence relations $\models^{n+}$ and $\models^{n-}$ are defined in accordance with (8.2) and (8.3).

Proposition 8.5 *For every $n \in \mathbb{N}$, the logic $B_{\sharp n}$ is refined.*[7]

Proof Obviously, $B_{\sharp n}$ is separated. That in $B_{\sharp n}$ the relations $\models^{n+}$ and $\models^{n-}$ are distinct, can again be seen by noticing that for every $n \in \mathbb{N}$, $(A \wedge_f B) \models^{n-} A$ but $(A \wedge_f B) \not\models^{n+} A$ (since for every $\mathscr{M}_3^n$ there may well exist $x, y \in \mathscr{P}^n(\mathbf{4})$ such that $x \sqcup_f y$ is t-positive, whereas either x or y is not). □

Proposition 8.6 *For every $n \in \mathbb{N}$, the logic $B_{\sharp n}$ is harmonious.*

Proof Again, we regard $\mathscr{L}_{tf}$ as being based on $\mathscr{C}^+ = \{\sim_t, \wedge_t, \vee_t\}$ and the set of connectives with matching arity $\mathscr{C}^- = \{\sim_f, \wedge_f, \vee_f\}$ so that $\mathscr{C} = \{\sim, \wedge, \vee\}$. It can easily be seen that the condition $f_{c_t} \neq f_{c_f}$ is satisfied for every $c \in \mathscr{C}$. We must show that in $B_{\sharp n}$ for all sets of $\mathscr{L}_t$-formulas Δ, Γ,

$$\Delta \models^{n+} \Gamma \text{ iff } \Delta^f \models^{n-} \Gamma^f.$$

We use the main result of Sect. 4.3, namely that for *every* $n \in \mathbb{N}$, the truth entailment relation $\models_t^n$ of $(\mathscr{L}_{tf}, \models_t^n, \models_f^n)$ restricted to $\mathscr{L}_t$ and the falsity entailment relation $\models_f^n$ of $(\mathscr{L}_{tf}, \models_t^n, \models_f^n)$ restricted to $\mathscr{L}_f$ can both be axiomatized as first-degree entailment. The proof systems differ only insofar, as every connective c_t is uniformly replaced by its "negative counterpart" c_f. Thus, given the completeness theorem for $\models_t^n$ and its counterpart for $\models_f^n$, it is enough to show that for all sets of $\mathscr{L}_t$-formulas $\Delta, \Gamma : \Delta \models^{n+} \Gamma$ in $B_{\sharp n}$ iff $\Delta \models_t^n \Gamma$, and for all sets of $\mathscr{L}_f$-formulas Δ, $\Gamma : \Delta \models^{n-} \Gamma$ in $B_{\sharp n}$ iff $\Delta \models_f^n \Gamma$. This follows from Corollary 4.1: For any $A, B \in \mathscr{L}_t$:

$$A \models_t^n B \text{ iff } \forall v^n (x \in v^n(A)^t \Rightarrow x \in v^n(B)^t)$$

and the analogous result for falsity entailment: For any $A, B \in \mathscr{L}_f$:

$$A \models_f^n B \text{ iff } \forall v^n (x \in v^n(B)^f \Rightarrow x \in v^n(A)^f).$$ □

8.6 Some Remarks on Generalizing Harmony

The point of departure for our considerations is the familiar and simple notion of an extensional n-valued propositional logic. We argued that the distinction between designated, antidesignated, and undesignated values and values that are both designated and antidesignated ought to be taken seriously. There is no reason to

[7] Note that $n \geq 1$.

privilege designation over antidesignation when it comes to defining entailment. Hence we suggested a slightly more general notion of an n-valued logic. Whereas the set $\mathscr{D}^+$ of designated values determines a positive entailment relation $\models^+$, the set $\mathscr{D}^-$ of antidesignated values determines a negative entailment relation $\models^-$. Once we have two entailment relations, it is natural to consider two languages. This consideration then led us to defining the notion of a harmonious n-valued logic.

As mentioned already, it was observed quite a while ago that one may consider n-valued logics in which the truth functions $f_\wedge$ and $f_\vee$ for conjunction and disjunction form a lattice on the underlying set of truth values, see, for example [209, 210]. Thus, given an n-valued propositional logic, one may wonder whether a lattice order can be defined from some given truth functions. However, in light of the research on bilattices and trilattices, the interest usually goes in the opposite direction. Given some natural partial orders on sets of semantical values, one may wonder whether these orderings form lattices on the underlying sets and thereby give rise to a conjunction (lattice meet), disjunction (lattice join), and hopefully also some sort of negation. In the bilattice $FOUR_2$, there is only one "logical" order, there referred to as the truth order. In a Belnap trilattice $\mathscr{M}_3^n$, there are two logical orderings, the truth order $\leq_t$ and the falsity order $\leq_f$. From the bi-consequence system of any Belnap trilattice, we obtained a harmonious finitely-valued logic. But neither did we consider the following relation induced by the information order $\models^i$, defined as

$$\Delta \models_i^n \Gamma \text{ iff } \forall v^n \ \sqcap_i \{v^n(A) \mid A \in \Delta\} \leq_i \sqcup_i \{v^n(A) \mid A \in \Gamma\},$$

nor did we consider in this chapter additional orderings or other sets of truth values with more than two lattice orderings on them. This would, however, be an interesting line of investigation to pursue, if one aims at finding examples of many-valued logics displaying a more general form of harmony. Thus, generalizations of the notion of a harmonious n-valued propositional logic can be obtained by replacing the set $\{\mathscr{D}^+, \mathscr{D}^-\}$ of distinguished subsets of the set of values $\mathscr{V}$ by a set $\{\mathscr{D}_1, \ldots, \mathscr{D}_n\}$ of distinguished subsets of $\mathscr{V}$. Moreover, the association of entailment relations with the distinguished subsets of $\mathscr{V}$ may vary.

A relation $\models \subseteq \mathscr{P}(\mathscr{L}) \times \mathscr{P}(\mathscr{L})$ is a Tarski–Scott multiple conclusion consequence relation iff it satisfies the following conditions:

1. For every $\Delta \subseteq \mathscr{L}, \Delta \models \Delta$ (reflexivity);
2. If $\Delta \models \Gamma \cup \{A\}$ and $\{A\} \cup \Theta \models \Sigma$, then $\Delta \cup \Theta \models \Gamma \cup \Sigma$ (transitivity);
3. If $\Delta \subseteq \Theta, \Gamma \subseteq \Sigma$, and $\Delta \models \Gamma$, then $\Theta \models \Sigma$ (monotony).

A Tarski–Scott multiple conclusion consequence relation is said to be structural iff it is closed under substitution.

In an n-valued logic which is refined in the sense of Definition 8.2, in addition to $\models^+$ and $\models^-$, several other possibly interesting semantical relations can be defined, though not all of them turn out to be Tarski–Scott multiple conclusion entailment relations. In order to refer to these relations in a compact way, we need a more systematic notation. Let $+$ stand for designated values, $-$ for antidesignated values, u for neither designated nor antidesignated values, and b for values that are both designated and antidesignated. Let $\overline{+}$, $\overline{-}$, $\overline{\mathsf{u}}$, and $\overline{\mathsf{b}}$ stand for the respective complements. Moreover, let $\Leftarrow$ indicate preservation from the conclusions to the premises, $\Rightarrow$ indicate preservation from the premises to the conclusions, and $\Leftrightarrow$ indicate preservation in both directions. Then $\models^+$ is denoted as $\models^{+\Rightarrow+}$ and $\models^-$ as $\models^{-\Leftarrow-}$. The relation $\models$ defined (for a given n-valued logic) by requiring that $\Delta \models \Gamma$ iff both $\Delta \models^+ \Gamma$ and $\Delta \models^- \Gamma$ (cf. the proof of Proposition 8.2) is denoted as $\models^{(+\Rightarrow+,-\Leftarrow-)}$, and the relation defined by requiring that $\Delta \models \Gamma$ iff ($\Delta \models^+ \Gamma$ or $\Delta \models^- \Gamma$) is denoted as $\models^{(+\Rightarrow+|-\Leftarrow-)}$. By way of example, we may consider, in addition to $\models^+$ and $\models^-$, the following "semantic consequence" relations:

- $\models^{+\Leftarrow+}$;
- $\models^{-\Rightarrow-}$;
- $\models^{\mathsf{u}\Rightarrow\mathsf{u}}$; $\models^{\mathsf{u}\Leftarrow\mathsf{u}}$;
- $\models^{\mathsf{b}\Rightarrow\mathsf{b}}$; $\models^{\mathsf{b}\Leftarrow\mathsf{b}}$;
- $\models^{\overline{+}\Rightarrow\overline{+}}$; $\models^{\overline{+}\Leftarrow\overline{+}}$;
- $\models^{\overline{-}\Rightarrow\overline{-}}$; $\models^{\overline{-}\Leftarrow\overline{-}}$;
- $\models^{\overline{\mathsf{u}}\Rightarrow\overline{\mathsf{u}}}$; $\models^{\overline{\mathsf{u}}\Leftarrow\overline{\mathsf{u}}}$;
- $\models^{\overline{\mathsf{b}}\Rightarrow\overline{\mathsf{b}}}$; $\models^{\overline{\mathsf{b}}\Leftarrow\overline{\mathsf{b}}}$;
- $\models^{\bullet\Rightarrow\circ}$; $\models^{\bullet\Leftarrow\circ}$;
  where $\bullet, \circ \in \{+, -, \mathsf{u}, \mathsf{b}, \overline{+}, \overline{-}, \overline{\mathsf{u}}, \overline{\mathsf{b}}\}$ and $\bullet \neq \circ$;
- $\models^{(\bullet\Rightarrow\circ,\blacklozenge\Leftarrow\Diamond)}$; $\models^{(\bullet\Leftarrow\circ,\blacklozenge\Rightarrow\Diamond)}$;
  where $\bullet, \circ, \blacklozenge, \Diamond \in \{+, -, \mathsf{u}, \mathsf{b}, \overline{+}, \overline{-}, \overline{\mathsf{u}}, \overline{\mathsf{b}}\}$ and $\bullet \neq \overline{\blacklozenge}$ or $\circ \neq \overline{\Diamond}$;
- $\models^{(\bullet\Rightarrow\circ,\blacklozenge\Rightarrow\Diamond)}$; $\models^{(\bullet\Leftarrow\circ,\blacklozenge\Leftarrow\Diamond)}$;
  where $\bullet, \circ, \blacklozenge, \Diamond \in \{+, -, \mathsf{u}, \mathsf{b}, \overline{+}, \overline{-}, \overline{\mathsf{u}}, \overline{\mathsf{b}}\}$ and $\bullet \neq \blacklozenge$ or $\circ \neq \Diamond$;
- $\models^{(\bullet\Rightarrow\circ|\blacklozenge\Leftarrow\Diamond)}$; $\models^{(\bullet\Leftarrow\circ|\blacklozenge\Rightarrow\Diamond)}$;
  where $\bullet, \circ, \blacklozenge, \Diamond \in \{+, -, \mathsf{u}, \mathsf{b}, \overline{+}, \overline{-}, \overline{\mathsf{u}}, \overline{\mathsf{b}}\}$ and $\bullet \neq \overline{\blacklozenge}$ or $\circ \neq \overline{\Diamond}$;
- $\models^{(\bullet\Rightarrow\circ|\blacklozenge\Rightarrow\Diamond)}$; $\models^{(\bullet\Leftarrow\circ|\blacklozenge\Leftarrow\Diamond)}$;
  where $\bullet, \circ, \blacklozenge, \Diamond \in \{+, -, \mathsf{u}, \mathsf{b}, \overline{+}, \overline{-}, \overline{\mathsf{u}}, \overline{\mathsf{b}}\}$ and $\bullet \neq \blacklozenge$ or $\circ \neq \Diamond$.

Since consequence relations normally are not required to be symmetric, relations like $\models^{+\Leftrightarrow+}$ are, perhaps, not of primary interest. But the relation $\models^{(+\Leftrightarrow+,\mathsf{u}\Rightarrow\overline{-})}$ $= \models^{+\Leftrightarrow+} \cap \models^{\mathsf{u}\Rightarrow\overline{-}}$, for example, might be of some interest. Investigating and applying such non-standard semantic consequence relations is not as exotic as it might seem at first sight, and, indeed, some such relations have been considered in the literature. We already noted that $\models^{(+\Rightarrow+,-\Leftarrow-)}$ is dealt with in [80], and there are other examples.

Example 8.1 G. Malinowski [163, 165, 166], emphasizing the distinction between accepted and rejected propositions, draws the distinction between designated and antidesignated values and uses it to generalize Tarski's notion of a consequence operation to the notion of a quasi-consequence operation (or just *q*-consequence operation). Also, single-conclusion q-consequence relations are defined; they relate not antidesignated assumptions to designated (single) conclusions. In our notation, multiple-conclusion q-consequence is the relation $\models^{\overline{-}\Rightarrow+}$.

Example 8.2 Another non-standard consequence relation has been presented in [16] where it is said *not* to be "overly outlandish or inconceivable", although it fails to be a Tarski–Scott multiple-conclusion consequence relation. In our notation, the "tonk-consequence" relation of [16] is the relation $\models^{(+\Rightarrow+|-\Leftarrow-)}$ on the set **4** with $\mathscr{D}^+ = \{\mathbf{T}, \mathbf{B}\}$ and $\mathscr{D}^- = \{\mathbf{F}, \mathbf{B}\}$. Since the logic is not transitive, sound truth tables for Prior's connective `tonk` are available such that this addition of `tonk` does not have a trivializing effect (but see also [269]).

Example 8.3 Formula-to-formula q-consequence, though not under this name, is also among the varieties of semantic consequence considered in Chaps. 3 and 4 of [251]. The types of consequence relations presented by Thijsse include the relations which, in our notation for the multiple conclusion case, are denoted as follows: $\models^{+\Rightarrow+}, \models^{\overline{-}\Rightarrow+}, \models^{\overline{-}\Rightarrow\overline{-}}, \models^{+\Rightarrow\overline{-}}$. Thijsse remarks that besides the familiar $\models^{+\Rightarrow+}$, the relation $\models^{\overline{-}\Rightarrow\overline{-}}$ "turns out to be interesting, both in theory and application".

Yet another way of defining a generalized notion of semantic consequence can be found in [38]. The two types of relation $\models^{+\Rightarrow+}$ and $\models^{-\Rightarrow-}$ are merged into a single *four-place* bi-consequence relation. Note also that n-place semantic sequents for n-valued logics have been considered by Schröter [218], see also [124, 125].

After these preparatory remarks, we are in a position to define a generalized notion of a harmonious n-valued propositional logic.

Definition 8.14 Let $\mathscr{L}$ be a language in a denumerable set of sentence letters and a finite non-empty set of finitary connectives $\mathscr{C}$. An n-valued propositional logic is a structure

$$\langle \mathscr{V}, \mathscr{D}_1, \ldots, \mathscr{D}_k, \{f_c : c \in \mathscr{C}\}\rangle$$

where $\mathscr{V}$ is a non-empty set of cardinality n $(2 \leq n)$, $2 \leq k$, every $\mathscr{D}_i$ $(1 \leq i \leq k)$ is a non-empty proper subset of $\mathscr{V}$, the sets $\mathscr{D}_i$ are pairwise distinct, and every f_c is a function on $\mathscr{V}$ with the same arity as c. The sets $\mathscr{D}_i$ are called distinguished sets. A valuation v is inductively extended to a function from the set of all $\mathscr{L}$-formulas into $\mathscr{V}$ by setting: $v(c(A_1, \ldots, A_m)) = f_c(v(A_1), \ldots, v(A_m))$ for every m-place $c \in \mathscr{C}$. For every set $\mathscr{D}_i$, two semantic consequence relation $\models_i^{\Rightarrow}$ and $\models_i^{\Leftarrow}$ are defined as follows:

1. $\Delta \models_i^{\Rightarrow} \Gamma$ iff for every valuation function v : (if for every $A \in \Delta, v(A) \in \mathscr{D}_i$, then $v(B) \in \mathscr{D}_i$ for some $B \in \Gamma$);
2. $\Delta \models_i^{\Leftarrow} \Gamma$ iff for every valuation function v : (if for every $A \in \Gamma, v(A) \in \mathscr{D}_i$, then $v(B) \in \mathscr{D}_i$ for some $B \in \Delta$).

Obviously, the relations $\models_i^{\Rightarrow}$ and $\models_i^{\Leftarrow}$ are inverses of each other.

Definition 8.15 Let $\Lambda = \langle \mathscr{V}, \mathscr{D}_1, \ldots, \mathscr{D}_k, \{f_c : c \in \mathscr{C}\}\rangle$ be an n-valued logic. Λ is said to be separated if for every $\mathscr{D}_i$, there exists no $\mathscr{D}_j$ such that $i \neq j$ and $\mathscr{V} \setminus \mathscr{D}_i = \mathscr{D}_j$. Λ is said to be refined if it is separated and the relations $\models_i^{\Rightarrow}$ are pairwise distinct (and hence also the relations $\models_i^{\circ}$ with $\circ \in \{\Rightarrow, \Leftarrow\}$ are distinct).

Definition 8.16 Let $\mathscr{C}$ be a finite non-empty set of finitary connectives, and let $\mathscr{L}_i$ be the language based on $\mathscr{C}_i = \{c_i \mid c \in \mathscr{C}\}$ for some $k \in \mathbb{N}$ such that $2 \leq i \leq k$. If $i \neq j, 2 \leq j \leq k$ and if A is an $\mathscr{L}_i$-formula, then let A_j be the result of replacing every connective c_i in A by c_j. If Δ is a set of $\mathscr{L}_i$ formulas, let $\Delta_j = \{A_j \mid A \in \Delta\}$. If the language $\mathscr{L}$ of a refined n-valued logic Λ with k distinguished sets is based on the set $\cup_{i \leq k} \mathscr{C}_i$, then Λ is said to be *harmonious* iff (i) for every i, j with $i \neq j$ and all sets of $\mathscr{L}_i$-formulas Δ, Γ the following holds: (i) $\Delta \models_i^{\Rightarrow} \Gamma$ iff $\Delta_j \models_j^{\Leftarrow} \Gamma_j$, (ii) $\Delta \models_i^{\Leftarrow} \Gamma$ iff $\Delta_j \models_j^{\Rightarrow} \Gamma_j$, and (iii) for every $c \in \mathscr{C}, f_{c_i} \neq f_{c_j}$.

The logics $B_{\sharp n}$ are harmonious in this generalized sense. We may set $k = 2$, $\mathscr{D}_1 = \mathscr{D}^+$, $\mathscr{D}_2 = \mathscr{D}^-, \mathscr{C}_1 = \{\sim_t, \wedge_t, \vee_t\}, \mathscr{C}_2 = \{\sim_f, \wedge_f, \vee_f\}, \models_1^{\Rightarrow} = \models^+, \models_1^{\Leftarrow} = (\models^+)^{-1}$ (the inverse of $\models^+$), $\models_2^{\Leftarrow} = \models^-$, and $\models_2^{\Rightarrow} = (\models^-)^{-1}$.

Another obvious generalization of the present considerations is giving up the restriction to finitely many values.

Chapter 9
Generalized Truth Values and Many-Valued Logics: Suszko's Thesis

[A] fundamental problem concerning many-valuedness is to know what it really is [57, p. 281].

Abstract According to Suszko's Thesis, there are but two logical values, *true* and *false*. In this chapter, we consider and critically discuss Roman Suszko's, Grzegorz Malinowski's, and Marcelo Tsuji's analyses of logical two-valuedness. Another analysis is presented which favors a notion of a logical system as encompassing possibly more than one consequence relation. Moreover, in light of these considerations, we will point out that the relation between the notion of a truth value and the notion of entailment is even more intimate than the connection emerging from the interaction between properties of entailment relations and truth values. In some cases it is possible to draw a strong analogy between them, namely to interpret entailment relations as a kind of truth value, and such an interpretation seems to be both natural and promising.

9.1 Introduction

Many-valued logic is one of the oldest branches of modern formal non-classical logic. It was developed in very influential works by Jan Łukasiewicz, Emil Post, Dmitry Bochvar, and Stephen Kleene in the 1920s and 1930s and is now a fully established and flourishing research program, see, for instance, [125, 165]. In the 1970s, however, the prominent logician Roman Suszko called into question the theoretical foundations of many-valued logic. Suszko distinguishes between the logical values *truth* and *falsity* (*true* and *false*) on the one hand, and algebraic values on the other hand. Whereas the algebraic values are just admissible referents of formulas, the logical values play another role. One of them, *truth*, is used to define valid semantic consequence: If every premise is true, then so is (at least one of) the conclusion(s). If being false is understood as not being true, then, by contraposition, also the other logical value can be used to explain valid semantic consequence: If the (every) conclusion is not true, then so is at least one of the

Y. Shramko and H. Wansing, *Truth and Falsehood*, Trends in Logic 36,
DOI: 10.1007/978-94-007-0907-2_9,

premises. The logical values are thus represented by a bi-partition of the set of algebraic values into a set of designated values (truth) and its complement (falsity).[1]

Essentially the same idea was taken up earlier by Dummett in his important paper [69], where he asks

> what point there may be in distinguishing between different ways in which a statement may be true or between different ways in which it may be false, or, as we might say, between degrees of truth and falsity [69, p. 153].

Dummett observes that, first,

> the sense of a sentence is determined wholly by knowing the case in which it has a designated value and the cases in which it has an undesignated one,

and moreover,

> finer distinctions between different designated values or different undesignated ones, however naturally they come to us, are justified only if they are needed in order to give a truth-functional account of the formation of complex statements by means of operators [69, p. 155].

Suszko's claim evidently echoes this observation by Dummett.

Suszko declared that "Łukasiewicz is the chief perpetrator of a magnificent conceptual deceit lasting out in mathematical logic to the present day" [246, p. 377], and he claimed that "there are but two logical values, true and false" [43, p. 169]. This challenging claim is now called *Suszko's Thesis*.[2] It has been given a formal content by the so-called Suszko Reduction, the proof that every structural Tarskian consequence relation and hence also every structural Tarskian many-valued propositional logic is characterized by a bivalent semantics.[3]

The Suszko Reduction calls for a careful analysis, not only because it seems to undermine many-valued logic, but also because it invites a re-consideration of the notion of a many-valued logic in particular and the notion of a logical system in general. According to Tsuji [253, p. 308], "Suszko thought that the key to logical two-valuedness rested in the *structurality* of the abstract logics (for Wójcicki's theorem was the cornerstone of his reduction method)". Wójcicki's Theorem states that every structural Tarskian consequence relation possesses a characterizing class of matrices.[4] As pointed out in [43, 44], Suszko's Reduction can, however, be carried out for *any* Tarskian consequence relation, and furthermore da

---

[1] This chapter brings together [233, 276].

[2] Sometimes Suszko's Thesis is stated in more dramatic terms. Tsuji [253, p. 299], for example, explains that "Suszko's Thesis maintains that many-valued logics do not exist at all".

[3] Bivalent interpretations of many-valued logics have also been presented by Urquhart [256], Scott [220], and da Costa and later Béziau, see [30, 57], and references therein. In [12], the Suszko Reduction provides the background for discussing the merits of many-valued non-deterministic matrices.

[4] Wójcicki proved his theorem for consequence operations, but the presentations in terms of consequence relations and operations are trivially interchangeable. Moreover, it should be

Costa et al. [57] emphasize that the assumption of structurality is not needed for a reduction to two-valuedness. Also, Béziau [30] noticed that a reduction to a bivalent semantics can be obtained even if a much weaker and less common notion of a logical system is assumed. In light of this observation, Tsuji [253] criticizes Grzegorz Malinowski's analysis of the Suszko Reduction [163, 164, 165, 166]. Malinowski takes up Suszko's distinction between algebraic many-valuedness and logical two-valuedness and highlights the bi-partition of the algebraic values into designated ones and values that are not designated, a division that plays a crucial role in the Suszko Reduction.

In fact, as we have noted in the previous chapter, in the literature on many-valued logics, sometimes an explicit distinction is drawn between a set $\mathcal{D}^+$ of *designated* algebraic values and a set $\mathcal{D}^-$ of *antidesignated* algebraic values, see, for example, [124, 125, 165, 207]. This distinction leaves room for values that are *neither* designated *nor* antidesignated and for values that are *both* designated *and* antidesignated. This therefore amounts to replacing logical two-valuedness (viewed as a bi-partition of the set of algebraic values) by logical four-valuedness in general and by logical three-valuedness if it is postulated that $\mathcal{D}^+ \cap \mathcal{D}^- = \emptyset$ or that $\mathcal{D}^+ \cup \mathcal{D}^-$ is the set of all algebraic values available. The first condition is imposed by Malinowski [163, 164, 165, 166] and Gottwald [125], and the second condition may be used to define systems of paraconsistent logic.[5]

In order to provide a counterexample to Suszko's Thesis, Malinowski defined the notion of a single-conclusion *quasi*-consequence (q-consequence) relation. The semantic counterpart of q-consequence is q-entailment. Single-conclusion q-entailment is defined by requiring that if every premise is not antidesignated, then the conclusion is designated. Malinowski [163] proved that for every structural q-consequence relation, there exists a characterizing class of q-matrices. These matrices comprise, in addition to a subset of designated values $\mathcal{D}^+$, a disjoint subset of antidesignated values $\mathcal{D}^-$.[6] Not every q-consequence relation has a bivalent semantics. Moreover, a q-consequence relation need not be reflexive, i.e., it may be the case that a formula is not a q-consequence of a set of formulas to which it belongs. Tsuji [253] observed that for *any* relation $\vdash \subseteq \mathcal{P}(\mathcal{L}) \times \mathcal{L}$ (where $\mathcal{L}$ may be seen as a propositional language and $\mathcal{P}(\mathcal{L})$ is the powerset of $\mathcal{L}$), i.e., for any abstract logical structure in the sense of Béziau's Universal Logic [28], reflexivity characterizes the existence of a set of bivaluations with respect to which $\vdash$ is complete.[7]

(Footnote 4 continued)
pointed out that the entailment relation defined by a matrix in this case takes into account only valuations which are homomorphisms, cf. Sect. 9.2.

[5] In [275] and Chap. 8 it is explained why it is not at all unreasonable to admit $\mathcal{D}^+ \cap \mathcal{D}^- \neq \emptyset$.

[6] Also Malinowski considers only valuations which are homomorphisms.

[7] It is assumed that an entailment relation $\models_X$ with respect to a set X of bivaluations from the set of $\mathcal{L}$-formulas into $\{0, 1\}$ is defined by

$$\Delta \models_X A \text{ iff for every } v \in X, v(A) = 1, \text{ whenever for every } B \in \Delta, v(B) = 1.$$

Completeness thus means that for every set of $\mathcal{L}$-formulas $\Delta \cup \{A\}$, $\Delta \models_X A$ implies $\Delta \vdash A$.

In this chapter, we argue that Malinowski's analysis *does* in a sense capture the central aspect of explaining the feasibility of the Suszko Reduction and the distinction between algebraic and logical values. If the idea of *logical* many-valuedness as opposed to a multiplicity of algebraic values is taken seriously, then Tsuji's [253] analysis is question-begging because it presupposes a notion of a logical system that admits at most logical two-valuedness if the single assumed consequence relation is reflexive. With another concept of a logical system, every such system is logically k-valued for some $k \in \mathbb{N}$ ($k \geq 2$), and, moreover, the definition of a logical system is such that every entailment relation in a logically k-valued logic is reflexive. By increasing the number of logical values (considered as separate subsets of the set of algebraic values) from two to three and showing that every q-consequence relation has an adequate trivalent semantics, Malinowski did a step toward logical many-valuedness, but in a way that gives up the idea of entailment as preservation of a logical value (from the premises to the conclusion, or vice versa) and thereby at the price of violating reflexivity.[8] In our analysis of Suszko's Thesis we will suggest taking a further step and to admit an increase not only in the number of logical values, but also in the number of entailment relations.

First, however, we will review the Suszko Reduction method in Sect. 9.2 and there also collect some well-known facts about consequence relations. In Sects. 9.2.1 and 9.2.2 we will present and briefly discuss Malinowski's and Tsuji's analyses of Suszko's Thesis, and in Sect. 9.3, we will present our own analysis. Moreover, in Sect. 9.4, we will briefly mention yet another analysis leading to higher-arity consequence relations, and we will draw some general conclusions from the previous discussion. In the remainder of this chapter, we will take up the discussion of matrices and sets of distinguished subsets of algebraic values. We will consider q-matrices and entailment relations which can be defined on q-matrices. In particular, we will observe an analogy between the generalized truth values of the bi-lattice $FOUR_2$ and the considered entailment relations by defining ordering relations not between truth values but between the entailment relations.

9.2 The Suszko Reduction

Suszko suggested the reduction to a bivalent semantics with respect to the standard notion of a structural Tarskian logic (structural Tarskian consequence relation) and the standard notion of an n-valued matrix (semantically defined n-valued logic). For the sake of uniformity of our account with the works by Polish logicians (in particular, Grzegorz Malinowski, Roman Suszko, Ryszard Wójcicki, and others), we use in this chapter the terminology of *logical matrices* rather than of *valuation*

8 The problem of lifting the Suszko Reduction method from a reduction to a trivalent semantics to an $n > 3$ is raised in [169], and Malinowski there presents some partial solutions to this technically as well as conceptually intricate problem.

systems utilized in preceding chapters. We occasionally reformulate some previous definitions needed for our analysis of Suszko's Thesis.

Let $\mathscr{L}$ be a propositional language in a denumerable set of sentence letters and a finite non-empty set of finitary connectives $\mathscr{C} = \{c_1, \ldots, c_m\}$. A Tarskian consequence relation on $\mathscr{L}$[9] is a relation $\vdash \subseteq \mathscr{P}(\mathscr{L}) \times \mathscr{L}$ such that for every $A, B \in \mathscr{L}$ and every $\Delta, \Gamma \subseteq \mathscr{L}$:

$$\Delta \cup \{A\} \vdash A \quad \text{(Reflexivity)} \tag{9.1}$$

$$\text{If } \Delta \vdash A \text{ then } \Delta \cup \Gamma \vdash A \quad \text{(Monotonicity)} \tag{9.2}$$

$$\text{If } \Delta \vdash A \text{ and } \Gamma \cup \{A\} \vdash B, \text{ then } \Delta \cup \Gamma \vdash B \quad \text{(Cut)} \tag{9.3}$$

A Tarskian consequence relation $\vdash$ on the language $\mathscr{L}$ is called *structural* iff for every $A \in \mathscr{L}$, every $\Delta \subseteq \mathscr{L}$, and every uniform substitution function σ on $\mathscr{L}$ (every endomorphism of the absolutely free algebra $(\mathscr{L}, c_1, \ldots, c_m)$) we have

$$\Delta \vdash A \text{ iff } \sigma(\Delta) \vdash \sigma(A) \quad \text{(Structurality)}, \tag{9.4}$$

where $\sigma(\Delta) = \{\sigma(B) \mid B \in \Delta\}$. It is standard terminology to call a pair $(\mathscr{L}, \vdash)$ a Tarskian (structural Tarskian) logic iff $\vdash$ is Tarskian (structural and Tarskian).

An n-valued matrix based on $\mathscr{L}$ is a structure $\mathfrak{M} = \langle \mathscr{V}, \mathscr{D}, \{f_c : c \in \mathscr{C}\}\rangle$, where $\mathscr{V}$ is a non-empty set of cardinality $n \geq 2$, $\mathscr{D}$ is a non-empty proper subset of $\mathscr{V}$, and every f_c is a function on $\mathscr{V}$ with the same arity as c. The elements of $\mathscr{V}$ are usually called *truth values* (or truth degrees), and the elements of $\mathscr{D}$ are regarded as the *designated* truth values. In Suszko's terminology, $\mathscr{V}$ is the set of algebraic values, whereas $\mathscr{D}$ and its complement represent the two logical truth values. A valuation function v in $\mathfrak{M}$ is a function from $\mathscr{L}$ into $\mathscr{V}$. Also, a structure $\mathfrak{M} = \langle \mathscr{V}, \mathscr{D}, \{f_c : c \in \mathscr{C}\}\rangle$ may be viewed as a logic because the set of designated truth values determines a Tarskian (semantical) consequence relation $\models_{\mathfrak{M}} \subseteq \mathscr{P}(\mathscr{L}) \times \mathscr{L}$ by defining $\Delta \models_{\mathfrak{M}} A$ iff for every valuation v in $\mathfrak{M}$, $v(\Delta) \subseteq \mathscr{D}$ implies $v(A) \in \mathscr{D}$, where $v(\Delta) = \{v(B) \mid B \in \Delta\}$. Usually, only truth-functional valuations are considered. A valuation v in $\mathfrak{M}$ is truth-functional (structural) iff it is a homomorphism from $(\mathscr{L}, c_1, \ldots, c_m)$ into $(\mathscr{V}, f_{c_1}, \ldots, f_{c_m})$. A pair $\mathscr{M} = \langle\langle \mathscr{V}, \mathscr{D}, \{f_c : c \in \mathscr{C}\}\rangle, v\rangle$, where $\mathfrak{M} = \langle \mathscr{V}, \mathscr{D}, \{f_c : c \in \mathscr{C}\}\rangle$ is an n-valued matrix and v a valuation in $\mathfrak{M}$, may be called an n-valued model based on $\mathfrak{M}$. A model $\langle \mathfrak{M}, v\rangle$ is structural iff v is a truth-functional valuation in $\mathfrak{M}$. Obviously, an n-valued model $\mathscr{M}$ determines a Tarskian (semantical) consequence relation $\models_{\mathscr{M}} \subseteq \mathscr{P}(\mathscr{L}) \times \mathscr{L}$ by defining $\Delta \models_{\mathscr{M}} A$ iff $v(\Delta) \subseteq \mathscr{D}$ implies $v(A) \in \mathscr{D}$.

A Tarskian logic $(\mathscr{L}, \vdash)$ is said to be characterized by an n-valued matrix $\mathfrak{M}$ iff $\vdash = \models_{\mathfrak{M}}$, $(\mathscr{L}, \vdash)$ is characterized by an n-valued model $\mathscr{M}$ iff $\vdash = \models_{\mathscr{M}}$, $(\mathscr{L}, \vdash)$

9 Often, consequence relations are defined for arbitrary sets. In this chapter we restrict our interest, however, to consequence relations defined on propositional languages.

is characterized by a class $\mathfrak{K}$ of n-valued matrices iff $\vdash = \bigcap\{\models_{\mathfrak{M}} | \mathfrak{M} \in \mathfrak{K}\}$, and, finally, $(\mathcal{L}, \vdash)$ is characterized by a class $\mathfrak{K}$ of n-valued models iff $\vdash = \bigcap\{\models_{\mathcal{M}} | \mathcal{M} \in \mathfrak{K}\}$. Wójcicki [284] showed that every structural Tarskian logic is characterized by its so-called Lindenbaum bundle:

$$\{\langle\langle \mathcal{L}, \{A \in \mathcal{L} \mid \Delta \vdash A\}, \mathcal{C}\rangle, v\rangle \mid \Delta \subseteq \mathcal{L}, v \text{ is an endomorphism of } \mathcal{L}\}.$$

Theorem 9.1 (Wójcicki) *Every structural Tarskian logic is characterized by a class of structural n-valued models, for some* $n \leq \aleph_0$.

Theorem 9.2 (Suszko [246], Malinowski [165])[10] *Every structural Tarskian logic is characterized by a class of two-valued models.*

Proof Let $\Lambda = (\mathcal{L}, \vdash)$ be a structural Tarskian logic. Then, by Wójcicki's Theorem, Λ is characterized by a class $\mathfrak{C}_\Lambda$ of structural n-valued models. For $\langle\langle \mathcal{V}, \mathcal{D}, \{f_c : c \in \mathcal{C}\}\rangle, v\rangle \in \mathfrak{C}_\Lambda$, the function t_v from $\mathcal{L}$ into $\{0, 1\}$ is defined as follows:

$$t_v(A) = \begin{cases} 1 & \text{if } v(A) \in \mathcal{D} \\ 0 & \text{if } v(A) \notin \mathcal{D} \end{cases}$$

The class $\{\langle\langle\{0,1\}, \{1\}, \{f_c : c \in \mathcal{C}\}\rangle, t_v\rangle \mid \langle\langle \mathcal{V}, \mathcal{D}, \{f_c : c \in \mathcal{C}\}\rangle, v\rangle \in \mathfrak{C}_\Lambda\}$ of two-valued models characterizes Λ. □

Malinowski [166, p. 79] highlights (though in other words) that the Suszko Reduction does not establish the existence of a characterizing class of *structural* two-valued models. Suszko was fully aware of this fact. In [246, p. 378] he explains that

> the logical valuations are morphisms (of formulas to the zero-one model) in some exceptional cases, only. Thus, the logical valuations and the algebraic valuations are functions of quite different conceptual nature. The former relate to the truth and falsity and, the latter represent the reference assignments.

This conception may explain why the sets $\mathcal{V}$ are regarded by Suszko as sets of "algebraic values". They are algebraic because they are assigned to formulas by valuations which are homomorphisms. If the elements of $\mathcal{V}$ need not be assigned by homomorphisms, they may be called *referential values*. The semantics which emerges as a result of the Suszko Reduction is referentially bivalent and hence logically bivalent in the sense that there exist only two distinct non-empty proper subsets of the set of algebraic values, see also Sect. 9.3.2. The fact that the valuations t_v need not be homomorphisms deprives the Suszko Reduction of much of its jeopardizing effect on many-valued logic if structurality is viewed as a

[10] As Caleiro et al. [54] point out, "there seems to be no paper where Suszko explicitly formulates (SR)[the Suszko Reduction] in full generality!".

defining property of a logic. It has been emphasized by da Costa et al. [57] and Caleiro et al. [43, 44], however, that the condition of structurality can be given up.

Theorem 9.3 *Every Tarskian logic is characterized by a class of n-valued models for some* $n \leq \aleph_0$.

Theorem 9.4 *Every Tarskian logic is characterized by a class of two-valued models.*

The proofs of Theorem 9.2 and 9.4 are non-constructive. Caleiro et al. [43], [45] considerably improve upon the Suszko Reduction by defining an effective method for associating an equivalent two-valued model to each structural finitely-valued model (under the assumption of effectively separable truth values).

9.2.1 Malinowski's Analysis of Suszko's Thesis

Malinowski [166, p. 80 f.] succinctly analyses Suszko Reduction as follows:

> [L]ogical two-valuedness ... is obviously related to the division of the universe of interpretation into two subsets of elements: distinguished and others. It also turned out, under the assumption of structurality, that Tarski's concept of consequence may be considered as a "bivalent" inference operation.

He then describes his response to the question of "whether logical many-valuedness is possible at all" as giving

> an affirmative answer to this question by invoking a formal framework for reasoning admitting rules of inference which lead from non-rejected assumptions to accepted conclusions.

This approach may be viewed as taking 'true' and 'false' to be expressions that give rise to contrary instead of contradictory pairs of sentences. As such, the pair 'true' versus 'false' is reflected by the contrary pairs 'designated' versus 'antidesignated' and 'accepted' versus 'rejected'. Admitting algebraic values that are neither designated nor antidesignated amounts to admitting, in addition to the logical values *true* and *false*, the third logical value *neither true nor false*. In other words, being false is distinguished from not being true. Whereas the algebraic values that are not designated are already given with the set of designated values $\mathscr{D}$ as its set-theoretical complement, the treatment of *true* and *false* as values that are independent of each other leads to distinguishing a non-empty set $\mathscr{D}^+$ of designated algebraic values from a non-empty set $\mathscr{D}^-$ of antidesignated algebraic values.

Let $\mathscr{L}$ be as above. An n-valued q-matrix (quasi-matrix) based on $\mathscr{L}$ is defined by Malinowski as a structure $\mathfrak{M} = \langle \mathscr{V}, \mathscr{D}^+, \mathscr{D}^-, \{f_c : c \in \mathscr{C}\}\rangle$, where $\mathscr{V}$ is a non-empty set of cardinality $n \geq 2$, $\mathscr{D}^+$ and $\mathscr{D}^-$ are distinct non-empty proper subsets of $\mathscr{V}$ such that $\mathscr{D}^+ \cap \mathscr{D}^- = \emptyset$, and every f_c is a function on $\mathscr{V}$ with the same arity as c. If it is not required that $\mathscr{D}^+ \cap \mathscr{D}^- = \emptyset$, we shall talk of generalized

q-matrices.[11] A valuation function v in $\mathfrak{M}$ is a function from $\mathscr{L}$ into the set of truth degrees $\mathscr{V}$, and Malinowski considers only valuations which are homomorphisms (from $(\mathscr{L}, c_1, \ldots, c_m)$ into $(\mathscr{V}, f_{c_1}, \ldots, f_{c_m})$).

To obtain a kind of entailment relation that does not admit a reduction to a bivalent semantics, Malinowski defines such a relation, called q-entailment, as depending on both sets $\mathscr{D}^+$ and $\mathscr{D}^-$. A q-matrix $\mathfrak{M} = \langle \mathscr{V}, \mathscr{D}^+, \mathscr{D}^-, \{f_c : c \in \mathscr{C}\}\rangle$ determines a q-entailment relation $\models_{\mathfrak{M}} \subseteq \mathscr{P}(\mathscr{L}) \times \mathscr{L}$ by defining $\Delta \models_{\mathfrak{M}} A$ iff for every (in Malinowski's case homomorphic) valuation v in $\mathfrak{M}$, $v(\Delta) \cap \mathscr{D}^- = \varnothing$ implies $v(A) \in \mathscr{D}^+$. A pair $\mathscr{M} = \langle\langle \mathscr{V}, \mathscr{D}^+, \mathscr{D}^-, \{f_c : c \in \mathscr{C}\}\rangle, v\rangle$, where $\mathfrak{M} = \langle \mathscr{V}, \mathscr{D}^+, \mathscr{D}^-, \{f_c : c \in \mathscr{C}\}\rangle$ is an n-valued q-matrix and v a valuation in $\mathfrak{M}$, may be called an n-valued q-model based on $\mathfrak{M}$. The relation $\models_{\mathscr{M}} \subseteq \mathscr{P}(\mathscr{L}) \times \mathscr{L}$ determined by such a model is defined by the following equivalence: $\Delta \models_{\mathscr{M}} A$ iff $v(\Delta) \cap \mathscr{D}^- = \varnothing$ implies $v(A) \in \mathscr{D}^+$. A model $\langle \mathfrak{M}, v\rangle$ is structural iff v is a truth-functional valuation in $\mathfrak{M}$. If $\mathfrak{M} = \langle \mathscr{V}, \mathscr{D}^+, \mathscr{D}^-, \{f_c : c \in \mathscr{C}\}\rangle$ is a q-matrix and $\mathscr{D}^+$ is not the complement of $\mathscr{D}^-$, there is no class of functions from $\mathscr{L}$ into $\{1, 0\}$ such that $\Delta \models_{\mathfrak{M}} A$ iff for every function v from that class, $v(\Delta) \subseteq \{1\}$ implies $v(A) = 1$.[12] Let $\mathscr{M} = \langle\langle \mathscr{V}, \mathscr{D}^+, \mathscr{D}^-, \{f_c : c \in \mathscr{C}\}\rangle, v\rangle$ be an n-valued q-model. Malinowski pointed out that an equivalent three-valued q-model $\mathscr{M}' = \langle\langle \{0, \frac{1}{2}, 1\}, \{1\}, \{0\}, \{f_c : c \in \mathscr{C}\}\rangle, t_v\rangle$ can be defined as follows:

$$t_v(A) = \begin{cases} 1 & \text{if } v(A) \in \mathscr{D}^+ \\ \frac{1}{2} & \text{iff } v(A) \in \mathscr{V} \setminus (\mathscr{D}^+ \cup \mathscr{D}^-) \\ 0 & \text{if } v(A) \in \mathscr{D}^- \end{cases}$$

A q-entailment relation $\models_{\mathfrak{M}}$ is a special case of what Malinowski calls a q-consequence relation. A q-consequence relation on $\mathscr{L}$ is a relation $\Vdash \subseteq \mathscr{P}(\mathscr{L}) \times \mathscr{L}$ such that for every $A \in \mathscr{L}$ and every Δ, $\Gamma \subseteq \mathscr{L}$:

$$\text{If } \Delta \Vdash A \text{ then } \Delta \cup \Gamma \Vdash A \quad \text{(Monotonicity)} \tag{9.5}$$

$$\Delta \cup \{B \mid \Delta \Vdash B\} \Vdash A \text{ iff } \Delta \Vdash A \quad \text{(Quasi-closure)} \tag{9.6}$$

A q-consequence relation on $\mathscr{L}$ is called *structural* iff for every $A \in \mathscr{L}$, every $\Delta \subseteq \mathscr{L}$, and every uniform substitution function σ on $\mathscr{L}$ we have

$$\Delta \Vdash A \text{ iff } \sigma(\Delta) \Vdash \sigma(A) \quad \text{(Structurality).} \tag{9.7}$$

---

[11] Basically, a generalized q-matrix is what has been introduced in the previous chapter as a "symmetric n-valued valuation system", see Definition 8.1.

[12] In [29, p. 120] it is stated that "Malinowski constructs (using an extended concept of a matrix) a consequence relation which has no two-valued logical semantics because it fails to obey the "identity" axiom of Tarski. However it has been shown (cf. 148) that we can adapt in some way two-valued logical semantics even in the case of such kind of consequence relation". Note that this "adaptation" is sketched in terms of *two* functions mod_1 and mod_2 each assigning to every formula and every set of formulas a class of models.

A pair $(\mathscr{L}, \Vdash)$ is said to be a q-logic, and it is structural iff $\Vdash$ is structural. A q-logic $(\mathscr{L}, \Vdash)$ is said to be characterized by an n-valued q matrix $\mathfrak{M}$ iff $\Vdash\, = \models_{\mathfrak{M}}$, $(\mathscr{L}, \Vdash)$ is characterized by an n-valued q-model $\mathscr{M}$ iff $\Vdash\, = \models_{\mathscr{M}}$, and $(\mathscr{L}, \Vdash)$ is characterized by a class $\mathfrak{K}$ of n-valued q-matrices (q-models) iff $\Vdash\, = \bigcap\{\models_{\mathfrak{M}} \mid \mathfrak{M} \in \mathfrak{K}\}$ ($\Vdash\, = \bigcap\{\models_{\mathscr{M}} \mid \mathscr{M} \in \mathfrak{K}\}$).

If $\Delta \subseteq \mathscr{L}$, let $\mathscr{D}^{+}_{\Delta} = \{A \in \mathscr{L} \mid \Delta \Vdash A\}$ and $\mathscr{D}^{-}_{\Delta} = \mathscr{L} \setminus (\Delta \cup \{A \in \mathscr{L} \mid \Delta \Vdash A\})$. Malinowski [163] showed that every structural q-logic is characterized by the following Lindenbaum bundle:

$$\{\langle\langle \mathscr{L}, \mathscr{D}^{+}_{\Delta}, \mathscr{D}^{-}_{\Delta}, \mathscr{C}\rangle, v\rangle \mid \Delta \subseteq \mathscr{L}, v \text{ is an endomorphism of } \mathscr{L}\}.$$

Theorem 9.5 (Malinowski) *Every structural q-logic is characterized by a class of structural n-valued q-models for some* $n \leq \aleph_0$.

By the above definition of three-valued q-models $\mathscr{M}'$ and by the Suszko Reduction for the case that $\mathscr{V} \setminus (\mathscr{D}^{+} \cup \mathscr{D}^{-}) = \emptyset$, it follows that q-logics are logically two-valued or three-valued.

Theorem 9.6 (Malinowski) *Every structural q-logic is characterized by a class of two-valued q-models or by a class of three-valued q-models.*

In Chap. 8 we have already seen a non-standard entailment relation, based on a certain generalized q-matrix, namely tonk-consequence. In [56] this relation is said not to be "overly outlandish or inconceivable". We noted that tonk-consequence fails to be a Tarskian consequence relation. The tonk-consequence relation for a given language $\mathscr{L}$ and set of connectives $\mathscr{C}$ is the relation $\models\, \subseteq \mathscr{P}(\mathscr{L}) \times \mathscr{L}$ defined for the generalized q-matrix $\mathfrak{M} = \langle\{\emptyset, \{T\}, \{F\}, \{T,F\}\}, \{\{T\}, \{F,T\}\}, \{\{F\}, \{T,F\}\}, \{f_c : c \in \mathscr{C}\}\rangle$ as follows: $\Delta \models A$ iff either for every truth-functional valuation v in $\mathfrak{M}$, $v(\Delta) \subseteq \mathscr{D}^{+}$ implies $v(A) \in \mathscr{D}^{+}$, or for every truth-functional valuation v in $\mathfrak{M}$, $v(\Delta) \cap \mathscr{D}^{-} = \emptyset$ implies $v(A) \notin \mathscr{D}^{-}$. Since tonk-entailment is not transitive, sound truth tables for Prior's connective `tonk` are available such that this addition of `tonk` does not have a trivializing effect (but also see [269]).

9.2.2 Tsuji's Analysis of Suszko's Thesis

Tsuji [253, p. 305] emphasizes that Malinowski's analysis "is not wrong—but that it misses the main point of logical two-valuedness." Taking up Béziau's idea of a Universal Logic [28], Tsuji assumes a very general notion of a logical system. An *abstract logical structure* is any pair $(S, \vdash)$, where S is an arbitrary set and $\vdash\, \subseteq \mathscr{P}(S) \times S$. We shall again restrict our attention to relations $\vdash\, \subseteq \mathscr{P}(\mathscr{L}) \times \mathscr{L}$, where $\mathscr{L}$ is a propositional language.

Theorem 9.7 (Tsuji [253]) *An abstract logical structure* $(\mathscr{L}, \vdash)$ *is complete with respect to a class of two-valued models iff* $\vdash$ *satisfies (Reflexivity).*

Proof Let $\vdash$ be reflexive and consider the class of two-valued models $\mathfrak{C} =$

$$\{\langle\langle\{0,1\},\{1\},\{f_c : c \in \mathscr{C}\}\rangle, v_\Delta\rangle \mid \Delta \subseteq \mathscr{L}, v_\Delta(A) = 1 \text{ iff } \Delta \vdash A\}.$$

Suppose $\Delta \nvdash A$. Then $v_\Delta(A) = 0$. By (Reflexivity), $v_\Delta(\Delta) \subseteq \{1\}$. Therefore $\Delta \not\models_{\mathfrak{C}} A$. Conversely, if $\vdash$ is not reflexive, there exist $\Delta \subseteq \mathscr{L}$ and $A \in \Delta$ with $\Delta \nvdash A$. But $\Delta \models_{\mathscr{M}} A$ for any two-valued model $\mathscr{M}$ since $A \in \Delta$. □

Tsuji [252, p. 308] takes this result to reveal that "the problem of logical two-valuedness has more to do with the "geometrical" properties of the" relation $\vdash$ of an abstract logical structure $\langle S, \vdash\rangle$, "than with the algebraic properties of it's set S or of it's Lindenbaum bundle". In the next section, we shall raise doubts about this analysis.

9.3 Logical *n*-Valuedness as Inferential Many-Valuedness

9.3.1 What is a Logical Value?

Suszko does not define the notion of a logical value except for stating that *true* and *false* are the only logical values, but he distinguishes the logical values from algebraic values and explains that the possibly many algebraic values are denotations of (propositional) formulas. Moreover, he claims that "any multiplication of logical values is a mad idea" [246, p. 378]. In any case the question arises by virtue of which properties *true* and *false* are to be considered as *logical* values. The logical value *true* is given with the specification of a set of distinguished algebraic values and the corresponding notion of entailment. A formula A is entailed by a set of premises Δ if and only if it is the case that if every premise is true (alias designated), then so is the conclusion. Thus, truth is what is preserved in a valid inference from the premises to the conclusion. Let us refer to this notion of entailment as t_m-entailment.[13] A formula A is logically true iff A is t_m-entailed by the empty set (iff for every assignment v of algebraic values to the formulas of the language under consideration, $v(A)$ is designated), and A is logically false iff A t_m-entails the empty set (iff for every assignment v, $v(A)$ is not designated). If we consider a bi-partition of the set of algebraic values, truth is identified with a

[13] Note the difference between the notion of t_m-entailment and the one of t-entailment introduced in Chaps. 4 and 8 by Definitions 4.1 and 8.12. Whereas t-entailment is defined in the context of truth-value lattices and is conceived as reflecting the corresponding lattice order (namely, the truth order), t_m-entailment is defined on a logical matrix, having the main characteristic to preserve designatedness of truth values. An analogous observation holds also for the notion of f_m-entailment introduced below. In fact, t_m-entailment and f_m-entailment are nothing else but the relations $\models^+$ and $\models^-$ defined in the previous chapter by (8.2) and (8.3).

non-empty subset $\mathscr{D}$ of designated algebraic values and falsity with the complement of $\mathscr{D}$. Now, it is characteristic for falsity to be preserved in the reverse direction, i.e., from the conclusion to at least one of the premises. One might wish to consider a notion of f_m-entailment understood as the backward preservation of values associated with falsity. Obviously, membership in the complement of $\mathscr{D}$ is preserved from the conclusion to the premises, but this gives *the very same* entailment relation. Since $\mathscr{D}$ is uniquely determined by its complement and vice versa, logical two-valuedness is, in fact, reduced to logical *mono-valuedness* if there is just one entailment relation defined as truth preservation form the premises to the conclusion.

However, if we treat 'falsity' *not* as a mere abbreviation for 'non-truth' and if we correspondingly distinguish not only a set $\mathscr{D}^+$ of designated algebraic values but also a set $\mathscr{D}^-$ of antidesignated algebraic values which are not obligatorily complements of each other, then f_m-entailment may well be different from t_m-entailment. Indeed, consider, for example, the *q*-matrix $\mathfrak{L}_3{*} := \langle\{T, \varnothing, F\}, \{T\}, \{F\}, \{f_c : c \in \{\sim, \wedge, \vee, \supset\}\}\rangle$, where the functions f_c are the usual operations of Łukasiewicz's three-valued logic (see, e.g., [167, 168]). Let us call $\mathfrak{L}_3{*}$ the Łukasiewicz *q*-matrix. It is easy to see (Proposition 8.1) that in $\mathfrak{L}_3{*}$ t_m-entailment and f_m-entailment do not coincide, for $A \wedge (A \supset B) \models^+ B$, but $A \wedge (A \supset B) \not\models^- B$. In this situation it would hardly be justifiable to prefer one entailment relation over the other and to deal, e.g., only with t_m-entailment and to completely disregard f_m-entailment. And if both entailment relations are *pari passu* (as the sets $\mathscr{D}^+$ and $\mathscr{D}^-$ are), why should one not conceive of logical systems with two (or perhaps even more) entailment relations?[14]

If logic is thought of as the theory of valid inferences, then a *logical value* may be seen as a value that is used to define in a canonical way an entailment relation on a set of formulas. By a canonical definition of entailment we mean a definition of entailment as a relation that (in the single conclusion case) preserves membership in a certain set of algebraic values, either from the premises to the conclusion of inferences or from the conclusion to the premises. Such a relation will be Tarskian (since preservation of a logical value from the conclusion to the premises means that if the conclusion possesses the value, then so does at least one of the premises, whereas preservation from the premises to the conclusion means that if every premise possesses the value, then so does the conclusion). Two logical values are independent of each other iff the canonically defined entailment relations associated with these values are distinct.

[14] The idea of a logical system comprising two consequence relations may be detected already in [58, Chap. 6], where H.B. Curry distinguishes between deducibility and refutability. Curry assumes a theory $\mathfrak{T}$ generated by axioms and a theory $\mathfrak{F}$ generated by *counteraxioms*. Moreover, it is assumed that a formula A is refutable, if a refutable formula is deducible from A. Curry then introduces a notion of negation as refutation by requiring (i) that the negation of every counteraxiom is provable and (ii) that the negation $\neg A$ of a formula A is provable, if the negation of a formula B is provable and B is deducible from A.

In addition to the term 'logical many-valuedness', Malinowski also uses the term 'inferential many-valuedness'. One might understand Malinowski's analysis as assuming (i) that *logical* values play a role in defining a single entailment (inference) relation and (ii) that this entailment relation need not be defined canonically. Truth and falsity are treated by Malinowski as logical values insofar as *both* sets of algebraic values are used to define q-entailment, which, however, is not defined canonically. If the idea of entailment as preservation of a logical value is given up, then entailment will not, in general, be a Tarskian relation. There are thus several issues involved: (i) the number of sets of designated values (as representing logical values), (ii) the relation of the algebraic values to the logical values, and (iii) the definition of entailment in terms of the logical values. As to the latter issue, we shall here join Jennings and Schotch [136, p. 89] in declaring that "we want inferability to preserve".

Therefore, on *our* conception of Malinowski's term 'inferential many-valuedness', it is distinctive of a logical value that it is used to *canonically* define an entailment relation.[15] A logic may then be said to be logically (or inferentially) k-valued if it is a language together with k canonically defined and pairwise distinct entailment relations on (the set of formulas of) this language. Each of these k entailment relations is Tarskian and hence, in particular, reflexive. Obviously, neither the logic coming with a q-entailment relation nor classical propositional logic is logically two-valued in this sense because these logics are defined by a language together with a single entailment relation.

If there is in fact more than one logical value, each determining a separate entailment relation, and if therefore it is reasonable to think of a logic as possibly containing more than just one entailment relation (or, syntactically speaking, more than just one consequence relation), our contribution to the discussion of Suszko's Thesis is this:

1. *Suszko's* notion of a many-valued logic as comprising a set of designated values $\mathscr{D}$ but not also a set of antidesignated values that may be distinct from the complement of $\mathscr{D}$ amounts to assuming logical mono-valuedness instead of logical two-valuedness when it is assumed that a logic comes with a single entailment relation.

---

[15] This is actually *not* Malinowski's understanding of inferential many-valuedness. In [168] he explains that the chief feature of a q-consequence operation is that the repetition rule:

$$\text{rep} = (\{A\}, A \mid A \in \mathscr{L})$$

in general is not a rule of the operation. Moreover, q-consequence is the central notion of a purely *inferential* approach in the theory of propositional logics in the sense that "[t]he principal motivation behind the quasi-consequence ... stems from the mathematical practice which treats some auxiliary assumptions as mere hypotheses rather than axioms". The fact that the repetition rule is not unrestrictedly valid allows Malinowski [168, Sect. 2] to define two congruence relations on $\mathscr{L}$, inferential extensionality and inferential intensionality, which in general are independent of each other.

2. *Malinowski's* conception of logical many-valuedness is that logical values contribute to the definition of a single entailment relation, but there are reasons to let *every* logical value give rise to an entailment relation.
3. *Tsuji's* characterization of completeness with respect to sets of bivaluations in terms of reflexivity, assuming Beziau's notion of an abstract logical structure containing a *single* consequence relation, is question-begging because, with another concept of a logical system, in a logically k-valued logic every entailment relation is reflexive.

Assuming three algebraic values and understanding logical truth as receiving one of these values under any valuation, da Costa et al. [57, p. 292 f.] raise the question "Can logical truth also be multivalent?" and reply:

> It seems that *a priori* there is no good philosophical argument to reject this possibility, and this is another reason why we can reject Suszko's thesis.

Can one provide evidence to the effect that logical truth is multivalent? More generally, is there more than one logical value, each of which may be taken to determine its own entailment relation? It seems that Suszko did not raise fundamental doubts about *algebraic* many-valuedness, that is, about assuming more than two algebraic values. One natural question then comes with the distinction between algebraic and logical values: Is *every* non-empty subset of a given set of algebraic values a logical value? We need not try to decide this question here because, for our present purposes, it is enough to insist on falsity as a logical value not uniquely determined by truth. Instead of assuming with Frege that logic is the science of the most general laws of being true, for our purposes it is enough to assume that logic is the science of the most general laws of being true and the most general laws of being false.[16]

9.3.2 Another Kind of Counterexample

Before developing further the concept of a logical system with more than one entailment relation, we wish to present another definition of an entailment relation, a relation which is, so to say, "essentially three-valued". The key point of Malinowski's reduction (the proof that every q-consequence relation has a two-valued or a three-valued semantics) is to presuppose a tri-partition of the set of algebraic values rather than a bi-partition. As already observed, Malinowski's q-entailment is not reflexive. It is nevertheless possible to define a reflexive entailment relation which is in a sense necessarily three-valued by employing

[16] Interpretations of distinguished sets of algebraic values need not appeal to truth or falsity. In a series of papers, Jennings, Schotch, and Brown have argued that paraconsistent logic can be developed as a logic that preserves a degree of incoherence from the premises to the conclusion of a valid inference, see [39, 136] and the references given there.

Malinowski's idea that entailment should depend on both designated and anti-designated values which do not exhaust the set of all values available. Let us return to the observation above that in Łukasiewicz's q-matrix $\mathfrak{L}_3*$, t_m-entailment and f_m-entailment are not coincident. In this context we may also consider the intersection of these two relations, which can be called *tf*-entailment. Dunn [80, p. 11] introduces these three entailment relations and shows that they are all distinct. The latter point is crucial for defining a three-valued entailment relation which cannot be provided with a logically two-valued semantics *within the given set of algebraic values*. Namely, for the logic based on Łukasiewicz's q-matrix $\mathfrak{L}_3*$ such a relation is represented by *tf*-entailment[17,18]:

$$\Delta \models^{+,-}_{\mathfrak{L}_3*} B \text{ iff } \forall\, v \text{ in } \mathfrak{L}_3* : \quad \begin{array}{l} (1)\ v(\Delta) \subseteq \mathcal{D}^+ \Rightarrow v(B) \in \mathcal{D}^+; \\ (2)\ v(B) \in \mathcal{D}^- \Rightarrow \exists A \in \Delta(v(A) \in \mathcal{D}^-). \end{array}$$

Incidentally, $\models^{+,-}_{\mathfrak{L}_3*}$ restricted to the so-called "first-degree consequences" (statements of the form $A \models^{+,-}_{\mathfrak{L}_3*} B$, where neither A nor B contains implications) represent the first-degree entailment fragment of Łukasiewicz's three-valued logic (cf. [80, p. 15]). Now, it is not difficult to show that every *tf*-entailment relation and hence $\models^{+,-}_{\mathfrak{L}_3*}$ is Tarskian. This means that, according to Theorem 9.4, $\models^{+,-}_{\mathfrak{L}_3*}$ has an adequate logically two-valued semantics. It turns out, however, that it is impossible to determine $\models^{+,-}_{\mathfrak{L}_3*}$ by a two-valued valuational function defined on the carrier of $\mathfrak{L}_3*$. Indeed, assume such a definition would be possible. Then we need to pick from the set $\{T, \varnothing, F\}$ a proper subset $\mathcal{D}$ of designated algebraic values. Moreover, $\mathcal{D}$ should contain T (because $\models^{+,-}_{\mathfrak{L}_3*}$ is truth-preserving). There are only three such subsets: $\{T\}$, $\{T, \varnothing\}$, and $\{T, F\}$. But in the first case, $\models^{+,-}_{\mathfrak{L}_3*}$ would coincide with $\models^{+}_{\mathfrak{L}_3*}$ (which is impossible), in the second case $\models^{+,-}_{\mathfrak{L}_3*}$ would coincide with $\models^{-}_{\mathfrak{L}_3*}$ (which again, is impossible), and in the third case the intended definition for $\models^{+,-}_{\mathfrak{L}_3*}$ would be simply inadequate (making it falsity preserving from the premises to the conclusion).

9.3.3 Examples of Natural Bi-consequence Logics

In Chap. 4 we have seen already a natural example of a logic with two entailment (or consequence) relations (which can be called a *bi-consequence logic*), namely

[17] That is, we disregard for a moment the t_m-entailment and f_m-entailment taken separately and concentrate solely on their intersection. Note that it is most important here to have a value which is neither designated nor anti-designated; otherwise *tf*-entailment would trivially collapse into t_m-entailment (and f_m-entailment).

[18] A generalization of this definition to obtain a *tf*-entailment relation for *any* q-matrix $\mathfrak{M}$ based on a three-element set of algebraic values using Malinowski's three-valued valuation t_v (as defined on p. 196) is straightforward.

the system $(\mathscr{L}_{tf}, \models_t, \models_f)$ of truth and falsity entailment induced by the trilattice $SIXTEEN_3$ of generalized truth values. Basically, the distinction between the two entailment relations is driven by treating truth and falsity equally while developing Belnap's idea of "told values" assigned to (atomic) propositions. In Chaps. 5–7 we have defined and investigated various extensions of $(\mathscr{L}_{tf}, \models_t, \models_f)$, in particular the system $(\mathscr{L}^*_{tf}, \models_t, \models_f)$ based on the language $\mathscr{L}_{tf} \cup \{\rightarrow_t, \rightarrow_f\}$. In Chap. 8 we have observed that $(\mathscr{L}_{tf}, \models_t, \models_f)$ induces the bi-consequence system $(\mathscr{L}_{tf}, \models^+, \models^-)$ in the language $\mathscr{L}_{tf}$, where the entailment relations $\models^+$ and $\models^-$ are defined with respect to a certain generalized q-matrix. Recall that the sets x^t and x^f are defined as follows:

$$x^t := \{y \in x \mid T \in y\}; \qquad x^f := \{y \in x \mid F \in y\}.$$

The structure $\mathfrak{B}_{16}$ has been defined as the generalized q-matrix $\langle \mathbf{16}, \{x \in \mathbf{16} \mid x^t$ is non-empty$\}, \{x \in \mathbf{16} \mid x^f$ is non-empty$\}, \{-_t, \sqcap_t, \sqcup_t, -_f, \sqcup_f, \sqcap_f\}\rangle$. That is, $\mathscr{D}^+ = \{x \in \mathbf{16} \mid x^t$ is non-empty$\}$, and $\mathscr{D}^- = \{x \in \mathbf{16} \mid x^f$ is non-empty$\}$. Moreover, for all sets of $\mathscr{L}_{tf}$-formulas Δ, Γ, semantic consequence relations $\models^+$ and $\models^-$ are canonically defined as follows:

1. $\Delta \models^+ \Gamma$ iff for every valuation function v: (if for every $A \in \Delta$, $v(A) \in \mathscr{D}^+$, then $v(B) \in \mathscr{D}^+$ for some $B \in \Gamma$);
2. $\Delta \models^- \Gamma$ iff for every valuation function v: (if for every $A \in \Gamma$, $v(A) \in \mathscr{D}^-$, then $v(B) \in \mathscr{D}^-$ for some $B \in \Delta$).

The logic B_{16}, viewed as the triple $(\mathscr{L}_{tf}, \models^+, \models^-)$, provides an example of what we have called a *harmonious* many-valued logic. Moreover, we have seen that there are denumerably many harmonious many-valued logics. We believe that the motivation for $(\mathscr{L}_{tf}, \models_t, \models_f)$, presented in Chaps. 3 and 4, and for $(\mathscr{L}_{tf}, \models^+, \models^-)$, presented in Chap. 8, provides sufficient motivation for assuming a generalized notion of a many-valued logic and hence also a generalized notion of a logical system. In the next section the concept of a logically k-valued logic consisting of a language $\mathscr{L}$ together with k entailment relations on $\mathscr{L}$ is introduced.

9.3.4 Logically n-*Valued Logics*

In this section, we suggest thinking of a (single conclusion) logic not as a *pair* $(\mathscr{L}, \vdash)$ consisting of a non-empty set of formulas $\mathscr{L}$ and a single binary derivability relation $\vdash \subseteq \mathscr{P}(\mathscr{L}) \times \mathscr{L}$ but as a $k+1$-tuple consisting of a non-empty set of formulas together with k single-conclusion binary derivability relations on $\mathscr{L}$ for some $k \geq 2$, $k \in \mathbb{N}$.

Definition 9.1 A *Tarskian k-dimensional logic* (Tarskian k-logic) is a $k+1$-tuple $\Lambda = (\mathscr{L}, \vdash_1, \ldots, \vdash_k)$ such that (i) $\mathscr{L}$ is a language in a denumerable set of sentence letters and a finite non-empty set $\mathscr{C}$ of finitary connectives, (ii) for every

$i \leq k$, $\vdash_i \subseteq \mathscr{P}(\mathscr{L}) \times \mathscr{L}$, and (iii) every relation $\vdash_i$ satisfies (Reflexivity), (Monotonicity), and (Cut). Λ is said to be structural iff every $\vdash_i$ satisfies (Structurality).

Definition 9.2 Let $\mathscr{L}$ again be a language in a denumerable set of sentence letters and a finite non-empty set of finitary connectives $\mathscr{C}$. An n-valued k-dimensional matrix (k-matrix) is a structure $\mathfrak{M} = \langle \mathscr{V}, \mathscr{D}_1, \ldots, \mathscr{D}_k, \{f_c : c \in \mathscr{C}\}\rangle$, where $\mathscr{V}$ is a non-empty set of cardinality n $(2 \leq n)$, $2 \leq k$, every $\mathscr{D}_i$ $(1 \leq i \leq k)$ is a non-empty proper subset of $\mathscr{V}$, the sets $\mathscr{D}_i$ are pairwise distinct, and every f_c is a function on $\mathscr{V}$ with the same arity as c.[19] The sets $\mathscr{D}_i$ are called *distinguished* sets. A function from $\mathscr{L}$ into $\mathscr{V}$ is called a valuation in $\mathfrak{M}$.

A pair $\mathscr{M} = \langle \mathfrak{M}, v\rangle$, where $\mathfrak{M} = \langle \mathscr{V}, \mathscr{D}_1, \ldots, \mathscr{D}_k, \{f_c : c \in \mathscr{C}\}\rangle$ is an n-valued k-matrix and v a valuation in $\mathfrak{M}$, is called an n-valued k-model based on $\mathfrak{M}$. If v is a homomorphism from $(\mathscr{L}, c_1, \ldots, c_m)$ into $(\mathscr{V}, f_{c_1}, \ldots, f_{c_m})$, $\mathscr{M}$ is called a structural n-valued k-model based on $\mathfrak{M}$. Given an n-valued k-model $\mathscr{M} = \langle\langle \mathscr{V}, \mathscr{D}_1, \ldots, \mathscr{D}_k, \{f_c : c \in \mathscr{C}\}\rangle, v\rangle$, for every set $\mathscr{D}_i$ the semantic consequence relations $\models^{\Rightarrow}_{i,\mathscr{M}} \subseteq \mathscr{P}(\mathscr{L}) \times \mathscr{L}$ and $\models^{\Leftarrow}_{i,\mathscr{M}} \subseteq \mathscr{P}(\mathscr{L}) \times \mathscr{L}$ are defined as follows:

1. $\Delta \models^{\Rightarrow}_{i,\mathscr{M}} A$ iff $v(\Delta) \subseteq \mathscr{D}_i$ implies $v(A) \in \mathscr{D}_i$;
2. $\Delta \models^{\Leftarrow}_{i,\mathscr{M}} A$ iff $v(A) \in \mathscr{D}_i$ implies $v(B) \in \mathscr{D}_i$ for some $B \in \Delta$.

Obviously, the relations $\models^{\Rightarrow}_{i,\mathscr{M}}$ and $\models^{\Leftarrow}_{i,\mathscr{M}}$ are definable from each other. We may therefore, without loss of generality, focus on the relations $\models^{\Rightarrow}_{i,\mathscr{M}}$. If we are interested in the preservation of membership in $\mathscr{D}_i$ from the conclusion to the premises, we may take the complement of $\mathscr{D}_i$ as a distinguished set of algebraic values. Moreover, the relations $\models^{\Rightarrow}_{i,\mathscr{M}}$ are all Tarskian semantic consequence relations, and accordingly the structure $(\mathscr{L}, \models^{\Rightarrow}_{1,\mathscr{M}}, \ldots, \models^{\Rightarrow}_{k,\mathscr{M}})$ is a Tarskian k-logic.

Unsurprisingly, given an n-valued k-matrix $\mathfrak{M} = \langle \mathscr{V}, \mathscr{D}_1, \ldots, \mathscr{D}_k, \{f_c : c \in \mathscr{C}\}\rangle$, for every set $\mathscr{D}_i$ the entailment relations $\models^{\Rightarrow}_{i,\mathfrak{M}}$ and $\models^{\Leftarrow}_{i,\mathfrak{M}}$ are defined as follows:

1. $\Delta \models^{\Rightarrow}_{i,\mathfrak{M}} A$ iff for every valuation v in $\mathfrak{M}$: $v(\Delta) \subseteq \mathscr{D}_i$ implies $v(A) \in \mathscr{D}_i$;
2. $\Delta \models^{\Leftarrow}_{i,\mathfrak{M}} A$ iff for every valuation v in $\mathfrak{M}$: $v(A) \in \mathscr{D}_i$ implies $v(B) \in \mathscr{D}_i$ for some $B \in \Delta$.

A Tarskian k-logic $(\mathscr{L}, \vdash_1, \ldots, \vdash_k)$ is said to be characterized by an n-valued k-model $\mathscr{M}$ (k-matrix $\mathfrak{M}$) iff for every $\vdash_i$, $\vdash_i = \models^{\Rightarrow}_{i,\mathscr{M}}$ ($\models^{\Rightarrow}_{i,\mathfrak{M}}$). $(\mathscr{L}, \vdash_1, \ldots, \vdash_k)$ is characterized by a class $\mathfrak{K}$ of n-valued k-model (k-matrices) iff for every relation $\vdash_i$, $\vdash_i = \bigcap\{\models^{\Rightarrow}_{i,\mathscr{M}} \mid \langle\langle \mathscr{V}, \mathscr{D}_1, \ldots, \mathscr{D}_i, \ldots, \mathscr{D}_k, \{f_c : c \in \mathscr{C}\}\rangle, v\rangle \in \mathfrak{K}\}$ ($\vdash_i = \bigcap\{\models^{\Rightarrow}_{i,\mathfrak{M}} \mid \langle \mathscr{V}, \mathscr{D}_1, \ldots, \mathscr{D}_i, \ldots, \mathscr{D}_k, \{f_c : c \in \mathscr{C}\}\rangle \in \mathfrak{K}\}$).

The following statements follow immediately from Wójcicki's Theorem and Theorem 9.3.

---

[19] The notion of a k-matrix is not entirely new. Every k-matrix is a ramified matrix in the sense of Wójcicki [287, p. 189]. Ramified matrices are also called generalized matrices, see [36, p. 410 ff.]. Note, however, that Wójcicki associates with a ramified matrix *a single* entailment relation, namely $\bigcap\{\models_{\mathfrak{M}} \mid \mathfrak{M} = \langle \mathscr{V}, \mathscr{D}_i, \{f_c : c \in \mathscr{C}\}\rangle, 1 \leq i \leq k\}$.

Theroem 9.8 *Every structural Tarskian k-logic is characterized by a class of structural n-valued k-models for some* $n \leq \aleph_0$.

Theroem 9.9 *Every Tarskian k-logic is characterized by a class of n-valued k-models for some* $n \leq \aleph_0$.[20]

To every relation $\vdash_i$ in a Tarskian k-logic, one may apply the Suszko Reduction or the reduction described in [30]. But this observation does not raise any deep philosophical doubts about many-valued logic. Caleiro et al. [44, p. 3] are quite right in explaining that "there is some metalinguistic bivalence that one will not easily get rid of: *either* an inference obtains *or* it does not, but *not both*". If falsity is taken seriously and not just dealt with as the complement of truth, we end up, in general, with inferential four-valuedness in the form of preservation of truth, preservation of falsity, preservation of being neither true nor false, and preservation of being both true and false. For this kind of logical four-valuedness we need at least three algebraic values. Given a set $\mathcal{V}$ of algebraic values, the set of all available logical values is $\mathcal{P}(\mathcal{V}) \setminus \varnothing$. But which non-empty subsets of a given set $\mathcal{V}$ should we view as logical values? This may depend on philosophical considerations, on the intuitive interpretation of the algebraic values,[21] and on the intended applications of a logical system. In Chap. 3 we emphasized the naturalness of the truth order $\leq_t$ and the falsity order $\leq_f$ on the set **16** of generalized truth values, which in Chap. 8 led us to the symmetric valuation system (generalized q-matrix) $\mathfrak{B}_{16}$. Moreover, we emphasized that if we consider $\mathcal{V} = \mathbf{4} = \mathcal{P}(\mathbf{2})$ and the famous bilattice $FOUR_2$ presented in Chap. 3, then truth and falsity are *not* dealt with as independent of each other, because in $FOUR_2$ it is assumed that T is at least as true as F. There is thus an interplay between choosing a set of algebraic values $\mathcal{V}$ and distinguished subsets $\mathcal{D}_1, \ldots, \mathcal{D}_k$ of $\mathcal{V}$.

Note also that even if one is not committed to the idea of entailment as preservation of semantical values from the premises to the conclusion(s) of an inference (or vice versa), it may still be quite natural to conceive of a logical system as comprising more than just one entailment relation. Malinowski's notion of q-entailment, for example, is in a sense one-sided. The requirement that if every premise is not antidesignated, then the conclusion is designated seems not to be privileged in comparison to the requirement that if every premise is designated,

[20] Note that since we defined the entailment relations from the *k-models*, the semantics of *k-matrices* leaves room for other definitions of entailment.

[21] For Suszko, the relation between the algebraic and the logical values is established via characteristic functions, and the original intuitive interpretation of the algebraic values is uncoupled from the understanding of the logical values as *truth* and *falsity*. Similarly, the understanding of the sets of designated values $\mathcal{D}_i$ is detached from the intuitive understanding of the values in $\mathcal{V}$. It may, of course, happen that there exists a bijection between $\mathcal{V}$ and $\{\mathcal{D}_i, \ldots, \mathcal{D}_k\}$. In the context of q-consequence relations, Malinowski [166, p. 83] explains that for some inferentially three-valued logics based on three-element algebras, referential assignments ... and logical valuations ... do coincide. This phenomenon delineates a special class of logics, which satisfy a "generalized" version of the Fregean Axiom identifying in a one–one way three logical values and three semantical correlates (or referents).

then the conclusion is not antidesignated. The latter notion of entailment has been studied by Frankowski [98, 99], who refers to it as p-entailment ("plausibility"-entailment). For a given interpreted language, p-entailment, need not coincide with q-entailment. Against the background of the q-matrix $\mathfrak{L}_3^*$ with the following truth table for conjunction:

$f_\wedge$	T	$\varnothing$	F
T	T	$\varnothing$	F
$\varnothing$	$\varnothing$	$\varnothing$	F
F	F	F	F

the formula $(A \wedge B)$ p-entails A, but it is obviously not the case that $(A \wedge B)$ q-entails A. A p-consequence relation on a language $\mathscr{L}$ is a subset of $\mathscr{P}(\mathscr{L}) \times \mathscr{L}$ which is reflexive and monotonic. Frankowski shows that for every structural p-consequence relation there exists a characterizing class of q-matrices, where characterization is defined in terms of p-entailment instead of q-entailment. Interestingly, a relation which is both a q-consequence relation and a p-consequence relation is Tarskian.

Mixed approaches to entailment are, of course, also possible, see Chap. 8. In [167], Malinowski considers an entailment relation which leads from non-accepted (alias undesignated) premises to rejected (alias false) conclusions.

9.4 Another Analysis and a Résumé

Instead of increasing the number of entailment relations by associating to every set of designated values an entailment relation that preserves membership in this set from the premises to the conclusion, one may increase the number of places of the entailment relation. To every q-matrix, for instance, one might associate a ternary relation $\models \subseteq \mathscr{P}(\mathscr{L}) \times \mathscr{P}(\mathscr{L}) \times \mathscr{L}$ by postulating $\models (\Delta, \Gamma, A)$ iff for every valuation v, if $v(\Delta) \subseteq \mathscr{D}^+$ and $v(\Gamma) \subseteq \mathscr{V} \setminus (\mathscr{D}^+ \cup \mathscr{D}^-)$, then $v(A) \in \mathscr{D}^+$. Da Costa et al. [57, p. 292], e.g., explain:

> The real n-dimensional logics ($n > 2$) have to be developed by breaking down the deepest root of the principle of bivalence. We can easily imagine, for instance, a rule of deduction with *three* poles or more (our emphasis).

Higher-arity sequent systems for many-valued logics were already considered in the literature in the 1950s by Schröter [218] and later by several other authors, see also [125] and [176]. Higher-arity sequents for modal logics have been investigated by Blamey and Humberstone [36]. The use of higher-arity sequents may have some technical and conceptual merits. However, we assume that a selection of logical values (alias non-empty subsets of some set $\mathscr{V}$) is particularly well-motivated if every logical value is associated with a dimension according to which $\mathcal{V}$ can be partially ordered (*truth*, *falsity*, *information*, *necessity*, *constructiveness*,[22] etc.). From this point of view, it is natural to let each logical value give rise to a *binary* entailment relation.

In our considerations on Suszko's Thesis, we argued that a logical value is a value that gives rise to an entailment relation in a canonical way. Although for each such entailment relation there exists a representation in terms of bivaluations, this does not show that it is unreasonable or unnecessary to assume more than one logical value. The observation that every entailment relation has a bivalent semantics *does*, however, cast doubt on the enterprise of providing a bivalent semantics seen as a research program that is not constrained by any further requirements, see also [26]. Indeed, as has been shown by Routley [211] using a canonical model construction, every logic based on a λ-categorical language has a characterizing bivalent possible worlds semantics, where by a logic Routley understands an axiomatic system consisting of a countable set of axioms and a countable set of derivation rules leading from a finite number of premises to a single conclusion. Routley [211, p. 331] also notes, "[I]f every logic on a λ-categorial language has a two-valued worlds semantics then ipso facto it has a three-valued ... semantics". This kind of "reduction" to three-valuedness certainly does not refute the claim that there are just two truth values, and likewise the Suszko Reduction does not prove that there are but two logical values.

9.5 A Bilattice of Four Entailment Relations

We will conclude this chapter by making some remarks concerning many-valuedness and entailment. Semantically, logic as a science of correct reasoning and valid arguments concentrates largely on the concept of entailment and investigations of its properties. Sometimes it is even said that the concept of entailment is *fundamental* to logic (see, e.g., [81, p. 55]). Usually, however, entailment is defined through another (philosophically perhaps even more) fundamental concept—the one of truth. The standard understanding is that a sentence A entails a sentence B if and only if B is true whenever A is true. If the underlying context is classical, and falsity is interpreted simply as the absence of truth (as non-truth) and truth as the absence of falsity (as non-falsity), then this is just another way of saying that B is not false whenever A is not false. Two other popular ways to express essentially the same idea is to say that always either A is false or B is true or that it is impossible that A is true and B is false.

Following Frege, truth and falsity are represented in logical semantics by the corresponding truth values, and in classical logic the only truth values that play a role in semantical constructions are the two truth values *the True* (T) and *the False*

22 See Chap. 3.

Fig. 9.1 Lattice *THREE*

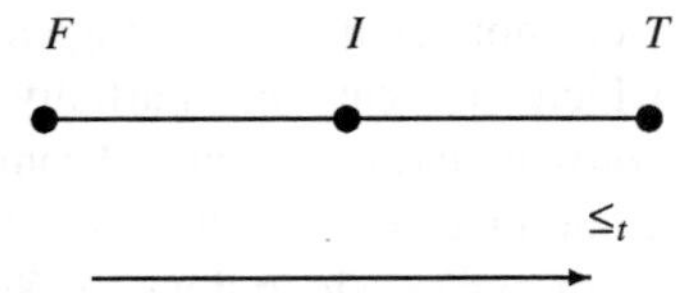

(F). As we have seen, the picture becomes more complex as soon as one brings into play the idea of a many-valued logic. It appears that the mere multiplication of semantical values not only affords room for defining various entailment relations, but this also allows one to define semantical relations differing in important respects from the familiar notion of semantic consequence. In light of our previous considerations, we will point out that the relation between the notion of a truth value and the notion of entailment is even more intimate than the connection emerging from the interaction between properties of entailment relations and truth values. In some cases it is possible to draw a strong analogy between truth values and entailment relations, namely to interpret entailment relations as a kind of truth values. Such an interpretation seems to be both natural and promising.

9.5.1 Definitions of Four Entailment Relations

We will first consider certain three-valued propositional logics. If, following [80, p. 11], we label the third value with I (as in a sense *intermediate* between classical truth T and falsity F), this notation may be seen to leave room for various more specific intuitive interpretations. We will be especially interested in two three-valued systems: Kleene's (strong) "logic of indeterminacy" K_3 and Priest's "logic of paradox" P_3 (alias *LP*), cf. the material from Sect. 8.3. If we restrict ourselves to a non-implicational language, then K_3 is determined by the *Kleene matrix* $\mathfrak{K}_3 = \langle\{T, I, F\}, \{T\}, \{f_c : c \in \{\sim, \wedge, \vee\}\}\rangle$, where the functions f_c are defined as follows:

$f_\sim$	
T	F
I	I
F	T

$f_\wedge$	T	I	F
T	T	I	F
I	I	I	F
F	F	F	F

$f_\vee$	T	I	F
T	T	T	T
I	T	I	I
F	T	I	F

The *Priest matrix* $\mathfrak{P}_3$ differs from $\mathfrak{K}_3$ only in that $\mathcal{D} = \{T, I\}$. Entailment in K_3 as well as in P_3 is defined by means of the standard definition[23]:

$$A \models B \text{ iff } \forall v : v(A) \in \mathcal{D} \Rightarrow v(B) \in \mathcal{D}. \tag{9.8}$$

It is also well-known that the truth values of both Kleene's and Priest's logic can be ordered to form a lattice (*THREE*), which is diagrammed in Fig. 9.1. Here

23 For the sake of simplicity, we consider entailment as a relation between (single) formulas while keeping in mind its possible generalization to sets of formulas.

T, I, and F are ordered by means of a "truth order" ($\leq_t$) so that the intermediate value I is "more true" than F but "less true" than T. The operations of meet and join with respect to $\leq_t$ are exactly the functions $f_\wedge$ and $f_\vee$ above, and $f_\sim$ is just the inversion of this order.

There are natural intuitive interpretations of I in K_3 and in P_3 as the *underdetermined* and the *overdetermined* values respectively (a truth-value gap and a truth-value glut). Formally these interpretations can be modelled by presenting the values as certain subsets of the set of classical truth values $\{T, F\}$. Then T turns into $\mathbf{T} = \{T\}$ (understood as "true only"), F into $\mathbf{F} = \{F\}$ ("false only"), I is interpreted in K_3 as $\mathbf{N} = \{\} = \varnothing$ ("*neither* true nor false"), and in P_3 as $\mathbf{B} = \{T, F\}$ ("*both* true and false").

If one combines all these new values into a joint framework, one obtains the four-valued logic B_4 introduced by Dunn and Belnap (see [22, 23, 75]), also known as the logic of first-degree entailment, or the logic of "tautological entailment" [3, § 15.2], cf. Chap. 2. This logic is determined by the *Belnap matrix* $\mathfrak{B}_4 = \langle\{\mathbf{N}, \mathbf{T}, \mathbf{F}, \mathbf{B}\}, \{\mathbf{T}, \mathbf{B}\}, \{f_c : c \in \{\sim, \wedge, \vee\}\}\rangle$, where, as we have seen already, the functions f_c are defined as follows:

$f_\sim$	
T	**F**
B	**B**
N	**N**
F	**T**

$f_\wedge$	**T**	**B**	**N**	**F**
T	**T**	**B**	**N**	**F**
B	**B**	**B**	**F**	**F**
N	**N**	**F**	**N**	**F**
F	**F**	**F**	**F**	**F**

$f_\vee$	**T**	**B**	**N**	**F**
T	**T**	**T**	**T**	**T**
B	**T**	**B**	**T**	**B**
N	**T**	**T**	**N**	**N**
F	**T**	**B**	**N**	**F**

Definition (9.8) applied to the Belnap matrix determines the entailment relation of B_4.

As we have seen in Chap. 3, the bilattice $FOUR_2$ is based on the set of truth values $\{\mathbf{N}, \mathbf{T}, \mathbf{F}, \mathbf{B}\}$ with its two partial ordering relations $\leq_i$ (the information order) and $\leq_t$ (the truth order), see Fig. 3.3. Lattice meet and join with respect to $\leq_t$ coincide with the functions $f_\wedge$ and $f_\vee$ in the Belnap matrix $\mathfrak{B}_4$, and $f_\sim$ turns out to be the truth order inversion. $FOUR_2$ is also of interest in that it allows another way of defining entailment. Namely, one can consider this relation as expressing agreement with the truth order. That is:

$$A \models B \text{ iff } \forall v : v(A) \leq_t v(B) \tag{9.9}$$

It is very well-known (see, e.g., [95, 239]) that (9.8) in B_4 and (9.9) in $FOUR_2$ define one and the same relation.

We have seen that if the set of truth values is *not* dichotomized, but in fact trisected (or even more), this has a significant impact on the very concept of logical entailment. For example, since "truth" (the set of designated values $\mathcal{D}^+$) does not generally coincide with "non-falsity" (the complement of the set of antidesignated values $\mathcal{D}^-$) any more, the expressions "B is true whenever A is true" and "B is not false whenever A is not false" need no longer mean the same. Moreover, the other usual characterizations of entailment: "in any case either A is false or B is true" and "it is impossible that A is true and B false", also become non-equivalent. Hence, the relations determined by these conditions may also be different.

We may approach this subject in a more systematic way. Given that sets of designated and antidesignated values are not obligatorily complementations of each other, in addition to the simple preservation of truth and the simple preservation of non-falsity from (in the present case) the single premise to the single conclusion, at least two other notions of entailment based on an obvious interplay between $\mathscr{D}^+$ and $\mathscr{D}^-$ which have a clear intuitive appeal (although these new notions are *not* defined in terms of preserving membership in some subset of the set of truth values $\mathscr{V}$) come to mind. That is, we obtain the following four primitive definitions[24]:

$$A \models^+ B \text{ iff } \forall v : v(A) \in \mathscr{D}^+ \Rightarrow v(B) \in \mathscr{D}^+ \quad (9.10)$$

$$A \models^- B \text{ iff } \forall v : v(A) \notin \mathscr{D}^- \Rightarrow v(B) \notin \mathscr{D}^- \quad (9.11)$$

$$A \models^q B \text{ iff } \forall v : v(A) \notin \mathscr{D}^- \Rightarrow v(B) \in \mathscr{D}^+ \quad (9.12)$$

$$A \models^p B \text{ iff } \forall v : v(A) \in \mathscr{D}^+ \Rightarrow v(B) \notin \mathscr{D}^- \quad (9.13)$$

As above, we will refer to these relations as t_m-entailment, f_m-entailment, q-entailment, and p-entailment, correspondingly. Whereas t_m-entailment is the standard truth-preserving relation, f_m-entailment incorporates the idea of non-falsity preservation (cf. [80, p. 10]). The relation of q-entailment can be seen as reflecting a reasoning from hypotheses (understood as statements that merely are taken to be non-refuted). Above, we pointed out that this relation has been defined by Malinowski (together with the underlying concept of a q-matrix) in order to define a notion of entailment which depends on both $\mathscr{D}^+$ and $\mathscr{D}^-$ and to provide thereby a counterexample to Suszko's Thesis. As already mentioned, p-entailment (p for "plausibility") has been studied by Frankowski [98, 99], who tried to explicate "reasonings wherein the degree of strength of the conclusion (i.e. the conviction it is true) is smaller th[a]n that of the premisses" [98, p. 41]. Usually q-entailment and p-entailment are firmly associated with the corresponding q- and p-matrices, but here we operate more generally, considering the definitions of q-entailment and p-entailment as such. The idea is that these definitions may give rise to various concrete entailment relations when brought into the context of a concrete matrix. And indeed, within a certain generalized four-valued setting, for example, some of these relations turn out to be indistinguishable. (In what follows, when we write $\models^x$ as a free-standing sign, we mean the respective entailment relation, i.e., the set of pairs (A, B) such that $A \models^x B$).

Proposition 9.1 *Let the Belnap generalized q-matrix* $\mathfrak{B}_4^*$ *be the structure* $\langle \{\mathbf{N}, \mathbf{T}, \mathbf{F}, \mathbf{B}\}, \{\mathbf{T}, \mathbf{B}\}, \{\mathbf{F}, \mathbf{B}\}, \{f_c : c \in \{\sim, \wedge, \vee\}\}\rangle$ *(functions* f_c *being defined as*

24 We say "primitive" and have in mind the possibility of considering further definitions obtained by combining the original conditions.

in the usual Belnap matrix), and let us introduce on this structure the above four entailment relations. Then $\models^+ = \models^-$ *and* $\models^q = \models^p$*. Moreover,* $\models^+$ *(alias* $\models^-$*)* $\neq \models^p$ *(alias* $\models^q$*).*

Proof Dunn has shown that in the logic of first-degree entailment $\models^+ = \models^-$ (see Proposition 9. 4 in [80, p. 10]). He defined for every valuation v its dual v^*, such that for any formula A: $v(A) = \mathbf{T} \Leftrightarrow v^*(A) = \mathbf{T}$; $v(A) = \mathbf{F} \Leftrightarrow v^*(A) = \mathbf{F}$; $v(A) = \mathbf{B} \Leftrightarrow v^*(A) = \mathbf{N}$; $v(A) = \mathbf{N} \Leftrightarrow v^*(A) = \mathbf{B}$. It is not difficult to see that in $\mathfrak{B}_4^*$ such a dual valuation does exist for every v. Note, that $v^{**} = v$.

Now, we show that $\models^p \subseteq \models^q$. Let $\forall v : v(A) = \mathbf{T}$ or $v(A) = \mathbf{B} \Rightarrow v(B) \neq \mathbf{F}$ and $v(B) \neq \mathbf{B}$. Assume $v(A) \neq \mathbf{F}$ and $v(A) \neq \mathbf{B}$. Then $v(A) = \mathbf{T}$ or $v(A) = \mathbf{N}$. Hence, $v^*(A) = \mathbf{T}$ or $v^*(A) = \mathbf{B}$. So, $v^*(B) \neq \mathbf{F}$ and $v^*(B) \neq \mathbf{B}$, and consequently, $v(B) \neq \mathbf{F}$ and $v(B) \neq \mathbf{N}$. Therefore $v(B) = \mathbf{T}$ or $v(B) = \mathbf{B}$.

Next, we show that $\models^q \subseteq \models^p$. First, observe that if $A \models^q B$, then there is no valuation v such that $v(A) = \mathbf{T}$ and $v(B) = \mathbf{B}$. Indeed, assume there exists such valuation v. Then $v^*(A) = \mathbf{T}$ and $v^*(B) = \mathbf{N}$. It follows that $v^*(B) = \mathbf{T}$ or $v^*(B) = \mathbf{B}$, a contradiction. Now, let $\forall v : v(A) \neq \mathbf{F}$ and $v(A) \neq \mathbf{B} \Rightarrow v(B) = \mathbf{T}$ or $v(B) = \mathbf{B}$. Assume $v(A) = \mathbf{T}$ or $v(A) = \mathbf{B}$. If $v(A) = \mathbf{T}$, then $v(B) = \mathbf{T}$ or $v(B) = \mathbf{B}$. But, the second case is impossible, and hence, $v(B) = \mathbf{T}$, i.e., $v(B) \neq \mathbf{F}$ and $v(B) \neq \mathbf{B}$. If $v(A) = \mathbf{B}$, then $v^*(A) = \mathbf{N}$ and thus, $v^*(B) = \mathbf{T}$ or $v^*(B) = \mathbf{B}$. Hence, $v(B) = \mathbf{T}$ or $v(B) = \mathbf{N}$. In both cases, again, $v(B) \neq \mathbf{F}$ and $v(B) \neq \mathbf{B}$, which is the required result.

Thus, in the generalized q-matrix $\mathfrak{B}_4^*$ the four entailment relations merge into two. To show that these two relations are distinct, we observe that in $\mathfrak{B}_4^*$: $A \models^+ A$ (as well as $A \models^- A$), but $A \not\models^p A$ (and also $A \not\models^q A$). (Otherwise we would have $v(A) \in \mathscr{D}^+ \Rightarrow v(A) \notin \mathscr{D}^-$ and $v(A) \notin \mathscr{D}^- \Rightarrow v(A) \in \mathscr{D}^+$, and this would mean that $\mathscr{D}^+ \cap \mathscr{D}^- = \varnothing$, which is not the case.)[25] □

9.5.2 Orderings on Entailment Relations

In our particular example of a *generalized four-valued* q-matrix, it turned out that $\models^+ = \models^-$ and $\models^q = \models^p$. Interestingly, in a non-generalized setting the picture may become even more complex. Let the (ordinary) *Kleene–Priest q-matrix* $\mathfrak{KP}_3^*$ be the structure $\langle\{T, I, F\}, \{T\}, \{F\}, \{f_c : c \in \{\sim, \wedge, \vee\}\}\rangle$, where the functions f_c are defined as in $\mathfrak{K}_3$ and $\mathfrak{P}_3$, and let us consider the entailment relations defined by (9.10)–(9.13) with respect to this matrix. Then, these relations are *all distinct*, $\models^+$ is the entailment relation of Kleene's logic, and $\models^-$ corresponds to the entailment

[25] Incidentally, this observation shows that in the context of $\mathfrak{B}_4^*$, $\models^p$ and $\models^q$ are not Tarskian *even if* these two relations coincide. (Recall that an entailment relation is Tarskian iff it satisfies (Reflexivity), (Monotonicity) and (Cut).) In this respect generalized q-matrices differ from non-generalized q-matrices since, as Frankowski has shown in [99, p. 198], if a relation defined on a non-generalized q-matrix is both a p- and a q-entailment relation, then it *is* Tarskian.

of Priest's logic (cf. [80]). Moreover, the following proposition exposes some simple facts about the interconnections between the entailments:

Proposition 9.2 *In the Kleene–Priest q-matrix:* (i) $\models^q \subseteq \models^+$; *(ii)* $\models^q \subseteq \models^-$; *(iii)* $\models^+ \subseteq \models^p$; *(iv)* $\models^- \subseteq \models^p$; (v) $\models^q \subset \models^+ \cap \models^-$.

Proof

(i) Let $\forall v : v(A) \notin \mathscr{D}^- \Rightarrow v(B) \in \mathscr{D}^+$. Assume $\exists v : v(A) \in \mathscr{D}^+$ and $v(B) \notin \mathscr{D}^+$. Then $v(A) \in \mathscr{D}^-$, a contradiction with the condition $\mathscr{D}^+ \cap \mathscr{D}^- = \varnothing$.

(ii)–(iv) The proof is analogous.

(v) Clauses (i) and (ii) state that $\models^q \subseteq \models^+ \cap \models^-$. That the set-inclusion is proper follows from the fact that in any proper q-matrix[26] (an hence in $\mathfrak{KP}_3^*$): $A \not\models^q A$, whereas both $A \models^+ A$ and $A \models^- A$.[27] □

Facts (i)–(iv) are mentioned by Devyatkin in [62]. Note, that they actually reveal that in $\mathfrak{KP}_3^*$ our four entailment relations are ordered by means of $\subseteq$ to form a lattice with $\models^q$ as the bottom and $\models^p$ as the top. At this place an analogy comes to mind with the information ordering in bilattice $FOUR_2$ which is defined exactly as set-inclusion. It turns out that it is not the only possible analogy one can draw here. Let $(\models^x)^T$ stand for the part of $\models^x$ which only preserves designated values and $(\models^x)^F$ for the part of $\models^x$ that only preserves non-antidesignated values:

$$(\models^x)^T := \{(A,B) \in \models^x \mid (\forall v : v(A) \in \mathscr{D}^+ \Rightarrow v(B) \in \mathscr{D}^+ \text{ and } \exists v : v(A) \notin \mathscr{D}^- \text{ and } v(B) \in \mathscr{D}^-)\} \tag{9.14}$$

$$(\models^x)^F := \{(A,B) \in \models^x \mid (\forall v : v(A) \notin \mathscr{D}^- \Rightarrow v(B) \notin \mathscr{D}^- \text{ and } \exists v : v(A) \in \mathscr{D}^+ \text{ and } v(B) \notin \mathscr{D}^+)\} \tag{9.15}$$

Intuitively $(\models^x)^T$ and $(\models^x)^F$ can be seen as representing correspondingly the "pure truth content" and the "pure falsity content" of $\models^x$. One can define then a "truth order" between the entailment relations:

$$\models^x \leq_t \models^y \text{ iff } (\models^x)^T \subseteq (\models^y)^T \quad \text{and} \quad (\models^y)^F \subseteq (\models^x)^F \tag{9.16}$$

The following proposition shows how the four entailment relations can be organized into a logical lattice with $\models^-$ as the bottom and $\models^+$ as the top:

Propostion 9.3 *In the Kleene–Priest q-matrix: (i)* $\models^- \leq_t \models^q$; *(ii)* $\models^- \leq_t \models^p$; *(iii)* $\models^q \leq_t \models^+$; *(iv)* $\models^p \leq_t \models^+$.

---

[26] We call a q-matrix *proper* iff $\mathscr{D}^+ \cup \mathscr{D}^- \neq \mathscr{V}$ holds for it.

[27] We leave as an open question whether in $\mathfrak{KP}_3^*$ $\models^p \subseteq \models^+ \cup \models^-$. An anonymous referee for [234] turned our attention to the fact that it is not difficult to define a particular q-matrix in which, e.g., $\models_p = \models_t = \models_f$. But it is also possible to define a q-matrix such that $\models^+ \cup \models^- \subset \models^p$. Indeed, add to $\mathfrak{KP}_3^*$ a unary truth function $f_\#$ such that $f_\#$(T) = I and $f_\#$(I) = F. Then obviously $A \models_p \#A$, but $A \not\models_t \#A$ and $A \not\models_f \#A$.

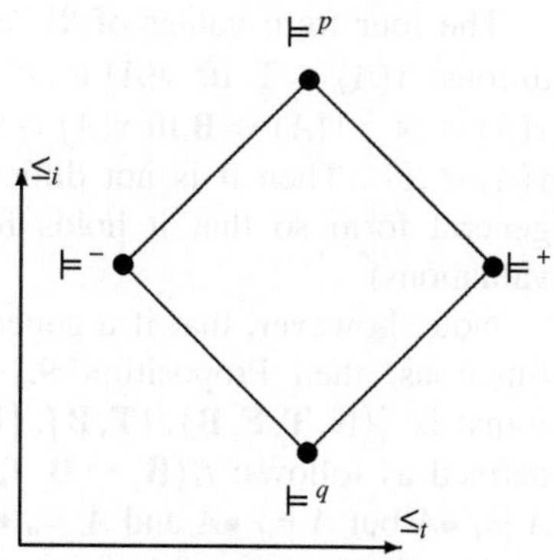

Fig. 9.2 The bilattice of entailment relations $FOUR_2^e$

Proof As a direct consequence of Proposition 9.2 (i–iv) and Definitions (9.14) and (9.15), we obtain: $(\models^+)^T = \models^+ \setminus \models^-$; $(\models^-)^T = \varnothing$; $(\models^q)^T = \varnothing$; $(\models^p)^T = (\models^+)^T$, as well as $(\models^-)^F = \models^- \setminus \models^+$; $(\models^+)^F = \varnothing$; $(\models^q)^F = \varnothing$; $(\models^p)^F = (\models^-)^F$. Hence the proposition. □

By combining Propositions 9.2 and 9.3, we immediately see that the entailment relations defined in the q-matrix $\mathfrak{K}\mathfrak{P}_3^*$ constitute a structure isomorphic to the bilattice $FOUR_2$, where $\models^+$ is analogous to **T**, $\models^-$ plays the role of **F**, $\models^q$ stands for **N**, and $\models^p$ is like **B**. This bilattice of four entailment relations $FOUR_2^e$ is presented in Fig. 9.2. In this way we obtain another representation of a four-valued logic whose values are formed by the entailment relations defined on the basis of a three-valued quasi-matrix.

9.5.3 Concluding Remarks

Throughout the preceding section, we constantly had in view the concrete q-matrix $\mathfrak{K}\mathfrak{P}_3^*$ and the concrete generalized q-matrix $\mathfrak{B}_4^*$. Nevertheless, one can observe that Proposition 9.1 can easily be extended to *any* generalized q-matrix, provided it is proper,[28] and its truth functions allow dualization, i.e., for any valuation v there exists a valuation v^* subject to the following conditions (for any formula A):

$$v(A) \in \mathscr{D}^+ \text{ and } v(A) \notin \mathscr{D}^- \Leftrightarrow v^*(A) \in \mathscr{D}^+ \text{ and } v^*(A) \notin \mathscr{D}^- \tag{9.17}$$

$$v(A) \notin \mathscr{D}^+ \text{ and } v(A) \in \mathscr{D}^- \Leftrightarrow v^*(A) \notin \mathscr{D}^+ \text{ and } v^*(A) \in \mathscr{D}^- \tag{9.18}$$

$$v(A) \in \mathscr{D}^+ \text{ and } v(A) \in \mathscr{D}^- \Leftrightarrow v^*(A) \notin \mathscr{D}^+ \text{ and } v^*(A) \notin \mathscr{D}^- \tag{9.19}$$

$$v(A) \notin \mathscr{D}^+ \text{ and } v(A) \notin \mathscr{D}^- \Leftrightarrow v^*(A) \in \mathscr{D}^+ \text{ and } v^*(A) \in \mathscr{D}^- \tag{9.20}$$

[28] Again, a generalized q-matrix is *proper* iff it satisfies the conditions: $\mathscr{D}^+ \cup \mathscr{D}^- \neq \mathscr{V}$ and $\mathscr{D}^+ \cap \mathscr{D}^- \neq \varnothing$.

The four truth values of $\mathfrak{B}_4^*$ can generally be modeled by the following conditions: $v(A) = \mathbf{T}$ iff $v(A) \in \mathscr{D}^+$ and $v(A) \notin \mathscr{D}^-$; $v(A) = \mathbf{F}$ iff $v(A) \notin \mathscr{D}^+$ and $v(A) \in \mathscr{D}^-$; $v(A) = \mathbf{B}$ iff $v(A) \in \mathscr{D}^+$ and $v(A) \in \mathscr{D}^-$; $v(A) = \mathbf{N}$ iff $v(A) \notin \mathscr{D}^+$ and $v(A) \notin \mathscr{D}^-$. Then it is not difficult to rewrite the proof of Proposition 9.1 in a general form so that it holds for any proper generalized q-matrix (with dual valuations).

Note, however, that if a generalized q-matrix includes "non-dualizable" truth functions, then Proposition 9.1 fails. Consider, for example, the generalized q-matrix $\langle\{\mathbf{N},\mathbf{T},\mathbf{F},\mathbf{B}\},\{\mathbf{T},\mathbf{B}\},\{\mathbf{F},\mathbf{B}\},\{f_\bullet\}\rangle$, where the unary truth function $f_\bullet$ is defined as follows: $f_\bullet(\mathbf{B}) = \mathbf{B}$, $f_\bullet(\mathbf{T}) = \mathbf{T}$, $f_\bullet(\mathbf{F}) = \mathbf{F}$, and $f_\bullet(\mathbf{N}) = \mathbf{F}$. Obviously, $A \models_t \bullet A$ but $A \not\models_f \bullet A$ and $A \models_p \bullet A$ but $A \not\models_q \bullet A$.

Also Propositions 9.2 (i–iv) and 9.3 can be extended to *any* (non-generalized) q-matrix (and Proposition 9.2 (v) to any proper non-generalized q-matrix), for in the corresponding proofs no specific use is made of any concrete feature of $\mathfrak{K}\mathfrak{P}_3^*$.

Thus, the main result of these preceding considerations can also be summarized as follows: We formulated a simple but fairly general method of constructing a generalized four-valued q-matrix by taking as its values the basic entailment relations defined on an arbitrary non-generalized q-matrix.

Chapter 10
Further Developments

Abstract In this concluding chapter, we will glance at some possible developments of our theory of generalized truth values. In particular, we will raise the issues of adding quantifiers and modal operators to the propositional languages considered in the previous chapters. Moreover, we will briefly touch on the idea of adverbially qualified truth values. Finally, we will look at further philosophical interpretations of Dunn and Belnap's four-valued logic and consider the possibilities of an extension of these interpretations beyond the four-valued case.

10.1 First-Order Trilattice Logics

A straightforward road to obtaining first-order trilattice logics consists of adding to some given axiom systems analogues of familiar axiom schemes that lead from classical propositional logic to classical first-order logic. If we again assume the language $\mathcal{L}^*_{tf}$ with its two separate sets of propositional connectives, consider the Hilbert-style proof systems HL^{t*} from Chap. 5, and adhere, e.g., to the axiomatization of classical first-order logic in Enderton's classical textbook [83], then the axioms for the universal quantifier should come in two versions:

- $\forall x(A \rightarrow_t B) \rightarrow_t (\forall xA \rightarrow_t \forall xB)$;
- $A \rightarrow_t \forall xA$, if there is no free occurrence of x in A;
- $\forall xA \rightarrow_t A(y/x)$, if y is free for x in A;
- $\forall x(A \rightarrow_f B) \rightarrow_f (\forall xA \rightarrow_f \forall xB)$;
- $A \rightarrow_f \forall xA$, if there is no free occurrence of x in A;
- $\forall xA \rightarrow_f A(y/x)$, if y is free for x in A.

These schemata are to be added to all universal closures of tautological schemata from HL^{t*}, and, of course, we keep *modus ponens* for both implications. Moreover, if we do not assume separate axioms for the existential quantifier, we

Y. Shramko and H. Wansing, *Truth and Falsehood*, Trends in Logic 36,
DOI: 10.1007/978-94-007-0907-2_10, © Springer Science+Business Media B.V. 2011

could introduce the existential quantifier as the dual of the universal quantifier with respect to the negations $\sim_t$ and $\sim_f$ by means of the usual definitions:

$$\exists xA :\Leftrightarrow \sim_t \forall x \sim_t A;$$
$$\exists xA :\Leftrightarrow \sim_f \forall x \sim_f A.$$

Another kind of expansion to first-order trilattice logics would be obtained by extending the separation between a truth vocabulary and a falsity vocabulary to the quantifiers and distinguishing between truth and falsity versions $\forall_t, \exists_t$ and $\forall_f, \exists_f$, respectively. It would then seem to be natural to postulate the following duality principles:

$$\exists_f xA :\Leftrightarrow \sim_f \forall_f x \sim_f A;$$
$$\exists_t xA :\Leftrightarrow \sim_t \forall_t x \sim_t A.$$

With this distinction between truth and falsity quantifiers, it seems natural to replace the above axiom schemes by the following axioms:

- $\forall_t x(A \rightarrow_t B) \rightarrow_t (\forall_t xA \rightarrow_t \forall_t xB)$;
- $A \rightarrow_t \forall_t xA$, if there is no free occurrence of x in A;
- $\forall_t xA \rightarrow_t A(y/x)$, if y is free for x in A;
- $\forall_f x(A \rightarrow_f B) \rightarrow_f (\forall_f xA \rightarrow_f \forall_f xB)$;
- $A \rightarrow_f \forall xA$, if there is no free occurrence of x in A;
- $\forall_f xA \rightarrow_f A(y/x)$, if y is free for x in A.

Another approach to first-order trilattice logics comes with the observation that "predicate logic can be seen as a special kind of propositional modal logic" [48, p. 95]. This idea has been taken up by van Benthem [27]. Let $\mathcal{M}$ be a first-order model and let $\alpha, \beta, \ldots$ range over variable assignments in that model. The familiar Tarskian truth definition for formulas $\forall xA$ is:

$$\mathcal{M} \models \forall xA[\alpha] \text{ iff for every assignment } \beta\text{: if } \alpha =_x \beta, \text{ then } \mathcal{M} \models A[\beta].$$

Here $\alpha =_x \beta$ means that β is the same function as α except for, possibly, the value assigned to x. Now, the variable assignments can be understood as states, and the concrete relations $=_x$ can be replaced by abstract two-place relations R_x of "update between states". In other words, to every individual variable x, there is an associated "accessibility relation" R_x. If we assume an interpretation of atomic formulas with free variables and denote the set of all states by W, we may write Tarski's evaluation clause in the style of Kripke's evaluation clause for the necessity operator:

$$\mathcal{M}, \alpha \models \forall xA \text{ iff for every state } \beta \in W\text{: if } \alpha R_x \beta, \text{ then } \mathcal{M}, \beta \models A.$$

The idea would thus be to develop first-order trilattice logics from modal trilattice logics. We will not enter into a development of first-order extensions of $(\mathcal{L}^*_{tf}, \models_t, \models_f)$, L^t, L^f, L^{t*}, L^{f*}, I_{16}, and IT_{16} along these lines here, but we will instead refer the reader to [27, 268] and [265, Chap. 12]. Besides, in the next section, we will outline some possible ways of extending propositional trilattice logics by means of standard modal operators.

Yet another approach to defining first-order trilattice logic would consist of using cylindrification operations from cylindric algebra, see, for example, [175].

10.2 Modal Trilattice Logics

There exist several avenues to defining modal trilattice logics, particularly avenues to extending the logics $(\mathscr{L}^*_{tf}, \models_t, \models_f), L^t, L^f, L^{t*}, L^{f*}, \mathrm{I}_{16}$, and IT_{16} to modal propositional logics, including, for example:

- the matrix approach,
- the axiomatic approach,
- Kripke model-based tableaux.

One could also investigate algebraic models for modal trilattice logics in order to generalize the characterization of normal modal logics by Boolean algebras with operators; see, for example [35]. We will not develop here any of these approaches in greater detail, but we will briefly comment on some possible directions for extending non-modal trilattice logics.

Nowadays, the least popular paradigm, perhaps, is the *matrix approach* to modal logics. If we try to pursue this approach, we are following the footsteps of Łukasiewicz and his enterprise of defining modalities by means of truth tables (and sets of designated truth values). The idea was that an adequate interpretation of modal expressions such as 'it is necessary that' and 'it is possible that' could be provided by many-valued valuation systems. This project is more or less dead, primarily because of Dugundji's Theorem [67] stating that none of the Lewis systems **S1–S5** has a finite characteristic matrix. Moreover, and more specifically, Łukasiewicz's attempt of defining modal logics by means of four-valued truth tables has been thoroughly analyzed and convincingly criticized by Font and Hájek [96], who emphasize that "apparently there is no satisfactory natural intuitive notion of necessity" for which the following valid formulas of Łukasiewicz's four-valued modal logic (notation adjusted) would be tautologies, where $\Diamond A$ ("it is possible that A") is understood as $\neg\Box\neg A$ ("it is not necessary that not A"):

$$(\Diamond A \wedge \Diamond B) \leftrightarrow \Diamond(A \wedge B),$$
$$(A \rightarrow B) \rightarrow (\Box A \rightarrow \Box B),$$
$$(A \rightarrow B) \rightarrow (\Diamond A \rightarrow \Diamond B),$$
$$\Box A \rightarrow (B \leftrightarrow \Box B),$$
$$\Box A \rightarrow (\Diamond B \leftrightarrow \Box B).$$

We hesitate, however, to declare Łukasiewicz's project a complete failure. On the one hand, finite matrices *are* useful in the investigation of modal logics, because these matrices provide a method for proving consistency and the independence of axioms. On the other hand, the trilattice *SIXTEEN*$_3$ and Belnap trilattices in general constitute much richer semantical structures than Łukaisewicze's four-valued matrix. Moreover, one may attempt to reconsider the very notion of a modal necessity or possibility operator in the context of truth value *lattices*. The definition of lattice operations (on multilattices of truth values) displaying properties that can arguably be seen as properties distinctive of modal operators might, in a sense, rehabilitate Łukasiewicz's approach.

One may also consider purely syntactic ways of defining modal trilattice logics. One could, for example, add sets of characteristic axioms that are adjoined to classical propositional logic **CPL** in order to obtain familiar and important modal extensions of **CPL**. If we focus on normal modal logics and, say, the axiom system HL^{t*} from Chap. 5 in the language $\mathscr{L}^*_{tf}$ as a non-modal base language, then in a first step one could consider analogues of the smallest normal modal propositional logic **K** by adding to HL^{t*} the necessitation rule

$$\vdash A / \vdash \Box A$$

and two versions of the so-called K axiom schema:

$$K^t\ \Box(A \to_t B) \to_t (\Box A \to_t \Box B); \quad K^f\ \Box(A \to_f B) \to_f (\Box A \to_f \Box B).$$

If we do not want to stipulate a separate rule and axioms for the possibility operator,[1] we could introduce $\Diamond$ as the dual of $\Box$ with respect to the negations $\sim_t$ and $\sim_f$ by means of the standard definitions:

$$\Diamond A :\Leftrightarrow\ \sim_t \Box \sim_t A, \quad \Diamond A :\Leftrightarrow\ \sim_f \Box \sim_f A.$$

Other well-known axiom schemes also come in pairs for truth and falsity implication:

$$\begin{array}{ll} D^t & \Box A \to_t \Diamond A, \\ D^f & \Box A \to_f \Diamond A, \\ T^t & \Box A \to_t A, \\ T^f & \Box A \to_f A, \\ 4^t & \Box A \to_t \Box\Box A, \\ 4^f & \Box A \to_f \Box\Box A, \\ 5^t & \Diamond A \to_t \Box\Diamond A, \\ 5^f & \Diamond A \to_f \Box\Diamond A. \end{array}$$

Since in $\mathscr{L}^*_{tf}$ all connectives arise in pairs, it seems natural to consider also pairs of modal operators, one pair for truth and the other pair for falsity: $\Box_t, \Diamond_t$ and $\Box_f, \Diamond_f$. The duality principles can then be adjusted as follows:

$$\Diamond_t A :\Leftrightarrow\ \sim_t \Box_t \sim_t A, \quad \Diamond_f A :\Leftrightarrow\ \sim_f \Box_f \sim_f A.$$

---

[1] The familiar $\Diamond$-version of the necessitation rules is $\vdash \neg A / \vdash \neg \Diamond A$, and the familiar $\Diamond$-version of the K axiom is the formula $(\neg \Diamond A \wedge \Diamond B) \to \Diamond(\neg A \wedge B)$, cf. [51, Chap. 4]. Since we have two versions of negation, conjunction, and implication in $\mathscr{L}^*_{tf}$, there are *several* options for formulating "$\Diamond$-versions" of the necessitation rule and the K-axiom.

If this is done, the K-axiom and the additional axiom schemes also need an adjustment:

$$K^{tt}\ \Box_t(A \to_t B) \to_t (\Box_t A \to_t \Box_t B),$$
$$K^{ff}\ \Box_f(A \to_f B) \to_f (\Box_f A \to_f \Box_f B),$$
$$D^{tt}\ \Box_t A \to_t \Diamond_t A,$$
$$D^{ff}\ \Box_f A \to_f \Diamond_f A,$$
$$T^{tt}\ \Box_t A \to_t A,$$
$$T^{ff}\ \Box_f A \to_f A,$$
$$4^{tt}\ \Box_t A \to_t \Box_t \Box_t A,$$
$$4^{ff}\ \Box_f A \to_f \Box_f \Box_f A,$$
$$5^{tt}\ \Diamond_t A \to_t \Box_t \Diamond_t A,$$
$$5^{ff}\ \Diamond_f A \to_f \Box_f \Diamond_f A.$$

Of course, mixed axioms like $\Diamond_f A \to_t \Box_f \Diamond_f A$ may be considered as well.

A purely syntactic treatment of modality in trilattice logics may be unsatisfactory, however, if it is not supplemented by a semantical interpretation of the modal operators. A relational semantics for the intuitionistic sequent calculus I_{16} and another Kripke semantics for the intuitionistic tableau calculus IT_{16} have been presented in Chap. 7. The Kripke models used in these characterizations can be extended to Kripke models for modal extensions of I_{16} and IT_{16}. Moreover, the extended models for IT_{16} can be used to develop a tableau calculus for modal extensions of IT_{16} along the lines of the tableau calculi for modal extensions of first-degree entailment logic presented by Priest in [198, 199]. In the semantics, the Kripke models introduced in Definition 7.13 may be extended by two modal accessibility relations $R_\Box$ and $R_\Diamond$, see [184]. The intuitionistic accessibility relation R and the modal accessibility relations $R_\Diamond$ and $R_\Box$ have to be correlated in a way that ensures the persistence of all formulas with respect to R, cf. Proposition 7.9. We then may use, for example, the following conditions for evaluating formulas $\Box A$ and $\Diamond A$ in a Kripke model $\mathcal{M} = \langle\langle M, R\rangle, \models_1, \models_2, \models_3, \models_4, v\rangle$:

$$x \models_1 \Box A \text{ iff } (\exists y \in M)(xR_\Box y \text{ and } y \not\models_1 A)$$
$$x \models_2 \Box A \text{ iff } (\exists y \in M)(xR_\Box y \text{ and } y \not\models_2 A)$$
$$x \models_3 \Box A \text{ iff } (\forall y \in M)xRy \text{ implies } (\forall z \in M)(\text{if } yR_\Box z \text{ then } z \models_3 A)$$
$$x \models_4 \Box A \text{ iff } (\forall y \in M)xRy \text{ implies } (\forall z \in M)(\text{if } yR_\Box z \text{ then } z \models_4 A)$$
$$x \models_1 \Diamond A \text{ iff } (\forall y \in M)xRy \text{ implies } (\forall z \in M)(\text{if } yR_\Diamond z \text{ then } z \not\models_1 A)$$
$$x \models_2 \Diamond A \text{ iff } (\forall y \in M)xRy \text{ implies } (\forall z \in M)(\text{if } yR_\Diamond z \text{ then } z \not\models_2 A)$$
$$x \models_3 \Diamond A \text{ iff } (\exists y \in M)(xR_\Diamond y \text{ and } y \models_3 A)$$
$$x \models_4 \Diamond A \text{ iff } (\exists y \in M)(xR_\Diamond y \text{ and } y \models_4 A)$$

The tableau rules for formulas $\Box A$ and $\Diamond A$ that can be read off from these above evaluation clauses are:

$$\begin{array}{cccc}
\Box A, j, \Delta_1 & \Box A, j, \Gamma_1 & \Box A, j, \Delta_2 & \Box A, j, \Gamma_2 \\
\downarrow & j\mathsf{r}_\Box k & \downarrow & j\mathsf{r}_\Box k \\
j\mathsf{r}_\Box k & \downarrow & j\mathsf{r}_\Box k & \downarrow \\
A, k, \Gamma_1 & A, k, \Delta_1 & A, k, \Gamma_2 & A, k, \Delta_2
\end{array}$$

$$\begin{array}{cccc}
\Box A, j, \Delta_3 & \Box A, j, \Gamma_3 & \Box A, j, \Delta_4 & \Box A, j, \Gamma_4 \\
j\mathsf{r}k & \downarrow & j\mathsf{r}k & \downarrow \\
k\mathsf{r}_\Box n & j\mathsf{r}k & k\mathsf{r}_\Box n & j\mathsf{r}k \\
\downarrow & k\mathsf{r}_\Box n & \downarrow & k\mathsf{r}_\Box n \\
A, n, \Delta_3 & A, n, \Gamma_3 & A, n, \Delta_4 & A, n, \Gamma_4
\end{array}$$

$$\begin{array}{cccc}
\Diamond A, j, \Delta_1 & \Diamond A, j, \Gamma_1 & \Diamond A, j, \Delta_2 & \Diamond A, j, \Gamma_2 \\
j\mathsf{r}k & \downarrow & j\mathsf{r}k & \downarrow \\
k\mathsf{r}_\Diamond n & j\mathsf{r}k & k\mathsf{r}_\Diamond n & j\mathsf{r}k \\
\downarrow & k\mathsf{r}_\Diamond n & \downarrow & k\mathsf{r}_\Diamond n \\
A, n, \Gamma_1 & A, n, \Delta_1 & A, n, \Gamma_2 & A, n, \Delta_2
\end{array}$$

$$\begin{array}{cccc}
\Diamond A, j, \Delta_3 & \Diamond A, j, \Gamma_3 & \Diamond A, j, \Delta_4 & \Diamond A, j, \Gamma_4 \\
\downarrow & j\mathsf{r}_\Diamond k & \downarrow & j\mathsf{r}_\Diamond k \\
j\mathsf{r}_\Diamond k & \downarrow & j\mathsf{r}_\Diamond k & \downarrow \\
A, k, \Delta_3 & A, k, \Gamma_3 & A, k, \Delta_4 & A, k, \Gamma_4
\end{array}$$

where $r_\Box$ and $r_\Diamond$ represent the modal accessibility relations, and the rules are to be understood as in Sect. 7.4.

10.3 Adverbially Qualified Truth Values

In developing a general theory of truth values, it also seems quite natural to take into account adverbial qualifications of ‘is true’ and ‘is false’ as giving rise to truth values. The values “logically true (false)” and “contingently true (false)” have been considered, for example, by Caton [49] and the values “necessarily true (false)” and “contingently true (false)” by Rescher [206]. In [231] a distinction is drawn between the truth values “constructive truth (falsity)” and “non-constructive truth (falsity)”; see also the modalized truth values in [31]. Moreover, in the literature, one also encounters uses of the expressions “epistemically true” [254], “morally true” [126], “provisionally true”, “eternally true” [42], etc.

One may then ask to what extent these adverbially qualified values are truth-functional. As to the modalized values “logically true (false)” and “contingently true (false)”, Caton suggested the application of a rewriting procedure to four-valued truth tables, and Rescher suggested four-valued tables motivated by the assumption that there exist exactly two incompatible states of affairs. It has been argued by MacIntosh [162] that under certain assumptions there is no *general* compositionality of adverbially qualified truth values. MacIntosh distinguishes between weak and strong adverbial qualifications of truth and falsity. The prime example of a "strong" (weak) qualification is "necessarily" (contingently). A sentence p is called strong iff p is strongly true or strongly false or its classical negation $\neg p$ is strongly true or strongly false. A sentence is weak iff it is not strong.

In addition to presupposing classical propositional logic, MacIntosh makes the following assumption:

It is impossible to soundly derive any weak sentence from a set consisting only of strong sentences.

Let t(f) stand for weak truth (falsity) and T(F) for strong truth (falsity). Neglecting non-qualified truth and falsity, and given the above assumptions, the value tables for negation, conjunction, and disjunction can be determined up to a small number of entries:

¬	
T	F
t	f
F	T
f	t

∧	T	t	F	f
T	T	t	F	f
t	t	t	F	–
F	F	F	F	F
f	f	–	F	–

∨	T	t	F	f
T	T	T	T	T
t	T	–	t	–
F	T	t	F	f
f	T	–	f	f

For instance, in view of the disjunction introduction inference $s \vdash (s \vee w)$, where s stands for a strong sentence and w for a weak one, it is clear that the disjunction of a sentence evaluated as T with any sentence at all is to be evaluated as T. For each of the non-gap entries, the functional values can be determined without any appeal to intuitions concerning specific adverbial qualifications. By turning to the modal case of necessity and contingency, however, it can be explained why the gaps in these tables are unavoidable. Whether a conjunction $(p \wedge q)$ of a contingently true formula p with a contingently false formula q, for example, is necessarily or contingently false depends on whether $\{p, q\}$ is unsatisfiable or not. There is thus no general result in this case. Modal considerations also explain the remaining gaps.

One may wonder what happens if the adverbial qualification of truth and falsity is internalized in the object language. MacIntosh emphasizes that there is no general answer to the question, whether for a strongly true sentence p, the sentence 'it is strongly true that p' is a strong or a weak sentence. He points out that in the modal logic **S4**, if p is strongly true, then so is $\Box p$ ("it is necessarily true that p") because of the **S4** axiom $\Box p \rightarrow \Box\Box p$. However, the axiom $\Box p \rightarrow \Box\Box p$ states that if $\Box p$ is true (simpliciter at a world in a model without any adverbial qualification), then $\Box\Box p$ is true. It seems thus that MacIntosh assumes that $\Box p$ is true iff p is strongly true. In this vein, in the absence of any reduction principles, iterated adverbial qualifications of truth and falsity suggest introducing infinitely many adverbially qualified values ("strongly strongly true", "strongly weakly true", "weakly strongly weakly false" etc., so that $\Box\Box p$ is true iff p is strongly strongly true).

10.4 Generalized Truth Values: Alternate Interpretations

Another promising line of research concerns possible applications of the machinery of generalized truth values to various philosophical interpretations of Dunn and Belnap's four-valued logic and exploring the possibilities of an extension of these interpretations beyond the four-valued case.

In this respect, it could be interesting to look, e.g., at the suggestion by David Lewis—to think of sentences in terms of *ambiguity*, especially of the ones regarded as both true and false under different readings, see [152]. Lewis claims that an ambiguous sentence cannot be considered true or false *simpliciter* until it is disambiguated in one way or another. Thus, the ambiguous sentence can only be said to be true (or false) *on some disambiguation* [152, p. 438]. The sentence is true (false) on some disambiguation *only*, Lewis tells us, if and only if it is true *on all* its disambiguations.

Let us formally reconstruct such an understanding as follows: if a sentence is true (false) on some disambiguation *simpliciter*, we mark it with the truth value $T(F)$; and if all the possible disambiguations for the given sentence have been carried out, we indicate this by putting the resulting truth value into parentheses. As Lewis explains, if one bars truth value gaps, then only three cases are possible: a sentence can be "true on all its disambiguations, false on all, or true on some and false on others" [152, p. 438]. More concretely, any ambiguous sentence is either:

- false on all disambiguations (in this case the sentence is *false only* and has the truth value $\{F\}$);
- true on all disambiguations (in this case the sentence is *true only* and has the truth value $\{T\}$);
- true on some disambiguations and false on all the others (in this case the sentence is *both true and false* and has the truth value $\{FT\}$).

In this way one obtains a certain intuitive interpretation (and justification) of the valuation system for Priest's Logic of Paradox or the "half-relevant" logic **R**-Mingle (see [76]).

Lewis also independently motivates the possibility for a sentence to be *neither true nor false* (gappy). Again, one can discriminate between cases when a sentence is gappy on *some* of its disambiguations or on *all* of its disambiguations (using the style of the notation adopted above, these cases can be marked by N and $\{N\}$, correspondingly). Lewis claims that by allowing truth-value gaps we have to envisage four more cases in which a sentence can be:

- gappy on all disambiguations (in our notation in this case the sentence has the truth value $\{N\}$);
- gappy on some disambiguations and false on all the other disambiguations (this can be marked as the truth value $\{NF\}$);
- gappy on some disambiguations and true on all the other disambiguations (this can be marked as the truth value $\{NT\}$);

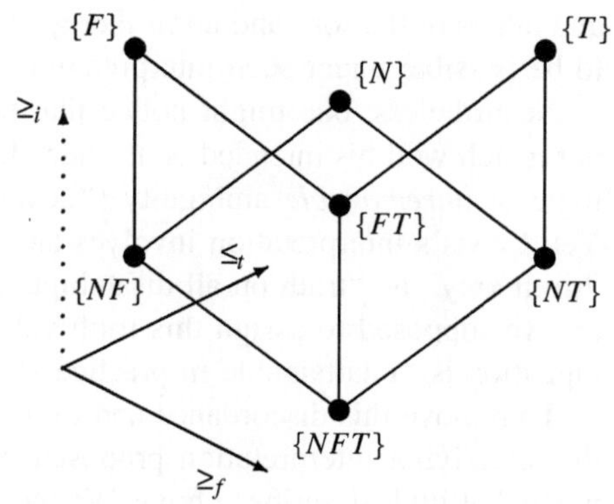

Fig. 10.1 Bi-and-a-half-lattice $SEVEN'_{2.5}$

- gappy on some disambiguations, true on some other disambiguations, and false on the rest of the disambiguations (the case for the truth value $\{NFT\}$).

These seven truth values taken together constitute a truth value (bi-and-a-half-) lattice $SEVEN'_{2.5}$ as presented in Fig. 10.1 (cf. lattice $SEVEN_{2.5}$, Fig. 4.1). Notice that the truth-value gap explicated in this way is not identical with Belnap's **N** and resembles much more Zaitsev's truth value for uncertainty **u**, see page 85.

In his considerations, Lewis concentrates on the three-valued version (without gaps) and argues to the effect that his so-conceived "logic for ambiguity" perfectly suits some "pessimistic" view on our language and reasoning. He recalls a standard logical requirement, according to which one has to "make sure that everything is fully disambiguated before one applies the methods of logic" [152, p. 439]. And then he gives the floor to a "pessimist" who might well complain that this requirement

> is a counsel of perfection, unattainable in practice. ... [A]mbiguity does not stop with a few scattered pairs of unrelated homonyms. It includes all sorts of semantic indeterminacy, open texture, vagueness, and whatnot, and these pervade all of our language. Ambiguity is everywhere. There is no unambiguous language for us to use in disambiguating the ambiguous language. So never, or hardly ever, do we disambiguate anything fully [152, p. 439].

Thus, to be more realistic, we have to adopt the pessimistic view and to weaken our logic "so that it tolerates ambiguity". Lewis's three (as well as seven) truth values and their informal interpretation are designed to provide the base for such a "logic for pessimists".

One may now ask why further combinations of Lewis's truth values are not considered and why cases where a sentence can be both true and false on some disambiguations and simultaneously true only, i.e., true and not false on some other disambiguations, are not admitted. It is easy to see, however, that the proposed interpretation entirely excludes the possibility of such combinations. Recall that for Lewis a sentence is true only if and only if it is true *on all* its disambiguations. In this case the sentence simply cannot be false (on any disambiguation). This precludes the construction of generalized truth values like $\{\{T\},\{FT\}\}$

and others of the sort, and no further generalization of Lewis's truth values appears to be possible under such interpretation.

Nevertheless, one might notice that the interpretation adopted by Lewis does not match well his intended goal. Indeed, the goal was to provide the basis for the logic of *unremovable* ambiguity ("never do we disambiguate anything fully"!). Yet, Lewis's interpretation involves an "absolutistic" understanding of the value "truth only" as "truth on all disambiguations", and similarly for falsity. But how are we supposed to assign this truth value if such a complete ("perfect") disambiguation is "unattainable in practice"?

To remove this discordance and to improve the situation, one has to reconsider the underlying interpretation proposed by Lewis to push it closer to the "pessimistic" attitude described above. We will sketch here some central points of such a possible reconsideration which can serve as the intuitive basis for a *generalized logic of* ambiguity.

Let us think of the disambiguation of a sentence as a certain procedure that can be accomplished in one way or another *by some person*. And the results of this procedure may well be different if the procedure is applied by several people. So, one cannot say that a sentence is true (false) on all disambiguations *simpliciter*, but merely that it is true (false) on all disambiguations accomplished by some (presumably "pessimistic") person. For the sake of simplicity, let us ignore truth value gaps. Then we arrive at the following reinterpretation of Lewis's initial three truth values:

$\{F\}$: a sentence is false on all disambiguations accomplished by some person;
$\{T\}$: a sentence is true on all disambiguations accomplished by some person;
$\{FT\}$: a sentence is false on some disambiguations accomplished by some person and true on other disambiguations accomplished by the same person.

Lewis admits what he calls "mixed disambiguations". These are "disambiguations of a compound sentence in which different occurrences of the same ambiguous constituent are differently disambiguated" [152, p. 439]. This understanding can be extended to the cases when *one and the same* occurrence of an ambiguous constituent is differently disambiguated by different people. Lewis's example of a mixed disambiguation is a disambiguation of the sentence 'Scrooge walked along the bank on his way to the bank' as 'Scrooge fancied a riverside stroll before getting to work with the money'. Consider just the sentence 'Scrooge walked along the bank'. Now imagine that Huey has disambiguated it as 'Scrooge fancied a riverside stroll', whereas Dewey takes it as 'Scrooge promenaded in front of the JPMorgan Chase Tower'. This would be a typical case of what can be called an *extended mixed disambiguation*.

Interestingly, Lewis [152, p. 440] recalls the "discussive logic"[2] developed by Jaśkowski [135], and he observes that Jaśkowski has conceived his logic inter alia

[2] Lewis uses the term "discursive" from the first English translation of Jaśkowsi's paper, but "discussive" seems to be the more appropriate translation of the Polish word "dyskusyjny", see the editorial note in [135, p. 55].

as a logic for ambiguity. In this connection, one can call into question the "discussiveness" of Lewis's logic itself. As it is formulated, it is, so to say, rather a logic for lonely (isolated) pessimists. But what if two or more pessimists have entered into a discussion? How should one evaluate a sentence on whose truth value they disagree or which is differently disambiguated by different disputants (extended mixed disambiguation)?

In exactly such a case, further combinations of Lewis's truth values may be helpful. And these combinations are directly supported by the above reconsideration that takes cognizance of a possible "inter-subjectivity" of the process of disambiguation. For example, the sentence 'Scrooge walked along the bank' in the case of an extended mixed disambiguation described above should take the generalized truth value $\{\{F\},\{T\}\}$. This value assumes then a natural interpretation as "the sentence is false on all disambiguations accomplished by some person and true on all disambiguations accomplished by some other person". And if the sentence has been disambiguated by three different people with three different outcomes, then $\{\{F\},\{T\},\{FT\}\}$ would seem to be the suitable generalized truth value for representing this situation.

Moreover, the procedure of disambiguation presumably can never be accomplished completely, i.e., once and for all. Disambiguation can be seen as a kind of a batch procedure that can be performed over and over again. An ambiguous sentence can be sent, again and again, into further rounds of disambiguation by different people. Then, e.g., the generalized truth value $\{\{\{T\}\}\}$ could mean that the sentence is true on all disambiguations accomplished by some person in three rounds. Clearly, truth value gaps can naturally arise in such an account of ambiguity.

It is not our intention here to elaborate in detail the proposed interpretation of ambiguity by means of generalized truth values. We believe, however, that this line of investigation can have manifold philosophical implications and is worthy of further development.

One can also try to combine the generalized logic of ambiguity with the conception of *confusion* presented by Camp in [46, pp. 125–160]. Camp has provided Belnap's four values with an interesting "confusion-based" intuitive motivation by developing what he called a "semantics of confused thought". Consider a rational agent who happens to mix up two very similar objects (say, *a* and *b*) and *ambiguously* uses one name (say, '*C*') for both of them. Now let such an agent assert some statement, saying, for instance, that *C* has some property. How should one evaluate this statement if *a* has the property in question, whereas *b* lacks it? Camp argues against ascribing truth values to such statements and puts forward an "epistemic semantics" in terms of "profitability" and "costliness" as suitable characterizations of sentences. A sentence *S* is said to be "profitable" if one would profit from acting on the belief that *S*, and it is said to be "costly" if acting on the belief that *S* would generate costs, for example as measured by failure to achieve an intended goal. If our "confused agent" asks some external observers whether *C* has the discussed property, the following four answers are possible:

- 'Yes' (mark the corresponding sentence with **Y**);
- 'No' (mark it with **N**);
- 'Cannot say' (mark it with **?**);
- 'Yes' and 'No' (mark it with **Y&N**).

Note that the external observers, who provide answers, are "non-confused" and have different objects in mind as to the referent of '*C*', in view of all the facts that may be relevant here. Camp conceives these four possible answers concerning epistemic properties of sentences as a kind of "semantic values", interpreting them as follows: the value **Y** is an indicator of profitability, the value **N** is an indicator of costliness, the value **?** is no indicator either way, and the value **Y&N** is both an indicator of profitability and an indicator of costliness. A strict analogy between this "semantics of confused reasoning" and Belnap's four valued logic is straightforward. Indeed, as [46, p. 157] observes, the set of implications valid according to his semantics is exactly the set of implications of the entailment system $\mathbf{E}_{fde}$.

Camp's interpretation of semantic values in terms of confusion has a strong intuitive appeal. But again, further generalization of these values is on the agenda. Indeed, what if the confused person, in his turn, would have to act as an expert ("of the second level") who is questioned about the object *C* and its properties? In this case an application of the methodology of generalized truth values naturally suggests itself. All the more so, if there are several such "confused experts" who provide information to some other person. Clearly, this other person should then be ready (and able) to deal with truth values such as {**N**, **Y&N**}, {**?**, **Y**}, etc. By combining the terminology employed by Priest and Camp, one can say that such truth values represent, so to say, *hyper-confusions*, and the person who happens to be unlucky enough to have "confused advisors" becomes then "hyper-confused". The application of the trilattice logics substantiated in the present book to investigations of hyper-confusions looks rather suggestive.

We conclude this section as well as the whole book by expressing our firm conviction that generalized truth values not only have a purely logical import for the analysis of various branches of modern non-classical logic, such as many-valued logic, paraconsistent logic, relevant logic, partial logic, constructive logic, etc., but may also be highly effective in elucidating significant philosophical problems connected to concepts such as truth, falsity, contradiction, information, ambiguity, vagueness, and many others.

References

1. Allwein, G., & MacCaull, W. (2001). A Kripke semantics for the logic of Gelfand quantales. *Sudia Logica, 68*, 173–228.
2. Almukdad, A., & Nelson, D. (1984). Constructible falsity and inexact predicates. *Journal of Symbolic Logic, 49*, 231–233.
3. Anderson, A. R., & Belnap, N. D. (1975). *Entailment: The logic of relevance and necessity* (Vol. I). Princeton, NJ: Princeton University Press.
4. Anderson, A. R., Belnap, N. D., & Dunn, J. M. (1992). *Entailment: The logic of relevance and necessity* (Vol. II). Princeton, NJ: Princeton University Press.
5. Anderson, D., & Zalta, E. (2004). Frege, Boolos, and logical objects. *Journal of Philosophical Logic, 33*, 1–26.
6. Arieli, O., & Avron, A. (1994). Logical bilattices and inconsistent data. In *Proceedings 9th IEEE annual symposium on logic in computer science* (pp. 468–476). IEEE Press.
7. Arieli, O., & Avron, A. (1996). Reasoning with logical bilattices. *Journal of Logic, Language and Information, 5*, 25–63.
8. Arieli, O., & Avron, A. (2000). Bilattices and paraconsistency. In D. Batens et al. (Eds.), *Frontiers of paraconsistent logic* (pp. 11–27). Baldock Hertfordshire: Research Studies Press.
9. Avigad, J., & Zach, R. (2008). The epsilon calculus, *The Stanford Encyclopedia of Philosophy (Fall 2008 Edition)*. Edward N. Zalta (Ed.). http://plato.stanford.edu/archives/fall2008/entries/epsilon-calculus/
10. Avron, A. (1996). The structure of interlaced bilattices. *Mathematical Structures in Computer Science, 6*, 287–299.
11. Avron, A. (1999). On the expressive power of three-valued and four-valued languages. *Journal of Logic and Computation, 9*, 977–994.
12. Avron, A. (2009). Multi-valued semantics: why and how. *Studia Logica, 92*, 163–182.
13. Avron, A., & Lev, I. (2005). Non-deterministic multi-valued structures. *Journal of Logic and Computation, 15*, 241–261.
14. Avron, A., & Zamansky, A. (2009). Non-deterministic semantics for logical systems - A survey. In D. Gabbay, & F. Guenthner (Eds.), *Handbook of philosophical logic*. Berlin: Springer-Verlag.
15. Baker, K. A. (1977). Finite equational bases for finite algebras in a congruence-distributive equational class. *Advances in Mathematics, 24*, 207–243.
16. Baaz, M., Fermüller, C., & Zach, R. (1994). Elimination of cuts in first-order many-valued logics. *Journal of Information Processing and Cybernetics, 29*, 333–355.
17. Baaz, M., Fermüller, C., Salzer, G., & Zach, R. (1998). Labeled calculi and finite-valued logics. *Studia Logica, 61*, 7–33.

Y. Shramko and H. Wansing, *Truth and Falsehood*, Trends in Logic 36, 227
DOI: 10.1007/978-94-007-0907-2, © Springer Science+Business Media B.V. 2011

18. Barwise, J., & Perry, J. (1981). Semantic innocence and uncompromising situations. *Midwest Studies in the Philosophy of Language, VI*, 387–403.
19. Barwise, J., & Perry, J. (1983). *Situations and attitudes*. Cambridge, MA: MIT Press.
20. Batyrshin, I., & Kaynak, O. (1999). Parametric classes of generalized conjunction and disjunction operations for fuzzy modeling. *IEEE Transactions on Fuzzy Systems, 7*, 586–596.
21. Beaney, M. (Eds. and transl.) (1997). *The Frege reader*. Oxford: Wiley.
22. Belnap, N. D. (1977). How a computer should think. In G. Ryle (Ed.), *Contemporary aspects of philosophy* (pp. 30–55). Stocksfield: Oriel Press Ltd.
23. Belnap, N. D. (1977). A useful four-valued logic. In J. M. Dunn, & G. Epstein (Eds.), *Modern uses of multiple-valued logic* (pp. 8–37). Dordrecht: D. Reidel Publishing Co.
24. Belnap, N. D. (2009). Truth values, neither-true-nor-false, and supervaluations. *Studia Logica, 91*, 305–334.
25. Belnap, N. D. (2010). How case-intensional semantics prevents the slingshot from hitting its target (Unpublished Manuscript) (22 pp.).
26. van Benthem, J. (1984). Possible worlds semantics: a research program that cannot fail? *Studia Logica, 43*, 379–393.
27. van Benthem, J. (1997). Modal foundations of predicate logic. *Logic jJurnal of the IGPL, 5*, 259–286.
28. Béziau, J.-Y. (1994). Universal logic. In T. Childers, & O. Majer (Eds.), *Proceedings Logica' 94* (pp. 73–93) Prague: Czech Academy of Sciences.
29. Béziau, J-Y. (1997). What is many-valued logic? In *Proceedings of the 27th international symposium on multiple-valued logic* (pp. 117–121). Los Alamitos CA: IEEE Computer Society Press.
30. Béziau, J.-Y. (1998). Recherches sur la logique abstraite: les logiques normales. *Acta Universitatis Wratislaviensis no. 2023, Logika, 18*, 105–114.
31. Béziau, J.-Y. (2006). Many-valued and Kripke semantics. In J. van Benthem et al. (Eds.), *The age of alternative logics* (pp. 89–101). Dordrecht: Springer-Verlag.
32. Bialynicki-Birula, A., & Rasiowa, H. (1957). On the representation of quasiboolean algebras. *Bulletin de l'Académie Polonaise des Sciences, 5*, 259–261.
33. Biedermann, K. (1999). An equational theory for trilattices. *Algebra Universalis, 42*, 253–268.
34. Birkhoff, G. (1967). *Lattice theory*. Rhode Island: Providence.
35. Blackburn, P., de Rijke, M., & Venema, Y. (2001). *Modal logic*. Cambridge University Press.
36. Blamey, S., & Humberstone, L. (1991). A perspective on modal sequent logic. *Publications of the Research Institute for Mathematical Sciences, Kyoto University, 27*, 763–782.
37. Bloom, S., & Suszko, R. (1972). Investigations into the sentential calculus with identity. *Notre Dame Journal of Formal Logic, 13*, 289–308.
38. Bochman, A. (1998). Biconsequence relations: A four-valued formalism of reasoning with inconsistency and incompleteness. *Notre Dame Journal of Formal Logic, 39*, 47–73.
39. Brown, B., & Schotch, P. (1999). Logic and aggregation. *Journal of Philosophical Logic, 28*, 265–287.
40. Burge, T. (1986). Frege on truth. In L. Haaparanta, & J. Hintikka (Eds.), *Frege synthesized* (pp. 97–154). Dordrecht: D. Reidel Publishing Co.
41. Burge, T. (1992). Frege on knowing the third realm. *Mind, 101*, 633–650.
42. Bush, W. T. (1908). Provisional and eternal truth. *The Journal of Philosophy, Psychology and Scientific Methods, 5*, 181–184.
43. Caleiro, C., Carnielli, W., Coniglio, M., & Marcos, J. (2005). Two's company: "The humbug of many logical values". In J.-Y. Beziau (Ed.), *Logica universalis* (pp. 169–189). Basel: Birkhäuser Verlag.
44. Caleiro, C., Carnielli, W., Coniglio, M., & Marcos, J. (2007). *Suszko's Thesis and dyadic semantics*. Preprint, http://www.cs.math.ist.utl.pt/ftp/pub/CaleiroC/03-CCCM-dyadic1.pdf.

45. Caleiro, C., Carnielli, W., Coniglio, M., & Marcos, J. (2007). *Dyadic semantics for many-valued logics*. Preprint, http://www.cs.math.ist.utl.pt/ftp/pub/CaleiroC/03-CCCM-dyadic1.pdf.
46. Camp, J. L. (2002). *Confusion: A study in the theory of knowledge*. Cambridge, MA: Harvard University Press.
47. Carnap, R. (1942). *Introduction to semantics*. Cambridge, MA: Harvard University Press.
48. Carnap, R. (1947). *Meaning and necessity. A study in semantics and modal logic*. Chicago: University of Chicago Press.
49. Caton, C. E. (1963). A stipulation of a modal propositional calculus in terms of modalized truth-values. *Notre Dame Journal of Formal Logic, 4*, 224–226.
50. Chagrov, A., & Zakharyaschev, M. (1997). *Modal logic*. Oxford: Clarendon Press.
51. Chellas, B. (1980). *Modal logic. An introduction.* Cambridge University Press.
52. Church, A. (1943). Review of Rudolf Carnap. *Introduction to Semantics. The Philosophical Review, 52*, 298–304.
53. Church, A. (1956). *Introduction to mathematical logic (Vol. I)* Revised and enlarged edition. Princeton: Princeton University Press.
54. Cignoli, R., D'Ottaviano, I., & Mundici, D. (2000). *Algebraic foundations of many-valued reasoning*. Dordrecht: Kluwer.
55. Czelakowski, J. (2001). *Protoalgebraic logics*. Dordrecht: Kluwer.
56. Cook, R. T. (2005). What's wrong with tonk (?). *Journal of Philosophical Logic, 34*, 217–226.
57. da Costa, N., Béziau, J.-Y., & Bueno, O. (1996). Malinowski and Suszko on many-valued logics: on the reduction of many-valuedness to two-valuedness. *Modern Logic, 6*, 272–299.
58. Curry, H. B. (1963). *Foundations of mathematical logic*. New York: McGraw-Hill.
59. D'Agostino, M., Gabbay, D., Hähnle, R., & Posegga, J. (Eds.). (1999). *Handbook of tableau methods*. Dordrecht: Kluwer.
60. Davidson, D. (1967). Truth and meaning. *Synthese, 17*, 304–323.
61. Davidson, D. (1969). True to the facts. *Journal of Philosophy, 66*, 748–764.
62. Devyatkin, L. (2007). Non-classical definitions of logical consequence (in Russian) (pp. 26–27). *Smirnov's readings in logic*. Moscow.
63. Devitt, M., & Sterelny, K. (1999). *Language and reality: An introduction to the philosophy of language*. Cambridge, MA: The MIT Press.
64. Dewitt, R. (2005). On retaining classical truths and classical deducibility in many-valued and fuzzy logics. *Journal of Philosophical Logic, 34*, 545–560.
65. Došen, K., & Schroeder-Heister, P. (Eds.). (1993). *Substructural logics*. Oxford: Oxford University Press.
66. Drai, D. (2002). The slingshot argument: an improved version. *Ratio, 15*, 194–204.
67. Dugundji, J. (1940). Note on a property of matrices for Lewis and Langford's calculi of propositions. *The Journal of Symbolic Logic, 5*, 150–151.
68. Dummett, M. (1977). *Elements of intuitionism.* Oxford: Oxford University Press.
69. Dummett, M. (1978). Truth. In *Truth and other enigmas* (pp. 1–24). Cambridge, MA: Harvard University Press (originally published In *Proceedings of the Aristotelian Society*, 1959, *59*, 141–162).
70. Dummett, M. (1981). *Frege: philosophy of language* (2nd ed.). London: Duckworth Publishers.
71. Dummett, M. (1991). *Frege and other philosophers.* Oxford: Oxford University Press.
72. Dummett, M. (2000). *Elements of intuitionism* (2nd ed.). Oxford: Clarendon Press.
73. Dunn, J. M. (1966). *The algebra of intensional logics.* Doctoral Dissertation. University of Pittsburgh, Ann Arbor (University Microfilms).
74. Dunn, J. M. (1971). An intuitive semantics for first degree relevant implications (abstract). *Journal of Symbolic Logic, 36*, 362–363.
75. Dunn, J. M. (1976). Intuitive semantics for first-degree entailment and 'coupled trees'. *Philosophical Studies, 29*, 149–168.

76. Dunn, J. M. (1976). A Kripke-style semantics for R-Mingle using a binary accessibility relation. *Studia Logica, 35*, 163–172.
77. Dunn, J. M. (1986). Relevance logic and entailment. In D. Gabbay, F. Guenter (Eds.), *Handbook of philosophical logic* (Vol. III, pp. 117–224). Dordrecht: D. Reidel Publishing Company.
78. Dunn, J. M. (1987). Relevant predication 1: The formal theory. *Journal of Philosophical Logic, 16*, 347–381.
79. Dunn, J. M. (1999). A comparative study of various model-theoretic treatments of negation: a history of formal negation. In D.M. Gabbay, & H. Wansing (Eds.), *What is negation?* Applied Logic Series 13 (pp. 23–51). Dordrecht: Kluwer.
80. Dunn, J. M. (2000). Partiality and its dual. *Studia Logica, 66*, 5–40.
81. Dunn, J. M., & Hardegree, G. M. (2001). *Algebraic methods in philosophical logic*. Oxford: Clarendon Press.
82. Dunn, J. M., & Zhou, C. (2005). Negation in the context of gaggle theory. *Studia Logica, 80*, 235–264.
83. Enderton, H. (2001). *A mathematical introduction to logic* (2nd ed.). New York: Academic Press.
84. Fidel, M. M. (1978). An algebraic study of a propositional system of Nelson. In *Mathematical logic, Proceedings of the first Brasilian conference, Campinas 1977*. Lecture Notes in Pure and Applied Mathematics, 39, (pp. 99–117). New York: Marcel Dekker Inc.
85. Fine, K. (1975). Vagueness, truth and logic. *Synthese, 30*, 265–300.
86. Fitting, M. (1989). Bilattices and the theory of truth. *Journal of Philosophical Logic, 18*, 225–256.
87. Fitting, M. (1990). Kleene's logic, generalized. *Journal of Logic and Computation, 1*, 797–810.
88. Fitting, M. (1990). Bilattices in logic programming. In G. Epstein (Ed.), *The twentieth international symposium on multiple-valued logic* (pp. 238–246). IEEE-Press.
89. Fitting, M. (1991). Bilattices and the semantics of logic programming. *Journal of Logic Programming, 11*, 91–116.
90. Fitting, M. (1994). Kleene's three-valued logic and their children. *Fundamenta Informaticae, 20*, 113–131.
91. Fitting, M. (1997). A theory of truth that prefers falsehood. *Journal of Philosophical Logic, 26*, 447–500.
92. Fitting, M. (2006). Bilattices are nice things. In T. Bolander, V. Hendricks, S. A. Pedersen (Eds.), *Self-reference* (pp. 53–77). Stanford: CSLI-Publications.
93. Fitting, M., & Mendelsohn, R. L. (1999). *First-order modal logic*. Dordrecht: Kluwer.
94. Føllesdal, D. (1983). Situation semantics and the 'slingshot' argument. *Erkenntnis, 19*, 91–98.
95. Font, J. M. (1997). Belnap's four-valued logic and De Morgan lattices. *Logic Journal of IGPL, 5*, 1–29.
96. Font, J. M., & H'ajek, P. (2002). On Łukasiewicz's four-valued modal logic. *Studia Logica, 70*, 157–182.
97. van Fraassen, B. (1966). Singular terms, truth-value gaps, and free logic. *Journal of Philosophy, 63*, 481–495.
98. Frankowski, S. (2004). Formalization of a plausible inference. *Bulletin of the Section of Logic, 33*, 41–52.
99. Frankowski, S. (2004). *P*-consequence versus *q*-consequence operations. *Bulletin of the Section of Logic, 33*, 197–207.
100. Frege, G. (1891). Function und Begriff. Vortrag, gehalten in der Sitzung vom 9. Januar 1891 der Jenaischen Gesellschaft für Medicin und Naturwissenschaft, Jena: H. Pohle, Jena, 31 pp. (Reprinted in [106]).
101. Frege, G. (1892). Über Sinn und Bedeutung. *Zeitschrift für Philosophie und philosophische Kritik, 100*, 25–50 (reprinted in [106]).

102. Frege, G. (1918). Der Gedanke. *Beiträge zur Philosophie des deutschen Idealismus, 1*, 58–77 (reprinted in [104]).
103. Frege, G. (1962). *Grundgesetze der Arithmetik, Bde. I und II* (2nd ed.). Darmstadt: Wissenschaftliche Buchgesellschaft.
104. Frege, G. (1967). *Kleine Schriften*. Ignacio Angelli (Ed.), Darmstadt: Wissenschaftliche Buchgesellschaft.
105. Frege, G. (1976). *Wissenschaftlicher Briefwechsel*. In G. Gabriel, H. Hermes, F. Kambartel, C. Thiel, & A. Veraart (Eds.), Hamburg: Felix Meiner Verlag.
106. Frege, G. (1986). Funktion, Begriff, Bedeutung. Fünf logische Studien. In G. Patzig (Ed.), Göttingen: Vandenhoeck & Ruprecht.
107. Frege, G. (1988). *Grundlagen der Arithmetik. Eine logisch-mathematische Untersuchung über den Begriff der Zahl*. Hamburg: Felix Meiner Verlag.
108. Frege, G. (1990). Einleitung in die Logik. In G. Frege (Ed.), *Schriften zur Logik und Sprachphilosophie* (pp. 74–91). Hamburg: Felix Meiner Verlag.
109. Gabbay, D. (1981). *Semantical investigations into Heyting's intuitionistic logic*. Dordrecht: Reidel.
110. Gabriel, G. (1984). Fregean connection: Bedeutung, value and truth-value. *The Philosophical Quarterly, 34*, 372–376.
111. Gabriel, G. (1986). Frege als Neukantianer. *Kant-Studien, 77*, 84–101.
112. Galatos, N., Jipsen, P., Kowalski, T., & Ono, H. (2007). *Residuated lattices: an algebraic glimpse at substructural logics*. Amsterdam: Elsevier.
113. Ganeri, J. (2002). Jaina logic and the philosophical basis of pluralism. *History and Philosophy of Logic, 23*, 267–281.
114. Ganter, B., & Wille, R. (1999). *Formal concept analysis: mathematical foundations*. Berlin: Springer.
115. Gargov, G. (1999). Knowledge, uncertainty and ignorance in logic: bilattices and beyond. *Journal of Applied Non-Classical Logics, 9*, 195–203.
116. Geach, P., & Black, M. (Eds.). (1952). *Translations from the philosophical writings of Gottlob Frege*. New York, NY: Philosophical Library.
117. Gentzen, G. (1935). Untersuchungen über das logische Schliessen. *Mathematische Zeitschrift, 39*, 176–210, 405–431.
118. Ginsberg, M. (1986). Multi-valued logics. In *Proceedings of AAAI-86, Fifth national conference on artificial intellegence* (pp. 243–247). Los Altos, CA: Morgan Kaufman Publishers.
119. Ginsberg, M. (1988). Multivalued logics: a uniform approach to reasoning in AI. *Computer Intelligence, 4*, 256–316.
120. Gödel, K. (1992). Zum intuitionistischen Aussagenkalkül. *Anzeiger der Akademie der Wissenschaften in Wien, Mathematisch-Naturwissenschaftliche Klasse, 69*, 65–66.
121. Gödel, K. (1944). Russell's mathematical logic. In P.A. Schilpp (Ed.), *The philosophy of Bertrand Russell* (pp. 125–153). Evanston and Chicago: Northwestern University Press.
122. Goguen, J. (1969). The logic of inexact concepts. *Synthese, 19*, 325–373.
123. Goré, R. (2000). Dual intuitionistic logic revisited. In R. Dyckhoff (Ed.), *Proceedings of the international conference on automated reasoning with analytic tableaux and related methods* (pp. 252–267). Berlin: Springer-Verlag.
124. Gottwald, S. (1989). *Mehrwertige Logik. Eine Einführung in Theorie und Anwendungen*. Berlin: Akademie-Verlag.
125. Gottwald, S. (2001). *A treatise on many-valued logic*. Baldock: Research Studies Press.
126. Gowans, C. (2004). Moral relativism, *The Stanford Encyclopedia of Philosophy* (Spring 2004 Edition). Edward N. Zalta (Ed.). http://plato.stanford.edu/archives/spr2004/entries/moral-relativism/
127. Grossmann, R. (1992). *The existence of the world*. London: Routledge.
128. Gurevich, Y. (1977). Intuitionistic logic with strong negation. *Studia Logica, 36*, 49–59.

129. Haack, S. (1996). *Deviant logic, fuzzy logic. Beyond the formalism.* Chicago: University of Chicago Press.
130. Hajek, P. (1998). *Metamathematics of fuzzy logic*. Dordrecht: Kluwer.
131. von Heusinger, K. (1997). *Salienz und Referenz Der Epsilonoperator in der Semantik der Nominalphrase und anaphorischer Pronomen.* Studia Grammatica 43. Berlin: Akademie Verlag.
132. Hilbert, D., & Bernays, P. (1970). *Grundlagen der Mathematik.* Bd. 2. Berlin: Springer.
133. Jain, P. (1997). *Investigating hypercontradictions.* (Unpublished Manuscript) (16 pp.).
134. James, W. (1956). The will to believe, 1897. In *Will to believe and other essays in popular philosophy* (pp. 21–31). New York, NY: Dover Publications.
135. Jaśkowski, S. (1999). A propositional calculus for inconsistent deductive systems. *Logic and Logical Philosophy, 7*, 35–56.
136. Jennings, R., & Schotch, P. (1984). The preservation of coherence. *Studia Logica, 43*, 89–106.
137. Kamide, N. (2002). Sequent calculi for intuitionistic linear logic with strong negation. *Logic Journal of the IGPL, 10*, 653–678.
138. Kamide, N. (2004). Quantized linear logic, involutive quantales and strong negation. *Studia Logica, 77*, 355–384.
139. Kamide, N. (2005). A cut-free system for 16-valued reasoning. *Bulletin of the Section of Logic, 34*, 213–226.
140. Kamide, N. (2005). Gentzen-type methods for bilattice negation. *Studia Logica, 80*, 265–289.
141. Kamide, N., & Wansing, H. (2009). Sequent calculi for some trilattice logics. *The Review of Symbolic Logic, 2*, 374–395.
142. Kamide, N., & Wansing, H. (2011). Completeness and cut-elimination theorems for trilattice logics. *Annals of Pure and Applied Logic, 162*, 816–835
143. Kamide, N., & Wansing, H. (2011). Proof theory of paraconsistent logic: A uniform perspective, (60 pp.).
144. Karpenko, A. (1983). Factor semantics for *n*-valued logics. *Studia Logica, 42*, 179–185.
145. Karpenko, A. (1989). Truth values: what are they? (in Russian). In V. Smirnov (Ed.), *Investigations in non-classical logics* (pp. 38–53). Moscow: Nauka.
146. Keefe, R. (2000). *Theories of vagueness*. Cambridge: Cambridge University Press.
147. Kracht, M. (1998). On extensions of intermediate logics by strong negation. *Journal of Philosophical Logic, 27*, 49–73.
148. Krause, D., & Béziau, J.-Y. (1997). Relativizations of the principle of identity. *Logic Journal of the IGPL, 5*, 1–12.
149. Kuhn, S. (1980). Quantifiers as modal operators. *Studia Logica, 39*, 145–158.
150. Lakshmanan, L. V. S., & Sadri, F. (1994). Probabalistic deductive databases. In M. Bruynooghe (Ed.), *Proceedings of 1994 international logic programming symposium* (pp. 254–268). MIT Press.
151. Lewis, C. I. (1943). The modes of meaning. *Philosophy and Phenomenological Research, 4*, 236–249.
152. Lewis, D. (1982). Logic for equivocators. *Noûs, 16*, 431–441.
153. Lowe, J. (1995). The metaphysics of abstract objects. *The Journal of Philosophy, 92*, 509–524.
154. Lowe, J. (1997). Objects and criteria of identity. In R. Hale, & C. Wright (Eds.), *A companion to the philosophy of language* (pp. 613–633). Oxford: Basil Blackwell.
155. Łukasiewicz, J. (1918). Farewell lecture by professor Jan Łukasiewicz, delivered in the Warsaw University Lecture Hall in March, 1918. In [157], 87–88.
156. Łukasiewicz, J. (1920). O logice trójwartościowej. *Ruch Filozoficny, 5*, 170–171. (English translation as "On three-valued logic." In [157], 87–88).
157. Łukasiewicz, J. (1921). Logika dwuwartościowa. *Przeglad Filosoficzny, 13*, 189–205 (English translation as "Two-valued logic." In [157], 89–109).

158. Łukasiewicz, J. (1970). Selected works. Amsterdam: North-Holland.
159. Łukasiewicz J. (1993) *Über den Satz des Widerspruch bei Aristoteles*. Hildesheim: Georg Olms Verlag.
160. MacFarlane, J. (2002). Review of Stephen Neale, *Facing Facts*. *Notre Dame Philosophical Reviews* (http://ndpr.nd.edu/review.cfm?id=1117).
161. MacFarlane, J. (2008). Truth in the garden of forking paths. In M. Kölbel, M. García-Carpintero (Eds.), *Relative Truth* (pp. 81–102). Oxford: Oxford University Press.
162. MacIntosh, J. J. (1991). Adverbially qualified truth values. *Pacific Philosophical Quarterly, 72*, 131–142.
163. Malinowski, G. (1990). *Q*-consequence operation. *Reports on Mathematical Logic, 24*, 49–59.
164. Malinowski, G. (1990). Towards the concept of logical many-valuedness. *Folia Philosophica, 7*, 97–103.
165. Malinowski, G. (1993). *Many-valued logics*. Oxford: Clarendon Press.
166. Malinowski, G. (1994). Inferential many-valuedness. In J. Woleński (Ed.), *Philosophical Logic in Poland* (pp. 75–84). Dordrecht: Kluwer.
167. Malinowski, G. (2001). Inferential paraconsistency. *Logic and Logical Philosophy, 8*, 83–89.
168. Malinowski, G. (2004). Inferential intensionality. *Studia Logica, 76*, 3–16.
169. Malinowski, G. (2009). Beyond three inferential values. *Studia Logic, 92*, 203–213.
170. Materna, P. (2000). Sense, denotation, reference: A terminological/philosophical chaos. Retrieved from http://www.phil.muni.cz/materna/sense.html. Cited 2000.
171. McGinn, C. (1976). A note on the Frege argument. *Mind, 85*, 422–423.
172. Mehlberg, H. (1958). *The reach of science*. Toronto: University of Toronto Press.
173. Meyer, R. K. (1978). *Why I am not a relevantist*, Research paper, no. 1, Australian National University, Logic Group, Research School of the Social Sciences, Canberra.
174. Miura, S. (1966). A remark on the intersection of two logics. *Nagoya Mathematical Journal, 26*, 167–171.
175. Monk, J. D. (2000). An introduction to cylindric set algebras. *Logic Journal of IGPL, 8*, 451–496.
176. Muskens, R. (1999). On partial and paraconsistent logics. *Notre Dame Journal of Formal Logic, 40*, 352–374.
177. Neale, S. (1995). The philosophical significance of Gödel's slingshot. *Mind, 104*, 761–825.
178. Neale, S. (2001). *Facing facts*. Oxford: Oxford University Press.
179. Negri, S., & von Plato, J. (2001). *Structural proof theory*. Cambridge: Cambridge University Press.
180. Nelson, D. (1949). Constructible falsity. *Journal of Symbolic Logic, 14*, 16–26.
181. Odintsov, S. (2003). Algebraic semantics for paraconsistent Nelson's logic. *Journal of Logic and Computation, 13*, 453–468.
182. Odintsov, S. (2008). *Constructive negations and paraconsistency*. Dordrecht: Springer.
183. Odintsov, S. (2009). On axiomatizing Shramko-Wansing's logic. *Studia Logica, 93*, 407–428.
184. Odintsov, S., & Wansing, H. (2004). Constructive predicate logic and constructive modal logic. Formal duality versus semantical duality. In V. Hendricks, et al. (Eds.), *First-order logic revisited* (pp. 269–286). Berlin: Logos Verlag.
185. Olson, K. R. (1987). *An essay on facts.* Lecture notes. No. 6, Stanford: CSLI Publications.
186. Omyła, M. (2003). Possible worlds in the language of non-Fregean logic. *Studies in Logic, Grammar and Rhetoric, 6*, 7–15.
187. Omyła, M. (2007). Remarks on non-Fregean logic. *Studies in Logic, Grammar and Rhetoric, 10*, 21–31.
188. Ono, H. (2003). Substructural logics and residuated lattices – an introduction. In V. Hendricks, & J. Malinowski (Eds.), *Trends in logic. 50 years of Studia Logica* (pp. 193–228). Dordrecht: Kluwer.

189. Oppy, G. (1997). The philosophical insignificance of Gödel's slingshot. Response to Stephen Neale. *Mind, 106*, 121–141.
190. Paoli, F. (2002). *Substructural logics: A primer*. Dordrecht: Kluwer.
191. Perry, J. (1996). Evading the slingshot. In A. Clark, J. Ezquerro, & J. Larrazabal (Eds.), *Philosophy and cognitive science. Categories, consciousness, and reasoning*. Dordrecht: Kluwer.
192. Poggiolesi, F. (2010). Display calculi and other modal calculi: a comparison. *Synthese, 173*, 259–279.
193. Popper, K. (1972). *Objective knowledge: An evolutionary approach*. Oxford: Oxford University Press.
194. Post, E. (1921). Introduction to a general theory of elementary propositions. *American Journal of Mathematics, 43*, 163–185.
195. Pottinger, G. (1979). On analysing relevance constructively. *Studia Logica, 38*, 171–185.
196. Priest, G. (1979). Logic of Paradox. *Journal of Philosophical Logic, 8*, 219–241.
197. Priest, G. (1984). Hyper-contradictions. *Logique et Analyse, 27*, 237–243.
198. Priest, G. (2008). *An introduction to non-classical logic. From if to is*. Cambridge: Cambridge University Press.
199. Priest, G. (2008). Many-valued modal logics: A simple approach. *Review of Symbolic Logic, 1*, 190–203.
200. Quine, W. V. O. (1953). Reference and modality. In *From a logical point of view* (pp. 139–159). Cambridge, MA: Harvard University Press.
201. Quine, W. V. O. (1960). *Word and object*. Cambridge, MA: Wiley and MIT Press.
202. Quine, W. V. O. (1969). *Ontological relativity and other essays*. New York, NY: Columbia University Press.
203. Rauszer, C. (1980). *An algebraic and Kripke-style approach to a certain extension of intuitionistic logic*, Dissertationes Mathematicae 167, Institute of Mathematics. Warsaw: Polish Academy of Sciences.
204. Rautenberg, W. (1979). *Klassische und nichtklassische Aussagenlogik*. Braunschweig: Vieweg.
205. Reck, E. (2007). Frege on truth, judgment, and objectivity. *Grazer Philosophische Studien, 75*, 149–173.
206. Rescher, N. (1965). An intuitive interpretation of systems of four-valued logic. *Notre Dame Journal of Formal Logic, 6*, 154–156.
207. Rescher, N. (1969). *Many-valued logic*. New York, NY: McGraw-Hill.
208. Restall, G. (2000). *An introduction to substructural logics*. London: Routledge.
209. Rose, A. (1950). A lattice-theoretic characterization of three-valued logic. *Journal of the London Mathematical Society, 25*, 255–259.
210. Rose, A. (1951). Systems of logics whose truth-values form lattices. *Mathematische Annalen, 123*, 152–165.
211. Routley, R. (1975). Universal semantics? *Journal of Philosophical Logic, 4*, 327–356.
212. Ruet, P. (2000). Non-commutative logic II: Sequent calculus and phase semantics. *Mathematical Structures in Computer Science, 10*, 277–312.
213. Ruffino, M. (2003). Wahrheit als Wert und als Gegenstand in der Logik Freges. In D. Greimann (Ed.), *Das Wahre und das Falsche. Studien zu Freges Auffassung von Wahrheit* (pp. 203–221). Hildesheim: Georg Olms Verlag.
214. Ruffino, M. (2004). Church's and Gödel's slingshot arguments. *Abstracta, 1*, 23–39.
215. Russell, B. (1919). *Introduction to mathematical philosophy*. London: George Allen and Unwin.
216. Ryan, M., & Sadler, M. (1992). Valuation systems and consequence relations. In S. Abramsky, D. Gabbay, T. Maibaum (Eds.), *Handbook of logic in computer science*. (Vol. 1, pp. 1–78). Oxford: Oxford University Press.
217. Schöter, A. (1996). Evidential bilattice logic and lexical inference. *Journal of Logic, Language and Information, 5*, 65–105.

218. Schröter, K. (1955). Methoden zur Axiomatisierung beliebiger Aussagen- und Prädikatenkalküle. *Zeitschrift für mathematische Logik und Grundlagen der Mathematik, 1*, 241–251.
219. Scott, D. (1973). Models for various type-free calculi. In P. Suppes, E. Nagel, A. Tarski (Eds.), *Logic, methodology and philosophy of science* (Vol. IV, pp. 157–187). Amsterdam: North-Holland.
220. Scott, D. (1973). Background to formalization. In H. Leblanc (Ed.), *Truth, syntax and modality* (pp. 244–273). Amsterdam: North-Holland.
221. Searle, J. (1995). Truth: a reconsideration of Strawson's view. In L.E. Hahn (Ed.), *The Philosophy of P. F. Strawson*. Chicago/Lasalle: Open Court Publishing.
222. Shramko, Y. (1989). On the problem of relevant entailment for intuitionistic logic (in Russian). In *Logico-philosophical investigations* (Vol. 1, pp. 165–175). Moscow.
223. Shramko, Y. (1998). A philosophically plausible modified Grzegorczyk semantics for first-degree intuitionistic entailment. *Logique et Analyse, 161-162-163*, 167–188.
224. Shramko, Y. (1999). *Intuitionismus und Relevanz*. Berlin: Logos-Verlag.
225. Shramko, Y. (2000). American plan for intuitionistic logic 1: an intuitive background. In T. Childers (Ed.), *The logica yearbook 1999* (pp. 53–64). Prague: Filosofia.
226. Shramko, Y. (2000). American plan for intuitionistic logic 2: generalized Kripke models (in Russian), *Logical Studies* (Online Journal), 5: http://logic.ru/en/node/160.
227. Shramko, Y. (2002). Generalized truth values: lattices and multilattices (in Russian), *Logical Investigations*, Moscow, 9: 264–291. (The On-line version see In *Logical Studies* (Online Journal), 8, 2002: http://logic.ru/en/node/185).
228. Shramko, Y. (2004). Die logische Wahrheitswerteontologie. In B. Christiansen, & U. Scheffler (Eds.), *Was folgt. Themen zu Wessel* (pp. 149–169). Berlin: Logos-Verlag.
229. Shramko, Y. (2005). Dual intuitionistic logic and a variety of negations: the logic of scientific research. *Studia Logica, 80*, 347–367.
230. Shramko, Y. (2009). Truth and Falsehood: what are truth values and what they are needed for (in Russian). *Logos, 70*, 96–121.
231. Shramko, Y., Dunn, J. M., & Takenaka, T. (2001). The trilattice of constructive truth values. *Journal of Logic and Computation, 11*, 761–788.
232. Shramko, Y., & Wansing, H. (2005). Some useful 16-valued logics: how a computer network should think. *Journal of Philosophical Logic, 34*, 121–153.
233. Shramko, Y., & Wansing, H. (2006). Hypercontradictions, generalized truth values, and logics of truth and falsehood. *Journal of Logic, Language and Information, 15*, 403–424.
234. Shramko, Y., & Wansing, H. (2007). Entailment relations and/as truth values. *Bulletin of the Section of Logic, 36*, 131–143.
235. Shramko, Y., & Wansing, H. (2009). The Slingshot-Argument and sentential idendity. *Studia Logica, 91*, 429–455.
236. Shramko, Y., & Wansing H. (2009). Truth Values. Part I, *Special issue of Studia Logica*, 91(3).
237. Shramko, Y., & Wansing H. (2009). Truth Values. Part II, *Special issue of Studia Logica*, v. 92, No 2.
238. Shramko, Y., & Wansing, H. (2010). Truth values, *The Stanford Encyclopedia of Philosophy* (Summer 2010 Edition), Edward N. Zalta (Ed.), URL = ⟨http://plato.stanford.edu/archives/sum2010/entries/truth-values/⟩.
239. Shramko, Y., & Zaitsev, D. (2004). Entailment and designated values (in Russian). *Logical Investigations, 11*, 126–137.
240. Simmons, K. (2002). Semantical and logical paradox. In D. Jacquette (Ed.), *A companion to philosophical logic* (pp. 115–130). Malden/Mass: Blackwell Publishes.
241. Sluga, H. (2002). Frege on the indefinability of truth. In E. Reck (Ed.), *From Frege to Wittgenstein: perspectives on early analytic philosophy* (pp. 75–95). Oxford: Oxford University Press.

242. Stelling, J. (2009). *Consequence relations, truth and logic*. University of Leipzig: Master thesis.
243. Suszko, R. (1968). Ontology in the Tractatus of L. Wittgenstein. *Notre Dame Journal of Formal Logic, 9*, 7–33.
244. Suszko, R. (1971). Identity connective and modality. *Studia Logica, 27*, 7–39.
245. Suszko, R. (1975). Abolition of the Fregean axiom. In R. Parikh (Ed.), *Logic colloquium*. Lecture notes in mathematics. 453 (pp. 169–239). Berlin: Springer.
246. Suszko, R. (1977). The Fregean axiom and Polish mathematical logic in the 1920's. *Studia Logica, 36*, 373–380.
247. Tarski, A. (1930). Über einige fundamentale Begriffe der Metamathematik. *Comptes Rendus des Séances de la Société des Sciences et des Lettres de Varsovie, XXIII* (Classe III), pp. 22–29.
248. Tarski, A. (1930). Fundamentale Begriffe der Methodologie der deduktiven Wissenschaften I. *Monatshefte für Mathematik und Physik, 37*, 361–404.
249. Tennant, N. (1987). Natural deduction and sequent calculus for intuitionistic relevant logic. *Journal of Symbolic Logic, 52*, 665–680.
250. Tennant, N. (1997). *The taming of The True*, Oxford University Press.
251. Thijsse, E. (1992). *Partial logic and knowledge representation*. PhD Thesis. Delft: Katholieke Universiteit Brabant Eburon Publishers.
252. Troelstra, A., & Schwichtenberg, H. (2000). *Basic proof theory*. (2nd ed.), Cambridge: Cambridge University Press.
253. Tsuji, M. (1998). Many-valued logics and Suszko's Thesis revisited. *Studia Logica, 60*, 299–309.
254. Tuomela, R. (1985). Truth and best explanation. *Erkenntnis, 22*, 271–299.
255. Urbas, I. (1996). Dual-intuitionistic logic. *Notre Dame Journal of Formal Logic, 37*, 440–451.
256. Urquhart, A. (1973). An interpretation of many-valued logic, *Zeitschrift für mathematische Logik und Grundlagen der Mathematik, 19*, 111–114.
257. Urquhart, A. (1986). Many-valued logic. In D.Gabbay, F. Guenther (Eds.), *Handbook of philosophical logic* (Vol. III, pp. 71–116). Dordrecht: D. Reidel Publishing Co.
258. Vakarelov, D. (1977). Notes on *N*-lattices and constructive logic with strong negation. *Studia Logica, 36*, 109–125.
259. Vardi, M. (1986). On epistemic logic and logical omniscience. In J.Y. Halpern (Ed.), *Theoretical aspects of reasoning about knowledge. Proceedings of the 1986 conference* (pp. 293–305). Los Altos, CA: Morgan Kaufmann Publishers.
260. Vasiliev, N. A. (1989). *Imaginary logic*. Moscow: Nauka.
261. Voishvillo, E. K. (1996). A theory of logical relevance. *Logique et Analyse, 155–156*, 207–228.
262. Voutsadakis, G. (2002). Poliadic concept analysis. *Order, 19*, 295–304.
263. Wagner, S. J. (1986). California semantics meets the Great Fact. *Notre Dame Journal of Formal Logic, 27*, 430–455.
264. Wansing, H. (1993). *The logic of information structures*. Lecture Notes in Artificial Intelligence 681. Berlin: Springer-Verlag.
265. Wansing, H. (1998). *Displaying modal logic*. Dordrecht: Kluwer.
266. Wansing, H. (1999). Predicate logics on display. *Studia Logica, 62*, 49–75.
267. Wansing, H. (2001). Negation. In L. Goble (Ed.), *The Blackwell guide to philosophical logic* (pp. 415–436). Cambridge, MA: Basil Blackwell Publishers.
268. Wansing, H. (2001). Short dialogue between M (Mathematician) and P (Philosopher) on multi-lattices. *Journal of Logic and Computation, 11*, 759–760
269. Wansing, H. (2006). Connectives stranger than tonk. *Journal of Philosophical Logic, 35*, 653–660.
270. Wansing, H. (2006). Logical connectives for constructive modal logic. *Synthese, 150*, 459–482.
271. Wansing, H. (2008). Constructive negation, implication, and co-implication. *Journal of Applied Non-Classical Logics, 18*, 341–364.

272. Wansing, H. (2010). The power of Belnap. Sequent systems for *SIXTEEN*$_3$. *Journal of Philosophical Logic, 39*, 369–393.
273. Wansing, H., & Belnap, N. (2010). Generalized truth values. A reply to Dubois. *Logic Journal of the Interest Group in Pure and Applied Logic*, 18, 921–935.
274. Wansing, H., & Kamide, N. (2010). Intuitionistic trilattice logics. *Journal of Logic and Computation*, 20, 1201–1229.
275. Wansing, H., & Shramko, Y. (2008). Harmonious many-valued propositional logics and the logic of computer networks. In C. Dégremont, L. Keiff & H. Rückert (Eds.), *Dialogues, logics and other strange things. Essays in honour of Shahid Rahman* (pp. 491–516), College Publications.
276. Wansing, H., & Shramko, Y. (2008). Suszko's Thesis, inferential many-valuedness, and the notion of a logical system, *Studia Logica, 88*, 89(147), 405–429.
277. Wansing, H., & Shramko Y. (2008). A note on two ways of defining a many-valued logic. In Pelis M. (Ed.), *Logica yearbook 2007* (pp. 255–266). Prague: Filosofia.
278. Widerker, D. (1983). The extensionality argument. *Nous, 17*, 457–468.
279. Wille, R. (1995). The basic theorem of triadic concept analysis. *Order, 12*, 149–158.
280. Williamson, T. (1990). *Identity and discrimination*. Oxford: Blackwell.
281. Williamsonm, T. (1994). *Vaguesness*. London: Routledge.
282. Wilson, G. (1978). On definite and indefinite descriptions. *Philosophical Review, 87*, 48–76.
283. Windelband, W. (1915). Präludien: Aufsätze und Reden zur Philosophie und ihrer Geschichte, 5. Aufgabe, Bnd. 1., Tübingen.
284. Wójcicki, R. (1970). Some remarks on the consequence operation in sentential logics. *Fundamenta Mathematicae, 68*, 269–279.
285. Wójcicki, R. (1984). R. Suszko's situational semantics. *Studia Logica, 43*, 326–327.
286. Wójcicki, R. (1986). Situation semantics for non-Fregean logic. *Journal of Non-Classical Logic, 3*, 33–67.
287. Wójcicki, R. (1988). *Theory of logical calculi. Basic theory of consequence operations*. Dordrecht: Kluwer.
288. Wójtowicz, A. (2005). The Slingshot Argument and non-Fregean logic. In A. Brozek, J. Jadacki, W. Strawinski (Eds.), *Logic, methodology and philosophy of science at Warsaw University* (Vol. 2, pp. 185–192). Warsaw: Wydawnictwo Naukowe Semper.
289. Wrigley, A. (2006). Abstracting propositions. *Synthese, 151*, 157–176.
290. Zach, R. (1993). *Proof theory of finite-valued logics*, Diploma Thesis, Technische Universität Wien, Vienna, Technical Report TUW-E185.2-Z.1-93. (Available at http://www.ucalgary.ca/rzach/papers/ptmvl.html)
291. Zadeh, L. (1965). Fuzzy sets. *Information and Control, 8*, 338–53.
292. Zadeh, L. (1975). Fuzzy logic and approximate reasoning. *Synthese, 30*, 407–425.
293. Zaitsev, D. (2009). A few more useful 8-valued logics for reasoning with tetralattice *EIGHT*$_4$. *Studia Logica, 92*, 265–280.
294. Zalta, E. (1983). *Abstract objects: an introduction to axiomatic metaphysics*. Dordrecht: D. Reidel Publishing Co.

Index

DOI: 10.1007/978-94-007-0907-2,

MIX
Papier aus verantwortungsvollen Quellen
Paper from responsible sources
FSC® C105338

If you have any concerns about our products,
you can contact us on
ProductSafety@springernature.com

In case Publisher is established outside the EU,
the EU authorized representative is:
Springer Nature Customer Service Center GmbH
Europaplatz 3, 69115 Heidelberg, Germany

Printed by Libri Plureos GmbH
in Hamburg, Germany